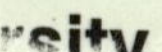

Lakhmi C. Jain, Vasile Palade and Dipti Srinivasan (Eds.)

Advances in Evolutionary Computing for System Design

Studies in Computational Intelligence, Volume 66

Editor-in-chief
Prof. Janusz Kacprzyk
Systems Research Institute
Polish Academy of Sciences
ul. Newelska 6
01-447 Warsaw
Poland
E-mail: kacprzyk@ibspan.waw.pl

Further volumes of this series
can be found on our homepage:
springer.com

Vol. 44. Nadia Nedjah, Luiza de Macedo Mourelle, Mario Neto Borges, Nival Nunes de Almeida (Eds.)
Intelligent Educational Machines, 2007
ISBN 978-3-540-44920-1

Vol. 45. Vladimir G. Ivancevic, Tijana T. Ivancevic
Neuro-Fuzzy Associative Machinery for Comprehensive Brain and Cognition Modeling, 2007
ISBN 978-3-540-47463-0

Vol. 46. Valentina Zharkova, Lakhmi C. Jain
Artificial Intelligence in Recognition and Classification of Astrophysical and Medical Images, 2007
ISBN 978-3-540-47511-8

Vol. 47. S. Sumathi, S. Esakkirajan
Fundamentals of Relational Database Management Systems, 2007
ISBN 978-3-540-48397-7

Vol. 48. H. Yoshida (Ed.)
Advanced Computational Intelligence Paradigms in Healthcare-1, 2007
ISBN 978-3-540-47523-1

Vol. 49. Keshav P. Dahal, Kay Chen Tan, Peter I. Cowling (Eds.)
Evolutionary Scheduling, 2007
ISBN 978-3-540-48582-7

Vol. 50. Nadia Nedjah, Leandro dos Santos Coelho, Luiza de Macedo Mourelle (Eds.)
Mobile Robots: The Evolutionary Approach, 2007
ISBN 978-3-540-49719-6

Vol. 51. Shengxiang Yang, Yew Soon Ong, Yaochu Jin Honda (Eds.)
Evolutionary Computation in Dynamic and Uncertain Environment, 2007
ISBN 978-3-540-49772-1

Vol. 52. Abraham Kandel, Horst Bunke, Mark Last (Eds.)
Applied Graph Theory in Computer Vision and Pattern Recognition, 2007
ISBN 978-3-540-68019-2

Vol. 53. Huajin Tang, Kay Chen Tan, Zhang Yi
Neural Networks: Computational Models and Applications, 2007
ISBN 978-3-540-69225-6

Vol. 54. Fernando G. Lobo, Cláudio F. Lima and Zbigniew Michalewicz (Eds.)
Parameter Setting in Evolutionary Algorithms, 2007
ISBN 978-3-540-69431-1

Vol. 55. Xianyi Zeng, Yi Li, Da Ruan and Ludovic Koehl (Eds.)
Computational Textile, 2007
ISBN 978-3-540-70656-4

Vol. 56. Akira Namatame, Satoshi Kurihara and Hideyuki Nakashima (Eds.)
Emergent Intelligence of Networked Agents, 2007
ISBN 978-3-540-71073-8

Vol. 57. Nadia Nedjah, Ajith Abraham and Luiza de Macedo Mourella (Eds.)
Computational Intelligence in Information Assurance and Security, 2007
ISBN 978-3-540-71077-6

Vol. 58. Jeng-Shyang Pan, Hsiang-Cheh Huang, Lakhmi C. Jain and Wai-Chi Fang (Eds.)
Intelligent Multimedia Data Hiding, 2007
ISBN 978-3-540-71168-1

Vol. 59. Andrzej P. Wierzbicki and Yoshiteru Nakamori (Eds.)
Creative Environments, 2007
ISBN 978-3-540-71466-8

Vol. 60. Vladimir G. Ivancevic and Tijana T. Ivacevic
Computational Mind: A Complex Dynamics Perspective, 2007
ISBN 978-3-540-71465-1

Vol. 61. Jacques Teller, John R. Lee and Catherine Roussey (Eds.)
Ontologies for Urban Development, 2007
ISBN 978-3-540-71975-5

Vol. 62. Lakhmi C. Jain, Raymond A. Tedman and Debra K. Tedman (Eds.)
Evolution of Teaching and Learning Paradigms in Intelligent Environment, 2007
ISBN 978-3-540-71973-1

Vol. 63. Wlodzislaw Duch and Jacek Mańdziuk (Eds.)
Challenges for Computational Intelligence, 2007
ISBN 978-3-540-71983-0

Vol. 64. Lorenzo Magnani and Ping Li (Eds.)
Model-Based Reasoning in Science, Technology, and Medicine, 2007
ISBN 978-3-540-71985-4

Vol. 65. S. Vaidya, L.C. Jain and Hiro Yoshida (Eds.)
Advanced Computational Intelligence Paradigms in Healthcare-2, 2007
ISBN 978-3-540-72374-5

Vol. 66. Lakhmi C. Jain, Vasile Palade and Dipti Srinivasan (Eds.)
Advances in Evolutionary Computing for System Design, 2007
ISBN 978-3-540-72376-9

Lakhmi C. Jain
Vasile Palade
Dipti Srinivasan
(Eds.)

Advances in Evolutionary Computing for System Design

With 134 Figures and 55 Tables

Prof. Lakhmi C. Jain
KES Centre
University of South Australia
Adelaide, Mawson Lakes Campus
South Australia
SA 5095
Australia
E-mail:- Lakhmi.jain@unisa.edu.au

Dr. Dipti Srinivasan
Department of Electrical
and Computer Engineering
National University of Singapore
Block E4-06-08
4 Engineering Drive 3
Singapore 117576
E-mail:- elesd@nus.edu.sg

Dr. Vasile Palade
Oxford University
Computing Laboratory
Wolfson Building
Parks Road
Oxford, OX1 3QD
England
E-mail:- Vasile.Palade@comlab.ox.ac.uk

Library of Congress Control Number: 2007926313

ISSN print edition: 1860-949X
ISSN electronic edition: 1860-9503
ISBN 978-3-540-72376-9 Springer Berlin Heidelberg New York

Springer is a part of Springer Science+Business Media
springer.com

Cover design: deblik, Berlin
Typesetting by the SPi using a Springer LaTeX macro package
Printed on acid-free paper SPIN: 11551430 89/SPi 5 4 3 2 1 0

Preface

In the last two decades, there has been a huge trend towards building intelligent systems that are inspired by biological systems from nature. Many nature-inspired algorithms and approaches have been developed and have become popular for designing intelligent systems with application to various real-world problems. Evolutionary computing paradigms, which are powerful search techniques inspired from natural genetics, proved to be at the forefront of this endeavour.

Undoubtedly, many authored and edited books have already revealed the field of evolutionary computing from many facets, which include deep theoretical analyses as well as more application-oriented studies. Edited volumes containing conference papers on the theory and application of evolutionary computing have also been very popular. In this book, the editors tried to put together a collection of extended and consistent chapters presenting original and latest research work from leading researchers on evolutionary computing. The focus of the book was to present challenging real-world applications of evolutionary computing paradigms as well as to provide an up-to-date reference manual for someone interested to use evolutionary techniques to system design.

This edited book consists of thirteen chapters covering a wide area of topics on evolutionary computing and applications. The application areas are diverse and include system control, bioinformatics, hardware optimization, traffic grooming, games theory, grid computing, and so on. Challenging and captivating theoretical issues are also introduced and discussed. These refer especially to the use of evolutionary computing techniques in conjunction with other machine learning approaches, and include: evolutionary neuro-fuzzy systems, fuzzy systems and genetic algorithms, evolutionary algorithms and immune learning for neural network design, evolutionary techniques for multiple classifier design, hybrid multi-objective evolutionary algorithms, evolutionary particle swarms.

Our hope is that this book will serve as a reference to researchers in evolutionary computing and to system designers and practitioners working in various application domains who are interested in using evolutionary computing paradigms in system design and implementation. This book can also be used by students and lecturers as an advanced reading material for courses on evolutionary computing.

The editors are grateful to the authors for their excellent contributions to the book. Thanks are due to the reviewers for providing precious feedback. The editorial support provided by Springer-Verlag is acknowledged. We hope that this book on evolutionary computing in system design will prove valuable to its readers.

Lakhmi C. Jain
University of South Australia
Australia

Vasile Palade
Oxford University
UK

Dipti Srinivasan
National University of Singapore
Singapore

Contents

1

Introduction to Evolutionary Computing in System Design

Lakhmi C. Jain[1], Shing Chiang Tan[2], Chee Peng Lim[3]

[1]School of Electrical and Information Engineering University of South Australia, Australia
[2]Faculty of Information Science and Technology Multimedia University, Malaysia
[3]School of Electrical and Electronic Engineering University of Science Malaysia, Malaysia

Summary. In this chapter, an introduction on the use of evolutionary computing techniques, which are considered as global optimization and search techniques inspired from biological evolutions, in the domain of system design is presented. A variety of evolutionary computing techniques are first explained, and the motivations of using evolutionary computing techniques in tackling system design tasks are then discussed. In addition, a number of successful applications of evolutionary computing to system design tasks are described.

1.1 Introduction

Evolutionary Computing (EC) [1] techniques have gained a lot of interests from researchers and practitioners in various fields as they provide robust and powerful adaptive search mechanisms that are useful for system design. In addition, the underlying biological concepts on which EC techniques are based on also contribute to their attractiveness in tackling system design problems. Although many different EC techniques have been proposed over the years, they, in general, share a common idea, i.e., evolving a population of potential solutions through the process of selection, recombination, and mutation. The search process of EC can be interpreted as a collective learning process in a population. Such interpretation is in agreement with the claims of some evolutionary models, such as the Baldwin evolution [2], that put emphasis on the role of the evolutionary learning process in a population.

On the other hand, the emergence of systems plays an important role in our daily life. According to [3], a system is defined as *a set of entities, real or abstract, comprising a whole where each component interacts with or is related to at least one other component and they all serve a common objective.* Researches in various types of systems are widespread and interdisciplinary, ranging from information and computer science, social and cognitive sciences,

L.C. Jain et al.: *Introduction to Evolutionary Computing in System Design*, Studies in Computational Intelligence (SCI) **66**, 1–9 (2007)
www.springerlink.com

to engineering and management research. For example, as stated in [3], in information and computer science, a system could also be a method or an algorithm. From the perspective of engineering, system engineering deals with the application of engineering techniques to form a structured development process that covers various stages of the process, from concept to production to operation and disposal. In social and cognitive sciences, systems comprise human brain functions and human mental processes as well as social/cultural behavioral patterns; while in management research, operations research, and organizational development, human organizations are viewed as conceptual systems of interacting components.

In this chapter, an introduction to EC in system designs is presented. In sections 1.2 and 1.3, what is EC, and why EC is useful in system design tasks are explained. In section 1.4, some successful applications of EC to a variety of system design problems are described. A description on each chapter in this book is given in Section 1.5, and a summary of this chapter is included in Section 1.6.

1.2 What is Evolutionary Computing

Evolutionary computing is referred to as computer methods that are inspired by biological evolutions [1]. From the literature, there are generally five basic types of EC, viz., genetic algorithms (GA), evolutionary programming (EP), evolution strategies (ES), genetic programming (GP), and learning classifier systems (LCS).

The basic idea behind all these EC variants is commonly the same from a unifying viewpoint. On availability of a population of individuals, natural selection, which is based on the principle of survival of the fittest following the existence of environmental pressures, is exercised to choose individuals that could better fit the environment. The operations involved are as follows. Given a fitness function to be optimized, a set of candidate solutions is randomly generated. The candidate solutions are assessed by the fitness function that indicates an abstract of fitness measure. Based on the fitness values, fitter candidate solutions have a better chance to be selected to seed the next generation through the application of recombination (crossover) and/or mutation. On one hand, the recombination operator is applied to two or more parent candidate solutions and one or more new candidate solutions (offspring) is/are produced. On the other hand, the mutation operator is applied to one parent candidate solution and one new candidate solution is produced. The new candidate solutions will compete among one another, based on the fitness values, for a place in the next generation. Even though candidate solutions with higher fitness values are favorable, in some selection schemes, the solution candidates with relatively low fitness values are included in the next generation so as to maintain diversity of the population. The processes of selection, recombination, and mutation are repeated from one generation of population to another until a terminating criterion is satisfied, i.e., either a candidate solution with sufficient quality has been obtained or the algorithm has been executed for a predefined number of generations.

Developed by Holland [4], the GA is the most popular and widely used EC paradigm. The GA is essentially a class of population-based search strategies that utilize an iterative approach to perform a global search on the solution space of a given problem. The three most popular search operators used in the GA are selection, crossover, and mutation. The selection operator chooses a subset group of candidate solutions (from a population) that are subsequently used to generate new candidate solutions with the crossover and mutation operators. In this regard, the crossover operator is applied to combine genetic information of the existing candidate solutions whereas the mutation operator is applied to create variations of a single candidate solution. While conventional GA models utilize binary codes to keep information of a candidate solution (i.e., a chromosome), real-valued chromosomes are now widely used to solve more difficult problems.

Introduced by Fogel, Owens, and Walsh [5], EP was aimed to simulate intelligent behavior by means of finite-state machines. In this regard, candidate solutions to a problem are considered as a population of finite-state machines. New candidate solutions (offspring) are generated from the existing candidate solutions (parents) through mutation. All candidate solutions are assessed by a fitness function. The crossover operator is not used in EP. Instead, the mutation operator is employed to provide variations in the population of candidate solutions.

ES [6, 7] was first developed to optimize parameters for aero-technology devices. This method is based on the concept of the *evolution of evolution*. Each candidate solution in the population is formed by genetic building blocks and a set of strategy parameters that models the behavior of that candidate solution in its environment. Both genetic building blocks and strategy parameters participate in the evolution. The evolution of the genetic characteristics is governed by the strategy parameters that are also adapted from evolution. Like EP, ES uses only the mutation operator.

Devised by Koza [8], GP was targeted at making the computer to solve problems without being explicitly programmed to do so. GP is an extension of conventional GA. However, the representation scheme of GP is different from that of conventional GA; the former represents individuals as executable programs (i.e., trees) whereas the latter uses string representation.

LCS [9] tackles the machine learning tasks as an evolutionary rule discovery component. The knowledge representation of LCS is a collection of production rules that are expressed in the *if-then* format. Each production rule, which is encoded as a binary string in Holland's standard approach, is considered as a classifier. The rules are updated according to some specific evolutionary procedure.

1.3 Why Evolutionary Computing

As explained in [10], one of the main advantages of EC is that it is conceptually simple and easy to use. An EC process normally covers random initialization of a population of candidate solutions, followed by some iterative variation and selection strategy, and by evaluation of candidate solutions in accordance with a performance index. This process is relatively simple, as compared with, for example, model-based approaches that deal with complicated mathematical formulation of the system under scrutiny.

The applicability of EC is broad and comprehensive, covering most problems that can be cast as an optimization task. In contrast to some numerical approaches that require continuous values or constraint formulation, the search space of EC can be disjoint, and the representation can be intuitive to humans [10]. Besides, EC techniques can be used to adapt solutions to changing environments. Indeed, a wide range of disciplines, including computer science, mathematics, medicine, engineering, and economics, to name a few, has been explored extensively by capitalizing EC techniques. Some successful application examples of EC are presented in section 1.4.

In solving complex, real-world problems, one may resort to methods that mimic the affinity from the nature. In this regard, the biologically inspired EC is cohesive with the idea of designing solutions that exploit a few aspects of natural evolutionary processes. In addition, recent trends in EC have geared towards devising hybrid models with other conventional optimization (e.g. gradient-based) techniques as well as other artificial intelligence paradigms. These hybrid models normally are highly automated and can give high efficiency, as evidenced in the work presented in other chapters of this book.

Another motivation of using EC stems from a technical standpoint. One trend from the growth of computer power has been the growing demand for problem-solving automation, which is vitally needed to cope with modern system design tasks of which their complexity is also ever increasing. EC techniques may turn out to be just as important because of the robustness and adaptation they can offer when dealing with modern computer-based applications. They are versatile enough and can normally provide satisfactory solutions within acceptance time with the advent of fast, powerful computing machines.

1.4 Applications of Evolutionary Computing to System Design

The applications of EC to system design are far reaching, especially from the perspective of meeting the requirements of software and hardware designs. A snapshot of successful applications of EC is presented, as follows.

EC can be deployed to solve problems that require any kind of optimization. Examples of optimization problems include operations-research tasks such as scheduling (e.g., [11, 12] and design optimization (e.g., [13, 14]). EC-based methods have been proposed for identifying an optimal broadcasting strategy in urban areas [15]. Automatic test data generation problems [16] can be formulated as an optimization problem and be solved using EC.

Classification and regression tasks are one of the fundamental intelligent processes in data analysis. Of many approaches, artificial neural networks (ANN) models, which are inspired from the functionality of biological brains, are a feasible method for tackling classification and regression tasks. Recently, various combinations between ANNs and EC have been investigated (e.g., [17]–[22]). The motivation of this synergistic machine learning lies in the retention of well-defined identity of the two entities, for which ANNs provide a framework of accurate and exact computation whereas EC provides a robust and efficient search, for better solution. Hybrid models combining EC and fuzzy systems (e.g., [23, 24]) as well as EC, ANNs, and fuzzy systems (e.g., [25–27]) have also been proposed, with the purpose of obtaining improved solutions.

Evolutionary robotics is an active field of research that utilizes EC to develop controllers for autonomous robots. In this regard, a population of candidate controllers is repeatedly evolved and tested with respect to a fitness function. The most feasible controller is identified for achieving an efficient robot learning application that could provide desired actions for the robot to perform. Successful applications of evolutionary robotics include research work by, to name a few, Paul et al. [28], Krasny and Orin [29], Reil and Husbands [30], Kodjabachian and Meyer [31], and Gallagher et al. [32].

Evolvable hardware is an emerging sub-specialty of EC that is aimed to automatically design and optimize electrical and mechanical structures, such as digital [33, 34]/analog [34–36] circuits, antennas [37], and robots [38–40] Research in this area has deployed various EC techniques including GA, GP, EP, adaptive GA [41], parallel GA [42]), compact GA [43], and Cartesian GP [44].

Recent advances in molecular biology and genomic technology have led to an explosive growth in the generation of biological data. Bioinformatics, as its name suggests, is a research area that utilizes computational methods to make discoveries from a huge biological database. The role of EC in bioinformatics, and data mining in general, is to assist in analyzing and, thus, extracting meaningful information from large data sets in a robust and computationally efficient way. EC has been applied to handle a broad classification of the various bioinformatics tasks, for instances, gene expression [45] and sequence alignment [46, 47], as well as for other data mining problems [48].

Other innovative application of EC to designing intelligent systems for solving a variety of practical problems include machine condition monitoring [49], medical diagnosis [50, 51], web-information search and retrieval [52, 53], and pedagogical educational learning system [54], as well as other problems in engineering and information systems [55]. In short, as computer technology continues to advance, the advantages of EC will become more profound, and

EC techniques will continue to be adopted as a practical problem-solving tool in diverse fields of system design.

1.5 Chapters Included in this Book

This book includes thirteen chapters. Chapter one introduces EC in system design. Chapter two by Castellano et al., is focused on evolutionary neuron-fuzzy systems and their applications. A scheme using neural learning and genetic optimization to learn a fuzzy model is presented. Chapter three, by Pratihar and Hui, describes the evolution of fuzzy controllers and applications. Some applications of the reported techniques are presented. Chapter four by Ranawana and Palade, is on neuro-genetic techniques for designing multi-classifier systems. An application for promoter recognition in DNA sequences is presented.

Chapter five by Xu and Liu is on evolutionary grooming of traffic in WDM optical networks. Future challenges and research directions are also included. Chapter six by Miranda presents evolutionary particle swarm optimization as an evolutionary meta-heuristic that implements a scheme of self adaptive recombination. Chapter seven by Tan and Wu is on the design of computationally efficient type-reduction strategies for type-2 fuzzy logic systems using genetic algorithm. Chapter eight by Ang et al. is on the design of neural network-based controller using an artificial immune system. Chapter nine by Srinivasan and Choy is on distributed problem solving using evolutionary learning in multi-agent systems.

Chapter ten by Tiwari et al. reports their work on evolutionary computing in grid-based environment. Chapter eleven by Vasilakos and Anastasopoulos is on the application of evolutionary game theory in wireless mesh networks. Chapter twelve by Garrett el al. is on the application of hybrid multiobjective evolutionary algorithms to the sailor assignment problem. The final chapter by Sanchez and Squillero presents the application of evolutionary computing techniques to hardware optimization problems related to the test and verification of advanced processors.

1.6 Conclusion

This chapter has presented an introduction to EC in handling system design problems. A variety of EC techniques, which include GA, EP, ES, GP, and LCS, have been described. The advantages of using EC techniques, especially for tackling optimization and search problems, as well as some successful applications of EC to different system design problems have also been elaborated. It is envisaged that EC techniques will ultimately play a significant role in solving practical system design tasks, especially those complex, nonlinear systems in various domains including computer and information sciences, engineering, management, and social and cognitive sciences, that have become more and more common in today's knowledge-based society.

References

1. Dumitrescu, D., Lazzerini, B., Jain, L.C., Dumitrescu, A., (2000), *Evolutionary Computation*, CRC Press, Boca Raton, FL, USA.
2. Suzuki, R., Arita, T., (2007), The dynamic changes in roles of learning through the Baldwin effect, *Artificial Life*, 13, 31–43.
3. http://en.wikipedia.org/wiki/System
4. Holland, J.H., (1962), Outline for a logical theory of adaptive systems, *J. ACM*, 3, 297–314.
5. Fogel, L.J., Owens, A.J., Walsh, M.J., (1966), *Artificial Intelligence through Simulated Evolution*, John Wiley, New York.
6. Rechenberg, I., (1994), Evolutionary strategy, *Computational Intelligence: Imitating Life*, Zurada, J.M., Marks II, R., Robinson, C., (Eds.), IEEE Press, 147–159.
7. Schwefel, H.-P., (1981), *Numerical Optimization of Computer Models*, John Wiley, Chichester, UK.
8. Koza, J.R., (1992), *Genetic Programming*, MIT Press, Cambridge, MA.
9. Holland, J.H., (1986), Escaping brittleness: The possibility of general-purpose learning algorithms applied to parallel rule-based systems, *Mach. Learn.*, 2, Michalski, R.S., Carbonell, J.G., Mitchell, T.M., (Eds.), Morgan Kaufmann, Los Altos, CA, 593–624.
10. Fogel, D.B. (1997), The advantages of evolutionary computation, *Bio-Computing and Emergent Computation,* Lundh, D., Olsson, B., Narayanan A., (Eds.), Skve, Sweden, World Scientific Press, Singapore, 1–11.
11. Lee, L.H., Lee, C.U., Tan, Y.P., (2007), A multi-objective genetic algorithm for robust flight scheduling using simulation, *Eur. J. Oper. Res.*, 177, 1948–1968.
12. Jozefowska, J., Mika, M., Rozycki, R., Waligora, G., Weglarz, J., (2002), A heuristic approach to allocating the continuous resource in discrete-continuous scheduling problems to minimize the makespan, *J. Scheduling*, 5, 487–499.
13. Greiner, H., (1996), Robust optical coating design with evolution strategies, *Appl. Opt.*, 35, 5477–5483.
14. Wiesmann, D., Hammel, U., Bäck, T., (1998), Robust design of multiplayer optical coatings by means of evolutionary algorithms, *IEEE Trans. Evol. Comput.*, 2, 162–167.
15. Alba, E., Dorronsoro, B., Luna, F., Nebro, A.J., Bouvry, P., Hogie, L., (2007), A cellular multi-objective genetic algorithm for optimal broadcasting strategy in metropolitan MANETs, *Comput. Commun.*, 30, 685–697.
16. Celso, C. Ribeiro, C.C., Martins, S.L., Rosseti, I., (2007), Metaheuristics for optimization problems in computer communications, *Comput. Commun.*, 30, 656–669.
17. Han, S.-J., Cho S.-B., (2006), Evolutionary neural networks for anomaly detection based on the behavior of a program, *IEEE Trans. Syst., Man Cybernet.—Part B*, 36, 559–570.
18. Fieldsend, J.E., Singh, S., (2005), Pareto Evolutionary Neural Networks, *IEEE Trans. Neural Netw.*,16, 338–354.

19. Bonissone, P.P., Chen, Y.-T., Goebel, K., Khedkar, P.S, (1999), Hybrid soft computing systems: industrial and commercial applications, *Proc. IEEE*, 87, pp. 1641–1667.
20. Yao, X., (1999), Evolving artificial neural networks, *Proc. IEEE*, 87, 1423–1447.
21. Vonk, E., Jain, L.C., and Johnson, R.P., (1997), *Automatic Generation of Neural Networks Architecture Using Evolutionary Computing*, World Scientific Publishing Company, Singapore.
22. Van Rooij, A., Jain, L.C., and Johnson, R.P., (1996), *Neural Network Training Using Genetic Algorithms*, World Scientific Publishing Company, Singapore.
23. Bonissone, P.P., Subbu, R., Eklund, N., Kiehl, T.R., (2006), Evolutionary algorithms + domain knowledge = real-world evolutionary computation, *IEEE Trans. Evol. Comput.*, 10, 256–280.
24. Hoffmann, F., (2001), Evolutionary algorithms for fuzzy control system design, *Proc. IEEE*, 89, 1318–1333.
25. Lin, C.J., Xu, Y.-J., (2006), A self-adaptive neural fuzzy network with group-based symbiotic evolution and its prediction applications, *Fuzzy Set Syst.*, 157, 1036–1056.
26. Oh, S.-K., Pedrycz, W., (2005), A new approach to self-organizing fuzzy polynomial neural networks guided by genetic optimization, *Phys. Lett. A*, 345, 88–100.
27. Jain, L.C. and Martin, N.M. (Eds), (1999), *Fusion of Neural Networks, Fuzzy Logic and Evolutionary Computing and their Applications*, CRC Press USA.
28. Paul, C., Valero-Cuevas, F.J., Lipson, H., 2006, Design and Control of Tensegrity Robots for Locomotion, *IEEE Trans. Robot.*, 22, 944–957.
29. Krasny, D.P., Orin, D.E., (2004), Generating High-Speed Dynamic Running Gaits in a Quadruped Robot Using an Evolutionary Search, *IEEE Trans. Syst., Man Cybernet.—Part B*, 34, 1685–1696.
30. Reil, T., Husbands, P., (2002), Evolution of Central Pattern Generators for Bipedal Walking in a Real-Time Physics Environment, *IEEE Trans. Evol. Comput.*, 6, 159–168.
31. Kodjabachian, J., Meyer, J.-A., (1998), Evolution and development of neural networks controlling locomotion, gradient following and obstacle avoidance in artificial insects, *IEEE Trans. Neural Netw.*, 9, 796–812.
32. Gallagher, J., Beer, R., Espenschiel, K., Quinn, R., (1996), Application of evolved locomotion controllers to a hexapod robot, *Robot. Autonomous Syst.*, 19, 95–103.
33. Hartmann, M., Haddow, P.C. (2004), Evolution of fault-tolerant and noiserobust digital designs, *Proc. Inst. Elect Eng.—Comput. Digit. Tech.*, 151, 287–294.
34. Higuchi, T., Iwata, M., Keymeulen, D., Sakanashi, H., Murakawa, H., Iajitani, I., Takahashi, E., Toda, K.,Salami, M., Kajihara, N., Oesu, N., (1999), Real-world applications of analog and digital evolvable hardware, *IEEE Trans. Evol. Comput.*, 220–235.
35. Lohn, J.D., (1999), Experiments on evolving software models of analog circuits, *Commun. ACM*, 42, 67–69.
36. Keymeulen, D., Zebulum, R.S., Jin, Y., Stoica, A., (2000), Fault-tolerant evolvable hardware using field-programmable transistor arrays, *IEEE Trans. Rel.*, 49, 305–316.
37. Hum, S.V., Okoniewski, M., Davies, R.J., (2005), An evolvable antenna platform based on reconfigurable reflect arrays, *Proc. NASA/DoD Conf. Evolvable Hardware*, Washington, DC, 139–146.

38. Lohn, J.D., Hornby, G.S., (2006), Evolvable hardware: using evolutionary computation to design and optimize hardware systems, *IEEE Computational Intelligence Magazine*, 1, 19–27.
39. Terrile, R.J., Aghazarian, H., Ferguson, M.I., Fink, W., Huntsberger, T.L., Keymeulen, D., Klimeck, G., Kordon, M.A., Seungwon, L., Allmen, P.V., (2005), Evolutionary computation technologies for the automated design of space systems, *Proc. NASA/DoD Conf. Evolvable Hardware*, Washington, DC, 131–138.
40. Yao, X., Higuchi, T., (1999), Promises and challenges of evolvable hardware, *IEEE Trans. Syst., Man Cybernet.—Part C,* 29, 87–97.
41. Ko, M.-S., Kang, T.-W., Hwang, C.-S., (1997), Function optimization using an adaptive crossover operator based on locality, *Eng. Appl. Artif. Intell.*, 10, 519–524.
42. Alba, E., Tomassini, M., (2002), Parallelism and evolutionary algorithms, *IEEE Trans. Evol. Comput.*, 6, 443–462.
43. Harik, G., Lobo, F., Goldberg, D., (1999), The compact genetic algorithm, *IEEE Trans. Evol. Comput.*, 3, 287–297.
44. Miller, J., (1999), An empirical study of the efficiency of learning Boolean functions using a Cartesian genetic programming approach, *Proc. Genetic Evol. Comput. Conf.*, Orlando, FL, 1, 1135–1142.
45. Tsai, H.-K., Yang, J.-M., Tsai, Y.-F., Kao, C.-Y., (2004), An evolutionary approach for gene expression patterns, *IEEE Trans. Inf. Technol. Biol.*, 8, 69–78.
46. Hung, C.-M., Huang, Y.-M., Chang, M.-S., (2006), Alignment using genetic programming with causal trees for identification of protein functions, *Nonlinear Analysis*, 65, 1070–1093.
47. Ngom, A., (2006), Parallel evolution strategy on grids for the protein threading problem, *J. Parallel Distrib. Comput.*, 66, 1489–1502.
48. Ghosh, A. and Jain, L.C. (Eds) (2005), *Evolutionary Computation in Data Mining*, Springer, Germany.
49. Guo, H., Jack, L.B., Nandi, A.K., (2005), Feature generation using genetic programming with application to fault classification, *IEEE Trans. Syst., Man Cybernet.—Part B*, 35, 89–99.
50. Toro, F.D., Ros E., Mota, S., Ortega, J., (2006), Evolutionary algorithms for multiobjective and multimodal optimization of diagnostic schemes, *IEEE Trans. Bio-Med. Eng.*, 53, 178–189.
51. Freitas, H.S., Bojarczuk, A.A., Lopes, C.C., (2000), Genetic programming for knowledge discovery in chest-pain diagnosis, *IEEE Eng. Med. Biol. Mag.*, 19, 38–44.
52. Gordon, M., Fan W.-G., Pathak, P., (2006), Adaptive web search: evolving a program that finds information, *IEEE Intell. Syst.,* 21, 72–77.
53. Kuo, R.J., Liao, J.L., Tu, C., (2005), Integration of ART2 neural network and genetic K-means algorithm for analyzing web browsing paths in electronic commerce, *Decis. Support Syst.*, 40, 355–374.
54. Bode'n, M., Bode'n, M., (2007), Evolving spelling exercises to suit individual student needs, *Applied Soft Computing*, 7, 126–135.
55. Jain, L.C. (Ed.), (2000), *Evolution of Engineering and Information Systems*, CRC Press USA.

2

Evolutionary Neuro-Fuzzy Systems and Applications

G. Castellano[1], C. Castiello[1], A.M. Fanelli[1], and L. Jain[2]

[1] Computer Science Department, University of Bari, Via Orabona 4, 70126 Bari, Italy [castellano, castiello, fanelli]@di.uniba.it
[2] School of Electrical and Information Engineering, University of South Australia, Adelaide, Mawson Lakes Campus, South Australia SA 5095, Australia
Lakhmi.Jain@unisa.edu.au

Summary. In recent years, the use of hybrid Soft Computing methods has shown that in various applications the synergism of several techniques is superior to a single technique. For example, the use of a neural fuzzy system and an evolutionary fuzzy system hybridises the approximate reasoning mechanism of fuzzy systems with the learning capabilities of neural networks and evolutionary algorithms. Evolutionary neural systems hybridise the neurocomputing approach with the solution-searching ability of evolutionary computing. Such hybrid methodologies retain limitations that can be overcome with full integration of the three basic Soft Computing paradigms, and this leads to evolutionary neural fuzzy systems. The objective of this chapter is to provide an account of hybrid Soft Computing systems, with special attention to the combined use of evolutionary algorithms and neural networks in order to endow fuzzy systems with learning and adaptive capabilities. After an introduction to basic Soft Computing paradigms, the various forms of hybridisation are considered, which results in evolutionary neural fuzzy systems. The chapter also describes a particular approach that jointly uses neural learning and genetic optimisation to learn a fuzzy model from the given data and to optimise it for accuracy and interpretability.

2.1 Introduction

Hybridisation of intelligent systems is a promising research field in the area of Soft Computing. It arises from an awareness that a combined approach may be necessary to efficiently solve real-world problems.

Each of the basic Soft Computing paradigms has advantages and disadvantages. Fuzzy systems do not have the automatic learning of neural networks, hence they can not be used in the absence of expert knowledge. Fuzzy systems have an advantage over neural networks and evolutionary strategies, as they can express knowledge in a linguistic form. This is familiar and forms part of the human reasoning process. Neural networks are developed specifically for learning, and are fundamental when a set of significant examples of the

G. Castellano et al.: *Evolutionary Neuro-Fuzzy Systems and Applications*, Studies in Computational Intelligence (SCI) **66**, 11–45 (2007)
www.springerlink.com

problem to be solved are available, rather than a solution algorithm for the problem. In general, neural networks can learn from examples. The way in which a knowledge base is extracted through a learning process is not easy for humans to understand. Evolutionary strategies, despite the learning speed being slower than neural networks, are able to use much more general functions than differentiable ones. This is because they do not require the computation of the gradient of the functions. Finally, as evolutionary algorithms explore in several directions the search space, they are less affected than neural learning by the problem of finding local minima.

A hybrid technique which makes use of a combination of the three Soft Computing paradigms is an interesting prospect. When these constituents are combined, they operate synergistically rather than competitively. Their mutual dependence may produce unexpected performance improvement.

In the last decade, several Soft Computing frameworks have been developed for a wide range of domains. Many of them solve a computational task by using a combination of different methodologies. The aim is to overcome the limitations and weakness of the several techniques.

This chapter gives an overview of different viewpoints of hybridisation among Soft Computing paradigms. The main features of principal Soft Computing paradigms are introduced. The attention is focused on all the possible ways of integrating the characteristics of two paradigms. This results in *neural fuzzy systems*, which is a hybrid of approximate reasoning method of fuzzy systems with the learning capabilities of neural networks, *evolutionary fuzzy systems* that use evolutionary algorithms to adapt fuzzy systems, and *evolutionary neural systems*, integrating the solution-searching ability of evolutionary computing with the neurocomputing approach. In conclusion, full hybridisation of the three paradigms is addressed. We summarise the research done on evolutionary neuro-fuzzy systems that integrate the solution-searching feature of evolutionary computing and the learning ability of neurocomputing with the explicit knowledge representation provided by fuzzy computing.

An example of full integration between Soft Computing paradigms is presented that makes use of neural learning and genetic algorithms to learn fuzzy models from data. This improves the accuracy and the ease of interpretation.

The chapter is organised as follows. Section 2.2 gives the fundamentals on the three main Soft Computing paradigms. That is, neural networks, fuzzy inference systems and evolutionary algorithms. Sections 2.3 to 2.5 provide a review of all possible forms of combination among the paradigms. In section 2.6 different examples of evolutionary-neuro-fuzzy hybridisation are considered, and some applications are reported. Section 2.7 presents a hybrid approach for the optimisation of neuro-fuzzy models based on an evolutionary algorithm. An example of the application is included in this section. Section 2.8 concludes the chapter.

2.2 Soft Computing Systems

The term Soft Computing [193] indicates a number of methodologies used to find approximate solutions for real-world problems which contain various kinds of inaccuracies and uncertainties. The guiding principle of Soft Computing is to develop a tolerance for imprecision, uncertainty, partial truth, and approximation to achieve tractability, robustness and low solution cost.

The underlying paradigms of Soft Computing are Neural Computing, Fuzzy Logic Computing and Evolutionary Computing. Systems based on such paradigms are Artificial Neural Networks (ANN's), Fuzzy Systems (FS's), and Evolutionary Algorithms (EA's).

Rather than a collection of different paradigms, Soft Computing is better regarded as a partnership in which each of the partners provides a methodology for addressing problems in a different manner. From this perspective, the Soft Computing methodologies are complementary rather than competitive. This relationship enables the creation of hybrid computing schemes which use neural networks, fuzzy systems and evolutionary algorithms in combination. Figure 2.1 shows their use in a Soft Computing investigation.

In the following, after a description of the three basic Soft Computing paradigms, some hybrid Soft Computing systems are overviewed. These show the rapid growth in the number and variety of combinations using ANN's, FS's and EA's.

2.2.1 Neural Networks

Artificial Neural Networks (ANN's) are computational models that are loosely modelled on biological systems and exhibit in a particular way some of the

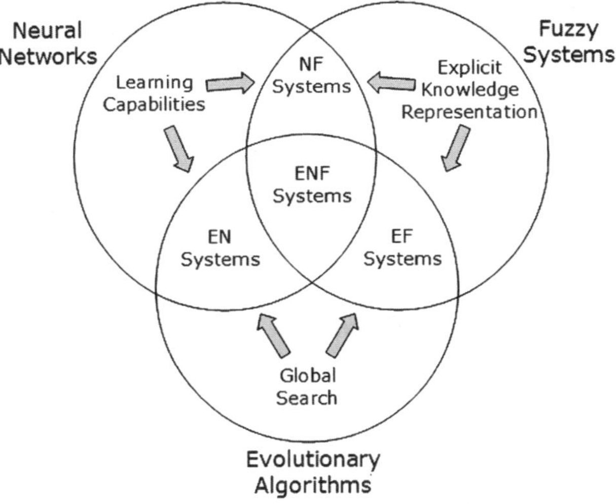

Fig. 2.1. Integration possibilities in Soft Computing

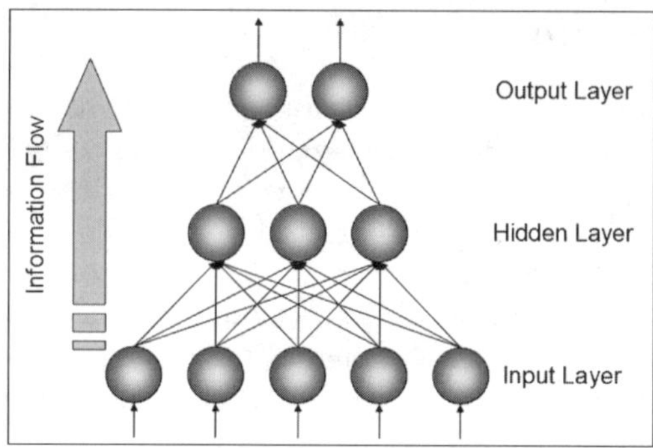

Fig. 2.2. The structure of three layers ANN

properties of the brain. They are composed by a number of simple processors (neurons) working in parallel, without any centralised control. The neurons are arranged in a particular structure which is usually organised in layers. A system of weighted connections determines the information flow through the network. In figure 2.2 a neural network is depicted with two layers, plus an input layer. The behaviour of an ANN is determined by the topology of the connections and by the properties of every processing unit, which typically evaluates a mathematical function. The numeric weights of the connections are modified in order to give ANN's a dynamic nature. ANN's possess also adaptive properties, since they are able to modify their output depending on the actual weight. The weight value is dependent on past experience. The learning of an ANN is based on the analysis of input data (training set) and the particular learning algorithms. The knowledge is eventually embedded into the weight configuration of the network.

Neural networks are commonly regarded as learning machines that work on the basis of empirical data. The only means of acquiring knowledge about the world in a connectionist system comes from observational instances. There are no a priori conceptual patterns that could lead to a learning process. The neurons composing the network only perform a very simple task. This consists of evaluating a mathematical function and transmitting a signal. This is similar to the neuronal cells of biological brains. The aim is to adapt to the surrounding data from the system. This rather simplified structure enables ANN's to perform quite complex tasks, endowing connectionist systems with the capability of approximating continuous functions to any required degree of accuracy. The task of a neural network is to realise a mapping of the type:

$$\phi : \mathbf{x} \in R^n \rightarrow \mathbf{y} \in R^m, \tag{2.1}$$

where $\mathbf{x}$ is a n-dimensional input vector and $\mathbf{y}$ is an m-dimensional output vector.

ANN's can be variously distinguished in terms of their *configurations*, *topology* and *learning algorithms*. Generally speaking, two main types of configurations can be identified: feedforward networks, these possess connections which propagate the information flow in one direction only. The other type is recurrent networks, whose nodes may have feedback connections, where the input for a neuron may be the output from another neuron. Here the input may be from the particular neuron in question. Each connectionist system is characterised by a particular topology. That is, the organisation of its own neurons. The most typical neural architecture is represented by the multilayered network, where the processing units are fully interconnected in three or more layers and the output is interpolated among all the nodes of the network. On the other hand, in competitive networks the nodes compete to respond to inputs. A particular scenario may have some units that are active, while all others are inhibited. Finally, learning algorithms play an important role in characterising the working mechanism of an ANN. The most commonly employed learning algorithm is the error back-propagation training. This involves adjusting the connection weights under a supervised learning scheme. Other types of learning algorithms include counter propagation mechanisms, Hebbian learning, Boltzmann algorithm and self-organising maps. Further details concerning the characterisation of neural networks, may be obtained from references which include [7,15,71,75,149].

2.2.2 Fuzzy Systems

Fuzzy mechanisms suitable for tasks involving reasoning have been proposed as an extension to classical formal logic. They were first introduced in set theory. The concept of a "fuzzy set" has been employed to extend classical sets, which are characterised by crisp boundaries. This addition permits a degree of flexibility for each object belonging to a particular set. This quality is realised by the definition of membership functions that give fuzzy sets the capacity of modelling linguistic, vague expression [192]. Fuzzy sets constitute the basis for fuzzy logic, a novel way for developing reasoning models by handling imprecise information, where truth can assume a continuum of values between 0 and 1. This kind of information is often referred to as *fuzziness*: it should be noted that fuzziness does not come from randomness, but from the uncertain and imprecise nature of abstract thoughts and concepts. Fuzzy reasoning realises a form of approximate reasoning that, using particular mathematical inferences, derives conclusions based on a set of fuzzy IF-THEN rules, where linguistic variables could be involved. In this way, fuzzy logic is suitable for describing the behaviour of systems which are either too complex or too ill-defined to be amenable to precise mathematical analysis. Classical systems cannot cope with inexact or incomplete information, because they do not provide any means of representing imprecise propositions and do not possess any mechanism that can make an inference from such propositions.

Fuzzy logic systems commonly contain expert IF-THEN rules and can be characterised in terms of their fundamental constituents: *fuzzification*, *rule base*, *inference*, *defuzzification*. Figure 2.3 is a schematic representation of such

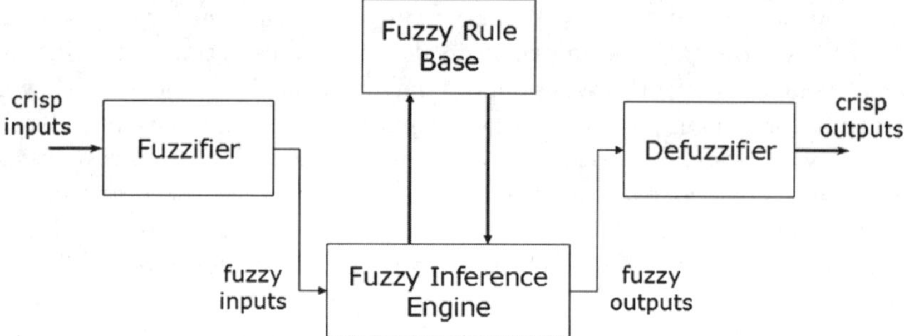

Fig. 2.3. The basic components of a fuzzy system

a fuzzy system. Fuzzification is a mapping from a crisp input space to fuzzy sets in a defined universe:

$$\mathrm{U} : x_i \in R \rightarrow X \in U \subset R^q. \tag{2.2}$$

Here x_i represents a crisp value and q is the number of fuzzy classes. The fuzzy sets are characterised by membership functions which portray the degree of belonging of x_i to the values in U, $\mu_F(x_i) : U \rightarrow [0, 1]$.

The rule base is constituted by an ensemble of fuzzy rules and the knowledge is expressed in the following form:

$$\begin{array}{c} \text{IF } x_1 \text{ is } A_1^k \text{ AND} \ldots \text{AND } x_n \text{ is } A_n^k \\ \text{THEN } y_1 \text{ is } b_1^k \text{ AND} \ldots \text{AND } y_m \text{ is } b_m^k, \end{array} \tag{2.3}$$

where the index $k = 1, \ldots, K$ indicates the k-th rule among the K rules in the rule base; A_i^k and b_j^k are fuzzy sets. These are defined over the input components x_i, $i = 1, \ldots, n$, and the output components y_j, $j = 1, \ldots, m$, respectively. The rule is a fuzzy implication that is usually represented by a Cartesian product of the membership functions of antecedents and consequents.

The fuzzy inference process can be described by starting with the definition of the membership functions $\mu_{ik}(\cdot)$ related to the k-th fuzzy rule and evaluated for each input component of a sample vector $\mathbf{x} = (x_1, \ldots, x_n)$. The most commonly employed membership functions are the triangular and the Gaussian functions. The values obtained by the fuzzification contribute to the AND conjunction of each rule, which is interpreted by a particular T-norm. This is most commonly the *min* or the *algebraic product* operators. After evaluating the degrees of satisfaction for the entire set of fuzzy rules, the K activation strengths are used in the OR disjunction represented by the alternative rules. These are interpreted by a particular S-norm, the *max* or the *algebraic sum* operators are most commonly used.

Finally, the defuzzification process is used to reconvert the fuzzy output values, deriving from the inference mechanism, into crisp values. These can then be eventually employed in different contexts. The most common strategy for defuzzification is to use the centre of area method which gives the centre of gravity of the output membership function.

Further details concerning the characteristics of fuzzy systems can be found in [22, 43, 147].

2.2.3 Evolutionary Algorithms

"Evolutionary Algorithms" is a general term for indicating a number of computational strategies that are based on the principle of evolution. Population-based generate-and-test algorithms are used. An evolutionary algorithm is an iterative probabilistic program, employed in optimisation problems. A population of individuals, $P(t) = \{p_1^t, \ldots, p_n^t\}$ is involved for each iteration t [52, 79, 115]. Each individual p_i^t represents a potential solution for the problem and is represented as some data structure $\mathcal{S}$. The metaphor underlying evolutionary computation enables us to review this encoding process as a translation from a phenotype representation to chromosomal representation or genotype encoding. The population members are successively evaluated in terms of a particular function. This function is specifically designed for the problem, in order to express the *fitness* of the individuals. In this way, at iteration $t+1$ a new population can be generated on the basis of the fittest elements. These are selected according to some prefixed rule to produce an offspring by means of "genetic" operators. They are usually represented by unary transformations, $m_i : \mathcal{S} \rightarrow \mathcal{S}$ (mutation), and higher order transformations, $c_j : \mathcal{S} \times \ldots \times \mathcal{S} \rightarrow \mathcal{S}$ called crossover, or recombination. The successive generations of several populations hopefully converge to a final offspring encoding an optimal solution. In figure 2.4 the general scheme of an evolutionary algorithm is shown.

The most common evolutionary computation techniques are genetic algorithms, so that the terms "Evolutionary Computation" and "GA-based methods" are used almost as synonyms. However, different evolutionary techniques should be distinguished. They can be characterised in terms of the data structures $\mathcal{S}$ employed to represent the chromosomes, the genetic operators and the individual selection schemes involved. In particular, Genetic Algorithms [59, 79, 115] codify chromosomes as binary strings. Genetic operators are applied only to the genotypical representations where the role of recombination is emphasised. Evolutionary Programming (EP) [52, 53] is a stochastic optimisation strategy similar to GA's, that gives emphasis on the behavioural linkage between parents and their offspring, rather than seeking to emulate specific genetic operators as observed in nature. The only genetic operator used in EP is mutation, applied to the phenotypical representations. Similarly to EP, Evolution Strategies [155, 156] make use of real-valued vectors to represent individuals. An individual comprises not only the specific object parameter set and its fitness function value, but usually also a set of endogenous (i.e. evolvable) strategy parameters that are used to control

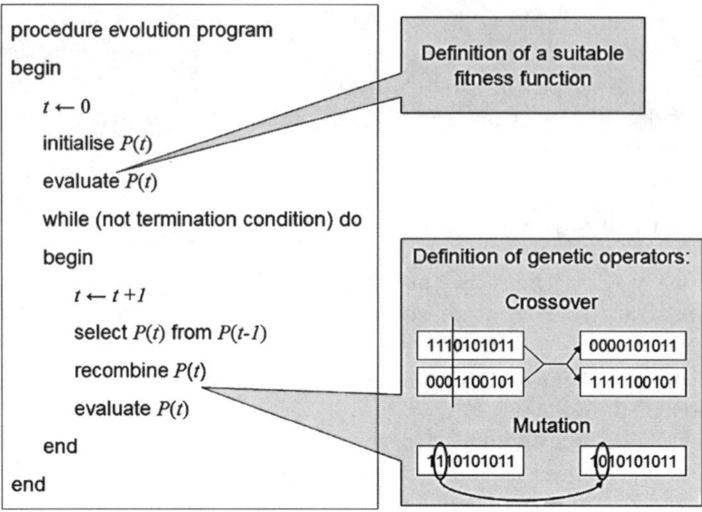

Fig. 2.4. The general structure of an evolutionary program

certain statistical properties of the genetic operators. Another evolutionary approach consists in Genetic Programming [5, 99, 101–104], where structures of possible computer programs (trees or even neural networks) evolve to represent the best solution to a problem.

2.3 Neuro-Fuzzy Systems

The neuro-fuzzy hybridisation represents by far the most fruitful and the most investigated strategy of integration in the context of Soft Computing. Both neural networks and fuzzy systems are dynamical, parallel processing systems that estimate input-output functions [118]. Fuzzy logic is capable of modelling vagueness, handling uncertainty and supporting human-type reasoning. Connectionist systems rely on numerical and distributed representation of knowledge and they are often regarded as black boxes, revealing little about their inductive engines. Also in terms of application requirements fuzzy logic and neural systems show very contrasting positions. Neural networks are capable of learning from scratch, without needing any a-priori intervention, provided that sufficient data are available or measurable. On the other hand, fuzzy systems make use of linguistic knowledge of the process, which can be supplied by human experts.

To a large extent, the key-points and the shortcomings of connectionist and fuzzy approaches appear to be complementary. It seems to be a natural practice to build up integrated strategies combining the concepts of the two paradigms [89,94,100,111,122,130,135]. It should be observed that neuro-fuzzy systems are the most prominent representatives of hybridisations in terms of the number of practical implementations. A number of successful applications can be observed in engineering and industrial applications [19].

An inspection of the multitude of hybridisation strategies proposed in literature which involve neural networks and fuzzy logic would be somewhat impractical. It is however straightforward to indicate the general lines underlying this kind of integration. This appears to be twofold. If we consider the fuzzy inference system as the main subject of the hybridisation, neural networks can add learning capabilities to an inference engine which reveals to be spoilt by some self-constructing feature. This approach is referred to as Neuro-Fuzzy System (NFS). However, when a connectionist system is regarded as the main subject of the hybridisation, fuzzy logic may assist by incorporating fuzziness into the neural framework. This may enable a better view of the black box. This argument is commonly referred to as Fuzzy-Neural System (FNS).

In the following we will briefly review some of the most prominent examples of neuro-fuzzy hybridisation proposed in literature.

2.3.1 Fuzzy-Neural Systems

Fuzziness could be incorporated into neural networks at different levels. Examples are given in literature concerning the fuzzification of input data, output results, learning procedures, error functions [69, 82, 95, 117, 119, 120, 129, 175]. One of the most intriguing examples of fuzzy-neural hybridisation is investigated by a field of research which constructs systems based on fuzzy neurons. These specific neurons are designed to realise the common operations of fuzzy set theory. That is, fuzzy union, intersection, aggregation. This is instead of the usual standard algebraic functions. Using this technique it may be possible to take advantage of the transparency and readability of the structures. This may result in improved interpretation for the overall system [76, 132, 136].

2.3.2 Neuro-Fuzzy Systems

A number of results have been presented to demonstrate the formal equivalence between fuzzy rule-based systems and neural networks [11, 20, 21, 70]. In particular, a functional equivalence has been shown to exist between radial basis function networks and fuzzy systems [88]. On the basis of this correspondence, various neuro-fuzzy integrations have been proposed to endow fuzzy inference engines with learning capabilities using the neural components. In practice, connectionist learning can be exploited to tune the parameters of an existing fuzzy system and/or to compile the structure of the rule base [34, 36, 50, 92, 96, 128, 166, 168, 170]. Among the most popular and pioneering neuro-fuzzy systems can be cited: GARIC [12], NEFCON [123], ANFIS [87]. All of these are models for successive hybridisation strategies [26, 98, 124–126, 167, 169].

2.4 Evolutionary-Fuzzy Systems

The effectiveness of evolutionary techniques in performing complex search processes over spaces has been successfully applied to provide learning capabilities to fuzzy systems. This kind of hybridisation led to the development of genetic fuzzy systems (GFS's) [6,42,134,152]. A GFS is therefore represented by a fuzzy rule based system which relies on genetic components to accomplish the main processes related to the system design: parameter optimisation and rule generation. Actually, the optimisation process usually concerns the tuning procedure applied over the variables of a fuzzy system. This may be already established in the structure. This configures the adaptation process. A more proper learning approach is realised when the genetic components are involved in the generation of the rule base, without reference to any pre-existing structure. It should be noted that, without loss of generality, we refer to the employment of genetic components. This is because genetic algorithms appear to be the most applied techniques inside this field of research.

The genetic search processes involved with GFS's are with a Knowledge Base constituted of:

- a Data Base, including the definitions of the parameters and variables of the fuzzy system;
- a Rule Base, comprising of a number of fuzzy rules.

As previously observed, the exploration of the different components of the Knowledge Base is correlated with the different objectives that can be attained. These range from parameter adaptation to pure learning activity. It should be observed that, similarly to any other search process, the application of evolutionary techniques to the Knowledge Base is subject to a trade-off between the dimension of the search space and the efficiency of the search. In this way, the smaller the search space will likely lead to a faster process yielding suboptimal solutions. A larger search space is more likely to contain optimal solutions, even if the search process would be less efficient.

For the purposes of our presentation, we are first going to describe the common applications of genetic procedures when used to create fuzzy rule-based systems, distinguishing different genetic adaptation and learning processes, and then some other kinds of hybridisation strategies proposed in literature.

2.4.1 Adaptation of the GFS Data Base

The tuning process applied to the Data Base of a GFS aims to optimise the membership functions in the fuzzy rule base. This is commonly predefined for the system.

Each individual may encode within its chromosomes the parameters of the Data Base. This codifies the particular shape of the membership functions or even the entire fuzzy partitions used to design the fuzzy rule base. This is especially true when dealing with descriptive fuzzy systems which involve linguistic variables. A number of references are addressed in literature as examples of this approach, and both real-valued and binary-valued chromosomes have been adopted [40,58,65,68,73,159].

2.4.2 Learning of the GFS Rule Base

The employment of genetic learning for the generation of fuzzy rules has been traditionally proposed in literature. There are three main approaches and these are the Pittsburgh [163], the Michigan [80] and the Iterative Rule Learning [173]. In the Pittsburgh approach an entire fuzzy rule base is encoded as a chromosome, and thus is one of the individuals of the candidate population. The Michigan approach, on the other hand, codifies each individual rule in a chromosome, so that the population represents the fuzzy rule base. The IRL approach encodes separately each rule and builds up the fuzzy rule base in an iterative fashion. This is done by adding a rule at every application of the genetic algorithm. In every case, when we consider approaches devoted to GFS Rule Base learning, we should be concerned with a particular scenario where the Data Base or the set of the involved membership functions is pre-configurated.

The most common way to codify a rule is to represent it in the disjunctive normal form. A set of rules is generally encoded as a list. In this way, the chromosomes can be given by the code of a single rule or by the assembly of several rules. Bibliographical references can be classified as belonging to the Pittsburgh approach [78,141], the Michigan approach [16,84] and the IRL approach [39,61].

2.4.3 Learning of the GFS Knowledge Base

The genetic component can be employed to determine the entire Knowledge Base pertaining to a GFS. Here both the parameter adaptation and the rule base learning are assessed by the genetic search process.

The fundamental strategies introduced in the previous section can be restated to characterise the approaches in literature. They distinguish between the Pittsburgh-based [8,23], the Michigan-based [131,172] and the IRL-based [40,41] learning methodologies.

2.4.4 Different Hybridisation Strategies

In order to complete the presentation, we mention a number of hybridisation strategies which integrate evolutionary techniques and fuzzy logic differently to the genetic adaptation and learning processes discussed in the previous sections.

First of all, it is pointed out how different approaches, other than genetic algorithms, have been employed to evolve fuzzy rule bases. In particular,

Genetic Programming can be utilised for this purpose, provided that the fuzzy rules are codified by coherent computer program structures, such as syntactic trees [2, 9, 35, 57, 77]. Recent sophisticated versions of genetic algorithms have been successfully adopted, and these include parallel GA's and coevolutionary algorithms [33, 45, 138, 143, 146].

Evolutionary algorithms have been also applied to solve peculiar questions connected with the management of a fuzzy rule base. It must be mentioned the employment of genetic techniques in high-dimensional problems to perform the selection of fuzzy rules [60,83,148], and the selection of features [24,62,108]. Recent works concern the growing interest in issues of comprehensibility. They should be regarded with the same attention as issues of accuracy arising from the application of fuzzy systems [83, 85, 91].

In a different way, GA's have found application in conjunction with fuzzy logic to improve the capabilities of clustering algorithms such as fuzzy C-means [14]. In particular, genetic strategies can be adopted to optimise the parameters of a FCM-type algorithm [66, 171], used to define the distance norm [191] and to directly produce the clustering results [18].

We point out that the evolutionary-fuzzy hybridisation can be analysed in the reverse direction, with respect to that so far considered. This is done even if it appears to be less profitable. Instead of considering the learning capabilities that genetic strategies could add to fuzzy system, *fuzzy evolutionary algorithms* focus on the management of GA parameters such as mutation rate or population size, using fuzzy systems [72, 137, 165].

To conclude our discussion, two bibliographical references assess the application possibility of GFS's [17, 46].

2.5 Evolutionary-Neural Systems

The combination of evolutionary algorithms and neural networks has led to the development of the so-called "Evolutionary Artificial Neural Networks" (EANN's) [181]. Using this term it is common to indicate a particular class of connectionist systems having augmented capabilities of adaptation to the external environment. If we regard the learning capabilities of a neural network as a form of adaptation based on the basis of data investigation, EANN's are able to show enhanced adaptability in a dynamic environment, due to the evolutionary components. This kind of fitting process may take place at several levels. These range from connection weights to network topology or even to learning rules. Strategies of evolution of connection weights are in many cases oriented to replace the classical back-propagation and conjugate gradient algorithms. It is hoped to overcome the drawbacks of gradient-descent techniques. Adaptation of architectures is also useful to avoid the tedious trial and error mechanism used by the human knowledge expert, which up to now remains the most common way of designing connectionist systems. With the evolution possibilities of learning rules a kind of "learning to learn" process could be developed. This would enable neural networks to become flexible and adjust to the problems to be solved. This could lead to an improvement of the learning performance, that is known to be dependent on the environment conditions and the task at hand. This would be instead of relying on individual training algorithms [179].

The adaptation process supported by the evolutionary components of an EANN is subjected to the usual cycle of computation. It is noted that some problems are inherently connected with the evaluation process of the individuals. The noisy fitness evaluation problem [182] compromises a fair estimate of the generated individuals. That is due mainly to the unavoidable one-to-many mapping between genotype and phenotype, whenever evolution of network topology is performed without consideration of the weight connections. The noise apparent during the evaluations of the fitness is related to the approximation of the genotype fitness value, that is the network encoding, with the phenotype fitness value. The network is evaluated by means of random initial weights during the evolutionary iterations. The permutation problem [67] could also occur when a many-to-one mapping from the genotypical to the phenotypical representation is done. This is due to the fact that a permutation of the hidden nodes of a network produces equivalent ANN's with different chromosome representations. Further analyses concerning the genotype/phenotype mapping can be found in [51, 52].

We will now briefly describe the different types of integration strategies of evolutionary algorithms and neural networks, as given in the literature.

2.5.1 Adaptation of the Connection Weights

The idea of applying evolutionary algorithms to adapt neural network weights could appear straightforward if we consider the network learning as a search for an optimal parameter configuration. Actually, EA's can be helpful in order to overcome the limitations of gradient descent methods. They may avoid the problem of them becoming trapped in local minima of the error function. They are then able to seek globally optimal solutions, without the need of referring to continuos and differentiable functions.

The employment of classical genetic algorithms is to determine a representation of connection weights via binary strings. Each individual can be reproduced by concatenating the binary code of the single weight configuration. This yields a general representation and allows a straightforward application of both crossover and mutation operators [90, 164, 178]. A more compact representation could be obtained by using real numbers which directly correspond to the weight values [63, 142, 153]. This approach can produce a more compact chromosome representation. Some difficulties can arise with the application of traditional mutation and crossover operators. For this reason, different strategies have been employed in this situation. These include evolutionary strategies implementing only mutation [154, 187].

The adoption of EA's for weight adaptation may be shown to be useful in enhancing the accuracy of neural networks. However, the most promising applications appear to be the development of cooperative training processes. In fact, the ability of EA's to find global basin of attraction may be coupled with some techniques of local search. The employment of evolutionary techniques to find near-optimal initial condition, followed by back-propagation local search has been successfully applied [107, 189].

2.5.2 Adaptation of the Architectures

In a similar way to the search for connection weights, the design of a network architecture can be regarded as a peculiar search process. In principle it can be done by means of evolutionary methods. The problem here can be embedded into the chromosome representation. The *direct encoding* approach aims at encoding all the parameters useful for realising a correspondence between the network topology and the employed data structure. This can be represented by a matrix recording the connections between the nodes [114, 116]. This kind of approach is likely to increase the dimension of the coding structure and therefore some *indirect encoding* strategies have been proposed. These make use only of a set of the most important factors needed to define the network architecture. Other details may be either disregarded, predefined or specified by means of deterministic rules [139, 174, 188].

In order to avoid the noisy fitness evaluation problem mentioned before, architecture and weight connections have sometimes evolved together. A fully specified neural network may be codified in a one-to-one mapping [48, 144, 182, 183, 185].

2.5.3 Adaptation of the Learning Rules

The ultimate application of EA's to neural network optimisation relies on the adaptation possibilities of the learning rules. The selection of an appropriate training algorithm, together with the weight initialisation and the topology design, plays a fundamental role in determining the behaviour of a connectionist system. It can be argued that the principle of applying the same learning rules for every task is impracticable. For these reasons, the help supplied by an evolutionary search in dynamically determining a proper learning strategy in different contexts would be valuable. The process of encoding a learning rule represents a challenging task. Some proposals can be found in literature, and these range from the plain evolution of the algorithmic parameters of back-propagation methods [97], to the effective evolution of learning rules, which have been assumed to be particular functions. These have been evolved in terms of their coefficients [10, 32].

2.5.4 Different Hybridisation Strategies

The hybridisation of evolutionary algorithms and neural networks can be done in ways which are quite different to those described in the previous sections. Instead of considering the support provided by EA's for designing a connectionist system in terms of weight configurations, topology and learning rules, a preprocessing activity can be used which is suitable for efficient neural training. Evolutionary methods are employed to develop an input feature selection process [64, 106].

The integration of the paradigms can also be considered in the use of an ANN to support evolving procedures. Some have suggested the employment of neural networks for controlling the parameters involved in fitness evaluation [37].

A fruitful field of research is represented by the use of ensembles of neural networks. An ensemble of ANN's is regarded as a population which may offer better generalisation opportunities when considered as an integrated system [113, 184].

2.6 Evolutionary-Neuro-Fuzzy Systems

An evolutionary neuro-fuzzy system (ENFS) is the result of adding evolutionary search procedures to systems, integrating fuzzy logic computing and neural learning. Using these three Soft Computing paradigms together can overcome some limitations of simpler forms of hybridisation.

One of the main problems with neuro-fuzzy systems is that the learning algorithm is typically based on a steepest descent optimisation technique minimising an error function. That is, back-propagation training is not guaranteed to converge. The algorithm may be trapped in a local minimum. It can never find the global solution. The tuning of the membership function parameters through neural learning is also not guaranteed. As previously stated (see section 2.5.1), experimental evidence has shown cases where evolutionary algorithms are inefficient for fine tuning solutions. They are better at finding global solutions. A hybrid learning scheme combining neural learning and evolutionary strategies may be able to solve this problem. A common approach involves using a genetic algorithm to rapidly locate a good region in the solution space and to initialise membership function parameters. The parameters are then fine tuned by applying a gradient descent learning algorithm that performs a local search in the good region to find a near optimal solution [194].

A problem of (neuro-)fuzzy modelling is the difficulty of determining the proper number of rules and the number of membership functions for each rule (see section 2.4). Evolutionary approaches can be integrated into neuro-fuzzy modelling to overcome this difficulty and to perform structure and parameter optimisation of the fuzzy rule base.

Several methodologies have been proposed in the literature to develop a form of evolutionary-neural-fuzzy hybridisation. We now review some of the existing approaches, grouping methodologies to reach the goal of optimisation. Some of them aim to learn parameters of a fuzzy rule base, whose structure is assumed to be fixed in advance. Other methods are developed to perform both structure and parameter learning of a fuzzy rule base.

2.6.1 Parameter Learning

A common approach given in the literature to assess an ENFS is by adding evolutionary learning capabilities to a neuro-fuzzy network. This network is usually a feed-forward multilayered network, which incorporates some fuzzy concepts. The result is a feed-forward multilayered network having both fuzzy and genetic characteristics.

In simplest approaches, a GA is used to learn or tune all the parameters of a neuro-fuzzy network which is assumed to have a fixed structure [112, 127, 145, 157, 177]. In these cases chromosomes of the GA encode all the parameters of the neuro-fuzzy model, including parameters of membership functions as defined in the antecedent and consequent of each fuzzy rule. In [49] the parameter identification of a fuzzy model is achieved in three separate learning phases. The first phase uses the Kohonen's self organising feature map algorithm to find the initial parameters of the membership functions. A maximum matching-factor algorithm is applied, in the second phase, to find the correct consequence part of each rule. After the fuzzy rules have been found and the whole network structure is established, the third learning phase fine-tunes the membership function parameters using a multiresolutional dynamic genetic algorithm (MRD-GA) that dynamically adapts the fuzzy-model configuration during the optimisation process.

2.6.2 Structure and Parameter Learning

Genetic algorithms may also be applied for structure optimisation of a fuzzy rule base. That is, to define the proper number of rules and the number of antecedents and consequents in each rule, as in [54].

Evolutionary strategies can be conveniently applied to simultaneously learning the parameters and the structure of a fuzzy model. Several approaches are given in the literature that involve the membership function parameters together with the fuzzy rule set, including the number of rules. In some cases the antecedent and consequent part of each rule are also included. As an example, the strategy proposed in [4] optimises the whole fuzzy system, represented as a neural network, in order to derive a minimum rule number fitting the given specifications, while training the network parameters. In [55] a GA-based method is used for a rough search for proper structures in the antecedent of fuzzy rules. The fine tuning of the parameters of the fuzzy model is done successively using neural learning. A complex hybrid genetic-neuro-fuzzy scheme is implemented in FuGeNeSys [150] that can learn a fuzzy rule base from the data. This uses the membership function parameters, the number of rules and the structure of each rule. These are simultaneously defined via a genetic algorithm that incorporates neural learning in the evolution procedure. Each individual in the population is made up of a set of rules, where each rule has an adjustable number of antecedents and consequents. Whenever a better individual is generated, it is transformed in a neuro-fuzzy network and trained with a back-propagation procedure until the error is reduced. Finally, the trained neuro-fuzzy network is retransformed in a genetic individual and used to replace the original individual.

In other approaches, such as the EvoNF (Evolving Neuro Fuzzy) model proposed in [1], all the adjustable parts of a fuzzy system, such as the membership function parameters, structure of the rule base are involved. This includes items such as the number of rules, representation of antecedents and consequents. The fuzzy operators are derived through an evolutionary strategy. Coding all such informations into a chromosome leads to individuals characterised by a complex layered structure. The optimisation strategy may become computationally expensive. Neural learning can be used to further tune the membership functions located by the evolutionary algorithm.

Finally, genetic algorithms can be used to evolve fuzzy neural networks. That is, neural networks that incorporate fuzzy numbers as weights, perform fuzzy operations in the nodes, and/or consider fuzzy nodes to represent membership functions. In this case, the learning process uses GA's to obtain the weights of the neural network, to adapt the transfer functions of the nodes, and/or to adapt the topology of the network, as in [105].

2.6.3 Applications

Hybrid systems using a combination of neural, fuzzy and evolutionary computing have been successfully employed in many applications, such as control [162], manufacturing [121], financial prediction [190], pharmacology [151], consumer products [161], telecommunications [13], modelling and decision making.

Some of these ENFS's application samples are reported in the literature. Our account does not claim to represent a complete overview.

In [161], the authors report an application of evolutionary computation in combination with neural networks and fuzzy systems for intelligent consumer products. The role of the evolutionary algorithm is to adapt the number of rules and to tune the membership functions to improve the performance of particular fuzzy systems. They are involved in predicting the number of dishes to be cleaned by a dish washer, estimating the amount of rice in a rice cooker and controlling a microwave oven. The paper also mentions evolutionary computation for fuzzy rule generation when applied to process control.

In [190] an evolutionary fuzzy neural network is proposed for financial prediction with hybrid input data sets from different financial domains.

In [3] a genetic-neuro-fuzzy system is applied to on-line recognition of on-line Arabic cursive handwriting. In this system, a genetic algorithm is used to select the best combination of characters recognised by a fuzzy neural network. The value returned from the fitness function for each gene represents the degree of match between the word represented by that gene and the real handwritten word. The calculation of the fitness value is based on the fuzzy values assigned to the recognition of each character by the fuzzy neural network.

In [121] the authors present a neuro-genetic-fuzzy system for computerised colour prediction, which is a challenging problem in paint production. A fuzzy knowledge base for predicting the pigment concentration of ten different colours for a given surface spectral reflectance is obtained by means of a neuro-fuzzy system. The fuzzy population generator uses this knowledge to seed the first generation of colour chromosomes. In addition, expert knowledge of the colour technician about the correct proportions of colours, the number of necessary colours and conflicts between complementary and similar colours is summarised in a fuzzy rule base that is introduced in the first generation of colour chromosomes. The GA calculates one component of colour chromosome fitness according to the compliance of the chromosome's colour with the fuzzy expert rules.

2.7 A Genetic-Based Approach to Optimise Neuro-Fuzzy Models

Research on NFS's and EFS's has for long been considered the objective of the learning process in terms of accuracy. Consequently, the error measure as minimised by neural learning, or the fitness function as maximised by the genetic algorithm was stated in terms of errors or distances from the target output. This causes a lack of interpretability and transparency in the resulting fuzzy rules. This is due to generation of meaningless and overlapping fuzzy sets.

Indeed, the most important motivation to use a fuzzy system is that the produced model is characterised by a clear reliance on linguistic terms that allows for easy understanding by humans. Hence, methods for training fuzzy models from data should not only find the best approximation of the data, but also and more importantly, they should extract knowledge from the data in the form of fuzzy rules that can be readily understood and interpreted.

Recently, the concepts of linguistic fuzzy modelling, interpretability, and similar ideas were considered to have qualities almost the opposite to accuracy. These have now been reconsidered [25] and today are viewed as an interesting part of the design process of a FS using neural learning and genetic algorithms. Several works have suggested both accuracy and interpretability as objectives in genetic-based learning systems [83,85,91]. This sets a situation for the learning process where several differing objectives have to be simultaneously considered. Some of the measures used to determine the level of interpretability of a fuzzy system are degree of compactness [83]. This may be regarded as the number of rules in the rule base, and rule simplicity [31]. This simplicity of rules may be evaluated through the number of input variables involved in each rule.

In this section we present an approach to identify accurate and interpretable fuzzy models based on a combination of neural learning and genetic algorithms [30]. The approach is aimed to optimise both the structure and the parameters of the fuzzy model extracted from data by a neural learning scheme through a multi-objective genetic algorithm. An iterative process is performed in each step using two optimisation procedures. A structure optimisation procedure that reduces the number of rules, and a multi-objective genetic algorithm that tunes membership function parameters. This is done by enforcing constraints on the fuzzy sets to ensure they result well-formed, leading to fuzzy rules that can be easily understood.

Returning to the fuzzy rule expressed in (2.3):

$$\begin{gathered}\text{IF } x_1 \text{ is } A_1^k \text{ AND} \ldots \text{AND } x_n \text{ is } A_n^k \\ \text{THEN } y_1 \text{ is } b_1^k \text{ AND} \ldots \text{AND } y_m \text{ is } b_m^k,\end{gathered}$$

we may suppose that the fuzzy model, generated by the neural learning algorithm, makes use of fuzzy singletons $b_j^k (j = 1, \ldots, m)$. These are defined over the output variables y_j and the fuzzy sets A_i^k are represented by Gaussian membership functions $\mu_{ik}(x_i) = \exp\left(-\frac{(x_i - c_{ik})^2}{2a_{ik}^2}\right)$, where c_{ik} and a_{ik} are the centre and the width of the Gaussian function, respectively.

Based on a set of K rules, the output of the fuzzy inference system for a given input $\mathbf{x}(0)$ is obtained as follows:

$$\hat{y}_j(0) = \frac{\sum_{k=1}^{K} \mu_k(\mathbf{x}(0)) b_j^k}{\sum_{k=1}^{K} \mu_k(\mathbf{x}(0))} \quad j = 1, \ldots, m. \tag{2.4}$$

Here $\mu_k(\mathbf{x}(0)) = \prod_{i=1}^{n} \mu_{ik}(x_i(0))$, where $k = 1, \ldots, K$ is the degree of fulfilment for the k-th rule, for $k = 1, \ldots, K$.

To better formalise the scheme of the GA-based approach, we denote by $FRB(\mathbf{w}, K)$ a Fuzzy Rule Base with parameter vector $\mathbf{w}$ and structure size K. That is, with a number of K fuzzy rules. We indicate by $FRB(\mathbf{w}_0, K_0)$ the rule base of the fuzzy model initially derived by neural learning. Firstly, the structure optimisation procedure is applied to $FRB(\mathbf{w}_0, K_0)$ to remove iteratively rules, until the difference between the accuracy of the reduced model and the accuracy of the initial model drops below a threshold ϵ. The result is a new fuzzy rule base $FRB(\mathbf{w}_S, K_S)$, with $0 \leq S \leq (K_0 - 1)$. Here $K_S \leq K_0$ and $ERR(\mathbf{w}_S) - ERR(\mathbf{w}_0) \leq \epsilon$, where $ERR(\cdot)$ is an error function used to evaluate the accuracy (generalisation ability) of the fuzzy model. Such a fuzzy

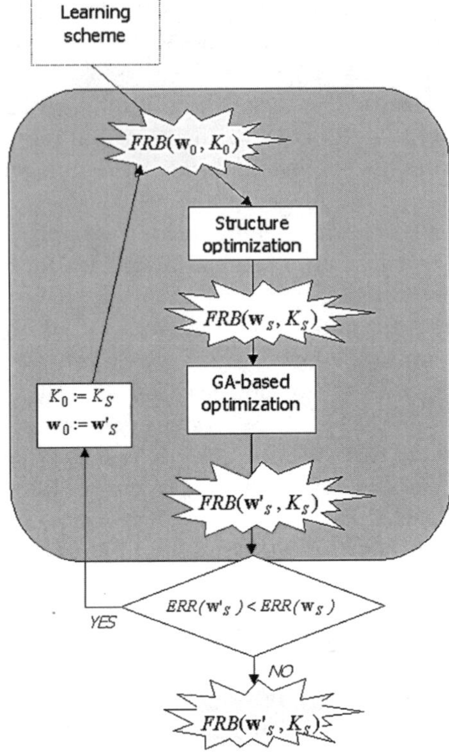

Fig. 2.5. Scheme of the GA-based optimisation approach

rule base is taken as starting point for the GA-based optimisation procedure. The result of this parameter optimisation is a rule base $FRB(\mathbf{w}'_S, K_S)$ that satisfies the following conditions:

$$ERR(\mathbf{w}'_S) \leq ERR(\mathbf{w}_S), \qquad INT(\mathbf{w}'_S) \geq INT(\mathbf{w}_S),$$

where $INT(\cdot)$ is a qualitative measure of the interpretability of the antecedent fuzzy sets, that will be detailed in Section 2.7.2. If the parameter optimisation process is able to improve the accuracy, that is, if $ERR(\mathbf{w}'_S) < ERR(\mathbf{w}_S)$, then the structure-parameter optimisation cycle can be reiterated, using the fuzzy rule base $FRB(\mathbf{w}'_S, K_S)$ as starting point. Otherwise, that is, if the local parameter tuning remains unchanged the accuracy, the whole optimisation algorithm is stopped. Providing the rule base $FRB(\mathbf{w}'_S, K_S)$ is the final rule base. The whole iterative algorithm ends when no further improvement in the accuracy and interpretability of the fuzzy rules can be observed. That is, when the best possible compromise between accuracy and interpretability of the fuzzy model is achieved. Figure 2.5 shows the flow chart of the whole optimisation approach.

We now describe both of the optimisation procedures.

2.7.1 Structure Optimisation

The structure optimisation procedure is referred to in [27,28] and is a systematic process that sequentially removes rules and investigates the model with fewer rules. At each step, rules to be removed are identified and the remaining rules are updated so the accuracy of the reduced model remains unchanged. This is achieved by updating only the consequent parameters, while premise parameters are left unchanged.

The removal of a rule involves the elimination of the fuzzy sets in its antecedent. Thus, after the structure simplification process, the number of fuzzy sets is also reduced, so as to produce a simple (hence easily interpretable) rule base. There is no guarantee that the remaining fuzzy sets still cover the entire input space. The use of Gaussian membership functions assures enough coverage and provides an acceptable accuracy when few rules are removed. As soon as many rules are removed, the coverage is no longer guaranteed. The remaining fuzzy sets may reveal a high degree of overlapping. That is, they may lack in distinguishability.

To deal with these problems, a multi-objective genetic algorithm is used to tune the premise parameters of reduced fuzzy rules. This is done to improve interpretability in the sense of readable fuzzy sets with no loss in accuracy of the model. In the following section we describe in detail the multi-objective genetic algorithm.

2.7.2 Multi-Objective GA for Parameter Optimisation

The parameters of membership functions in the antecedents of each fuzzy rule are encoded into an individual (chromosome). Since we have nK premise membership functions, each function being represented by two real values (i.e. c_{ik} and a_{ik}), the length of each individual is $2nK$. To limit the length of individuals, we adopt a coding with real-valued genes in the chromosome which directly represent the premise parameters of a fuzzy rule base. Thus, the i-th chromosome is a string of the form:

$$\mathbf{s}_i = (\underbrace{c_{11}^{(i)}, a_{11}^{(i)}, \cdots, c_{n1}^{(i)}, a_{n1}^{(i)}}_{premise\ of\ rule\ 1}, \cdots, \underbrace{c_{1K}^{(i)}, a_{1K}^{(i)}, \cdots, c_{nK}^{(i)}, a_{nK}^{(i)}}_{premise\ of\ rule\ K}).$$

The first individual of the initial population is generated as a copy of the premise parameters of the fuzzy rules produced by the structure optimisation procedure. The remaining individuals are initialised with random values constrained within certain permissible ranges, that is:

$$c_{ik} = rand \cdot range\{x_i(t)\} + \min_{t=1,\dots,N}\{x_i(t)\}$$

and

$$a_{ik} = rand \cdot \frac{2}{3}\frac{range\{x_i(t)\}}{K_S} + a_{i,min},$$

where *rand* is a random number in $[0,1]$, $range\{x_i\}$ is defined as:

$$range\{x_i\} = \left[\max_{t=1,\ldots,N}\{x_i(t)\} - \min_{t=1,\ldots,N}\{x_i(t)\}\right]$$

and $a_{i,min}$ is the minimum width among all the widths in the first individual. By doing so, the centres of membership functions are constrained within the corresponding input range, while the widths are greater than zero and not wider than the whole input range.

The reproduction probability of an individual is defined from the values of the fitness function according to: $p_R(\mathbf{s}_i) = \frac{fit(\mathbf{s}_i)}{\sum_{j=1}^{N_I} fit(\mathbf{s}_j)}$. To speed up the convergence, the best individual in the population is always selected and kept unchanged in the next generations, according to the *elitist strategy* [59].

The simplest form of crossover, which is the single point crossover, is adopted. Individuals are mutated by adding a random value generated within the permissible ranges, and defined by:

$$\epsilon = N(\alpha \cdot range\{x_i\}, (\beta \cdot range\{x_i\})^2),$$

where α and β assume different values depending whether the gene is the centre or the width of a Gaussian function. For each input variable x_i, genes representing the centres of extreme membership functions are fixed to the extremes of $range\{x_i\}$ and are never changed by mutation. A readable fuzzy partition for x_i requires the leftmost (or the rightmost) membership function to attain its maximum value at the extreme point defined by the minimum (or the maximum) of $range\{x_i\}$. Dynamic crossover and mutation probability rates are used, since they provide faster convergence [160]. The crossover probability is set high at the beginning of the evolution and decreases exponentially with the increasing number of generations. The mutation probability increases exponentially with each succeeding generation.

The expression for the fitness function was derived to model the two objectives of the optimisation process. These are the interpretability and the accuracy of the fuzzy model. To define the fitness term with respect to interpretability, the concept of interpretability is translated into the concepts of completeness and distinguishability of the fuzzy sets, that are expressed through a fuzzy similarity measure. If the similarity of two neighbouring fuzzy sets is zero or too small, this indicates that either the fuzzy partitioning is incomplete or they do not overlap enough. If the similarity value is too large, then it means that the two fuzzy sets overlap too much and the distinguishability between them is lost. The following measure [158] has been used to define the similarity between the two fuzzy sets A and B:

$$S(A,B) = \frac{M(A \bigcap B)}{M(A) + M(B) - M(A \bigcap B)}, \tag{2.5}$$

where $M(A) = \int_{+\infty}^{-\infty} \mu_A(x)dx$ is the size of fuzzy set A. This is computed as the area of the triangle approximating the Gaussian function $\mu_A(\cdot)$. To keep the fuzzy sets in a proper shape, the fuzzy similarity measure of any two neighbouring membership functions must satisfy the following condition:

$$S_{low} \leq S(A_l, A_{l+1}) \leq S_{up}, \tag{2.6}$$

where A_l and A_{l+1} are the two neighbouring fuzzy sets, S_{low} and S_{up} are the desired lower and upper bound of the fuzzy similarity measure, respectively. For each input variable x_i, the quantity:

$$\beta_{il} = \begin{cases} 1 & if\ S_{low} \leq S(A_{il}, A_{i,l+1}) \leq S_{up} \\ 0 & otherwise \end{cases} \tag{2.7}$$

is defined.

We can derive the following term which must be inserted into the fitness function of an individual:

$$fit_{int}(\mathbf{s}) = \frac{\beta}{n(K_S - 1)}, \tag{2.8}$$

where K_S is the number of fuzzy sets defined on each input variable. This is equal to the number of rule units in the network and β is defined as:

$$\beta = \sum_{i=1}^{n} \sum_{l=1}^{K_S - 1} \beta_{il}. \tag{2.9}$$

The value $fit_{int}(\mathbf{s})$ is maximum (that is, equal to 1) when all adjacent fuzzy sets coded into the individual $\mathbf{s}$ satisfy the condition (2.6). Conversely, if this condition is not satisfied by most of adjacent fuzzy sets, the corresponding fitness term will be assigned a very low value which means that the corresponding individual is unlikely to survive.

To preserve and, hopefully, improve the accuracy of the fuzzy model, the modelling error $ERR(\mathbf{w}_i)$ should be minimised. The value $\mathbf{w}_i$ is obtained by joining the premise parameters coded into the individual $\mathbf{s}_i$ and the vector of consequent parameters. This vector is taken from the rule base $FRB(\mathbf{w}_S, K_S)$ resulting from the structure optimisation procedure, where $\mathbf{w}_i = [\mathbf{s}_i, \mathbf{b}_S]$. The corresponding fitness term is defined as:

$$fit_{acc}(\mathbf{s}_i) = \exp\left(-\lambda ERR(\mathbf{w}_i)^2\right), \tag{2.10}$$

where λ is a fixed parameter with a large value. A high value of $fit_{acc}(\mathbf{s}_i)$ corresponds to a very low error, or high accuracy, of the fuzzy model coded into the individual $\mathbf{s}_i$.

To summarise, the analytical expression of the fitness function for an individual $\mathbf{s}_i$ is given by:

$$fit(\mathbf{s}_i) = fit_{acc}(\mathbf{s}_i) + \gamma fit_{int}(\mathbf{s}_i), \tag{2.11}$$

where γ is a factor that controls the influence of the term $fit_{int}(\mathbf{s}_i)$ during the whole GA evolution. It is made less relevant during the first generations, and has more and more influence as the evolution proceeds.

2.7.3 Illustrative Example

In this section, we will examine the application of this approach when used to derive fuzzy models for a problem of medical diagnosis. In particular, the Heart Disease data set from the University of California, Irvine, is considered. The database contains 270 samples. Each sample is represented by 13 attributes. There are two classes: presence or absence of heart-disease.

By using the neuro-fuzzy learning method proposed in [29] and performing a 10-fold cross validation strategy, we derived 10 different fuzzy models with a number of rules varying between 10 and 16. The application of the GA-based approach gave the results summarised in Table 2.1. It can be seen that a great improvement can be achieved in terms of complexity reduction and accuracy on the test set. The final number of rules was between 2 and 4. Almost in all trials, the structure-parameter optimisation required two iterations to improve the neuro-fuzzy model. As an illustrative example of such a behaviour, Table 2.2 gives the detailed results obtained for the model no. 6 where, starting from a 15-rule approximate model with low classification rate

Table 2.1. Results obtained after the optimisation of neuro-fuzzy models with 10-fold cross validation

model	before optimisation		after optimisation	
	number of rules	class. rate	number of rules	class. rate
1	16	59.25	4	70.37
2	12	70.37	3	70.37
3	13	70.37	2	81.48
4	10	74.07	2	74.07
5	11	74.07	3	70.37
6	15	70.37	3	74.07
7	12	55.56	2	55.56
8	10	74.07	4	88.89
9	13	74.07	2	85.18
10	11	59.26	4	88.89
average	12.3	68.15	2.9	75.92

Table 2.2. Results obtained at each stage of the GA-based optimisation approach in the case of the model no. 6

model	number of rules	class. rate	
		train. set	test set
initial model	15	70.37	70.37
	iteration no. 1		
structure opt.	10	69.96	66.67
GA-based opt.	10	66.67	74.07
	iteration no. 2		
structure opt.	3	66.67	70.37
GA-based opt.	**3**	**65.43**	**74.07**

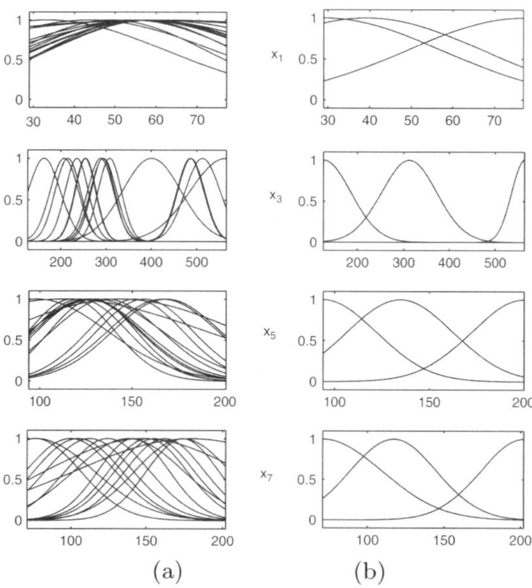

Fig. 2.6. Fuzzy sets of some input variables before (a) and (b) after the application of the GA-based optimisation process

on the test set, a final fuzzy model with only 3 rules and improved classification rate was obtained. Also, the interpretability of the final model is clearly improved, as shown by the well-formed final fuzzy sets in fig. 2.6.

2.8 Conclusions

In this chapter we have discussed the integration possibilities among the three paradigms that normally constitute the key components of the Soft Computing field of research. These are artificial neural networks, fuzzy logic and evolutionary algorithms. We reviewed the hybridisation mechanisms at the basis of the coupling approaches. The intent was to address the issues related to the intrinsic benefits and difficulties connected with their implementation. We considered the more complex integration of the three paradigms, which gives rise to the development of Evolutionary Neuro-Fuzzy Systems. We proposed a technique concerning design techniques used for the realisation of EFNF's, together with their application feasibility.

In addition, we have presented a particular approach of evolutionary-neuro-fuzzy integration, devoted to data-driven learning of fuzzy inference models. Particular attention has been paid in organising a hybrid system which could prove to be effective in constructing an accurate and comprehensible fuzzy rule base.

We hope that this work could trigger the increase in the research efforts devoted to progress the Soft Computing field of investigation. To further motivate such an endeavour, we offered an elaborated series of bibliographical references which could serve as a directory for the inclined reader to be oriented among both classical and novel literature contributes.

References

1. Abraham A (2002) EvoNF: A framework for optimization of fuzzy inference systems using neural network learning and evolutionary computation. In: Proc. 2002 IEEE International Symposium on Intelligent Control (ISIC'02), Canada, IEEE Press, 327–332
2. Alba E, Cotta C, Troya JM (1999) Evolutionary design of fuzzy logic controllers using strongly-typed GP. Mathware Soft Comput. 6(1):109–124
3. Alimi AM (1997) An evolutionary neuro-fuzzy approach to recognize on-line Arabic handwriting. In: Proceedings of the Fourth International Conference on Document Analysis and Recognition, 382–386
4. Alpaydin G, Dandar G, Balkir S (2002) Evolution-based design of neural fuzzy networks using self-adapting genetic parameters. IEEE Trans. Fuzzy Systems 10(2):211–221
5. Angeline PJ, Kinnear KE (eds) (1996) Advances in Genetic Programming II. MIT Press, Cambridge, MA
6. Angelov PP (2002) Evolving Rule-Based Models. A Tool for Design of Flexible Adaptive Systems. Physica-Verlag, Wurzburg
7. Arbib MA (ed) (2002) The Handbook of Brain Theory and Neural Networks (2nd Edition). MIT Press, Cambridge, MA
8. Baron L, Achiche S, Balazinski M (2001) Fuzzy decision support system knowledge base generation using a genetic algorithm. Int. J. Approximate Reasoning 28(1):125–148
9. Bastian A (2000) Identifying fuzzy models utilizing genetic programming. Fuzzy Sets and Systems 113:333–350
10. Bengio S, Bengio Y, Cloutier J, Gecsei J (1992) On the optimization of a synaptic learning rule. In: Preprints Conf. Optimality in Artificial and Biological Neural Networks
11. Benitez JM, Castro JL, Requena I (1997) Are artificial neural networks black boxes?. IEEE Trans. Neural Networks 8:1156–1164
12. Berenji HR, Khedkar P (1992) Learning and tuning fuzzy logic controllers through reinforcements. IEEE Trans. Neural Networks 3:724–740
13. Beritelli F, Casale S, Russo M (1995) Robust phase reversal tone detection using soft computing. In: Proc. of ISUMA - NAFIPS '95 The Third International Symposium on Uncertainty Modeling and Analysis and Annual Conference of the North American Fuzzy Information Processing Society, 589–594
14. Bezdek JC (1981) Pattern Recognition with Fuzzy Objective Function Algorithms. Plenum Press, New York
15. Bishop CM (1995) Neural Networks for Pattern Recognition. Oxford University Press
16. Bonarini A (1996) Evolutionary learning of fuzzy rules: competition and cooperation. In: [133]

17. Bonissone PP, Khedkar PS, Chen Y (1996) Genetic algorithms for automated tuning of fuzzy controllers: a transportation application. In: Proc. Fifth IEEE Int. Conf. on Fuzzy Systems (FUZZ-IEEE96), pp. 674–680
18. Buckles BP, Petry FE, Prabhu D, George R, Srikanth R (1994) Fuzzy clustering with genetic search. In: Proc. 1st IEEE Conf. on Evolutionary Computation (ICEC94), pp. 46–50
19. Buckley JJ, Feuring T (1999) Fuzzy and Neural: Interactions and Applications. Studies in Fuzziness and Soft Computing. Physica-Verlag, Heidelberg, Germany
20. Buckley JJ, Hayashi Y (1993) Numerical relationship between neural networks, continuous functions and fuzzy systems. Fuzzy Sets Syst. 60(1):1–8
21. Buckley JJ, Hayashi Y, Czogala E (1993) On the equivalence of neural nets and fuzzy expert systems. Fuzzy Sets Syst. 53(2):129–134
22. (2003) Advanced Fuzzy Systems Design and Applications. Studies in Fuzziness and Soft Computing (112), Physica-Verlag
23. Carse B, Fogarty TC, Munro A (1996) Evolving fuzzy rule based controllers using genetic algorithms. Fuzzy Sets and Systems 80:273–294
24. Casillas J, Cordon O, del Jesus MJ, Herrera F (2001) Genetic feature selection in a fuzzy rule-based classification system learning process for high dimensional problems. Inform. Sci. 136:169–191
25. Casillas J, Cordon O, Herrera F, Magdalena L (2003) Interpretability Issues in Fuzzy Modeling, Springer
26. Castellano G, Castiello C, Fanelli AM, Mencar C (2005) Knowledge Discovery by a Neuro-Fuzzy Modeling Framework. Fuzzy Sets and Systems 149:187–207
27. Castellano G, Fanelli AM (1996) Simplifying a neuro-fuzzy model. Neural Proc. Letters 4:75–81
28. Castellano G, Fanelli AM (1997) An approach to structure identification of fuzzy models. In: Proc. of the Sixth IEEE International Conference on Fuzzy Systems (FUZZ-IEEE 1997) 531–536
29. Castellano G, Fanelli AM (2000) Fuzzy inference and rule extraction using a neural network. Neural Network World Journal 3:361–371
30. Castellano G, Fanelli AM, Gentile E, Roselli T (2002) A GA-based approach to optimize fuzzy models learned from data. Proc. of Workshop on Approximation and Learning in Evolutionary Computing, part of Genetic and Evolutionary Computation Conference (GECCO 2002), New York, USA, 5–8
31. Castillo L, Gonzalez A, Perez R (2001) Including a simplicity criterion in the selection of the best rule in a fuzzy genetic learning algorithm. Fuzzy Sets and Systems 120(2):309–321
32. Chalmers DJ (1990) The evolution of learning: An experiment in genetic connectionism. In: Proc. 1990 Connectionist Models Summer School, pp. 81–90
33. Cheong F, Lai R (2000) Constraining the optimization of a fuzzy logic controller using an enhanced genetic algorithm. IEEE Trans. Systems Man Cybernet. B 30:31–46
34. Chen J, Xi Y (1998) Nonlinear system modeling by competitive learning and adaptive fuzzy inference system. IEEE Trans. Syst., Man, Cybern. 28:231–238
35. Chien B-C, Lin JY, Hong T-P (2002) Learning discriminant functions with fuzzy attributes for classification using genetic programming. Expert Systems Appl. 23(1):31–37

36. Cho KB, Wang BH (1996) Radial basis function based adaptive fuzzy systems and their applications to system identification and prediction. Fuzzy Sets Syst. 83:325339
37. Chou L-D, Wu J-LC (1995) Parameter adjustment using neural-network-based genetic algorithms for guaranteed QOS in atm networks. IEICE Trans. Commun. E78-B(4):572–579
38. Chung I-F, Lin CJ, Lin CT (2000) A GA-based fuzzy adaptive learning control network. Fuzzy Sets and Systems 112(1):65–84
39. Cordon O, del Jesus MJ, Herrera F, Lozano M (1999) MOGUL: a methodology to obtain genetic fuzzy rule-based systems under the iterative rule learning approach. Int. J. Intelligent Systems 14(11):1123–1153
40. Cordon O, Herrera F (1997) A three-stage evolutionary process for learning descriptive and approximate fuzzy logic controller knowledge bases from examples. Int. J. Approximate Reasoning 17(4):369–407
41. Cordon O, Herrera F (2001) Hybridizing genetic algorithms with sharing scheme and evolution strategies for designing approximate fuzzy rule-based systems, Fuzzy Sets and Systems 118(2):235–255
42. Cordon O, Herrera F, Hoffmann F, Magdalena L (2001) Genetic Fuzzy Systems - Evolutionary Tuning and Learning of Fuzzy Knowledge Bases. World Scientific, Singapore
43. Czogala E, Leski J (2000) Fuzzy and Neuro-Fuzzy Intelligent Systems. Studies in Fuzziness and Soft Computing (47), Physica-Verlag
44. de Oliveira JV (1999) Towards Neuro-Linguistic Modeling: Constraints for Optimization of Membership Functions. Fuzzy Sets and Systems 106:357–380
45. Delgado MR, Von Zuben F, Gomide F (2004) Coevolutionary genetic fuzzy systems: a hierarchical collaborative approach. Fuzzy Sets and Systems 141:89–106
46. Dote Y, Ovaska SJ (2001) Industrial applications of soft computing: a review. Proc. IEEE 89(9):1243–1265
47. Eiben AE, Back T, Schoenauer M, Schwefel HP (eds) (1998) Parallel Problem Solving from Nature (PPSN) V. Lecture Notes in Computer Science, vol. 1498 Springer-Verlag, Berlin, Germany
48. Fang J, Xi Y (1997) Neural network design based on evolutionary programming. Artificial Intell. Eng. 11(2):155–161
49. Farag WA, Quintana VH, Lambert-Torres G (1998) A Genetic-Based Neuro-Fuzzy Approach for Modeling and Control of Dynamical Systems. IEEE Trans. on Neural Networks 9(5):756–767
50. Feng JC, Teng LC (1998) An Online Self Constructing Neural Fuzzy Inference Network and its Applications. IEEE Trans. on Fuzzy Systems 6(1):12–32
51. Fogel D (1995) Phenotypes, genotypes, and operators in evolutionary computation. In Proc. 1995 IEEE Int. Conf. Evolutionary Computation (ICEC95), pp. 193–198
52. Fogel D (1999) Evolutionary Computation: Towards a New Philosophy of Machine Intelligence (2nd edition). IEEE Press
53. Fogel LJ, Owens AJ, Walsh MJ (1966) Artificial Intelligence Through Simulated Evolution. John Wiley & Sons, Chichester, UK
54. Fukuda T, Ishigami H, Shibata T, Arai F (1993) Structure Optimization of Fuzzy Neural Network by Genetic Algorithm, Fifth JFSA World Congress, 964–967

55. Furuhashi T, Matsushita S, Tsutsui H (1997) Evolutionary Fuzzy Modeling Using Fuzzy Neural Networks and Genetic Algorithm, IEEE International Conference on Evolutionary Computation, 623–627
56. Furuhashi T (2001) Fusion of fuzzy/neuro/evolutionary computing for knowledge acquisition, Proceedings of the IEEE 89(9):1266–1274
57. Geyer-Schulz A (1995) Fuzzy Rule-Based Expert Systems and Genetic Machine Learning. Physica-Verlag, Heidelberg
58. Glorennec PY (1997) Coordination between autonomous robots. Int. J. Approximate Reasoning 17(4):433–446
59. Goldberg DE (1989) Genetic Algorithms in Search, Optimization and Machine Learning. Addison Wesley, Reading, MA
60. Gomez-Skarmeta AF, Jimenez F (1999) Fuzzy modeling with hybrid systems, Fuzzy Sets and Systems 104:199–208
61. Gonzalez A, Perez R (1999) SLAVE: a genetic learning system based on an iterative approach, IEEE Trans. Fuzzy Systems 7(2):176–191
62. Gonzalez A, Perez R (2001) An experimental study about the search mechanism in SLAVE learning algorithm: hill-climbing methods versus genetic algorithm. Inform. Sci. 136(14):159–174
63. Greenwood GW (1997) Training partially recurrent neural networks using evolutionary strategies. IEEE Trans. Speech Audio Processing 5:192–194
64. Guo Z, Uhrig RE (1992) Using genetic algorithms to select inputs for neural networks. In: Proc. Int. Workshop Combinations of Genetic Algorithms and Neural Networks (COGANN-92), pp. 223–234
65. Gurocak HB (1999) A genetic-algorithm-based method for tuning fuzzy logic controllers. Fuzzy Sets and Systems 108(1):39–47
66. Hall LO, Bezdek JC, Boggavarapu S, Bensaid A (1994) Genetic fuzzy clustering. In: Proc. NAFIPS94, pp. 411–415
67. Hancock PJB (1992) Genetic algorithms and permutation problems: A comparison of recombination operators for neural net structure specification. In Proc. Int. Workshop Combinations of Genetic Algorithms and Neural Networks (COGANN-92), pp. 108–122
68. Hanebeck UD, Schmidt GK (1996) Genetic optimization of fuzzy networks. Fuzzy Sets and Systems 79 (1):59–68
69. Hayashi Y (1994) Neural expert system using fuzzy teaching input and its application to medical diagnosis. Inform. Sci. Applicat. 1:47–58
70. Hayashi Y, Buckley JJ (1994) Approximations between fuzzy expert systems and neural networks. Int. J. Approx. Reas. 10:63–73
71. Haykin S (1999) Neural Networks - A Comprehensive Foundation (2nd edition). Prentice Hall
72. Herrera F, Lozano M (1996) Adaptation of genetic algorithm parameters based on fuzzy logic controllers. In: [74]
73. Herrera F, Lozano M, Verdegay JL (1995) Tuning fuzzy controllers by genetic algorithms. Int. J. Approx. Reasoning 12:299–315
74. Herrera F, Verdegay JL (eds) (1996) Genetic Algorithms and Soft Computing. Physica-Verlag, Wurzburg
75. Hertz J, Krogh A, Palmer RG (1991) Introduction to the theory of neural computation. Addison Wesley, Reading, MA
76. Hirota K, Pedrycz W (1994) OR/AND neuron in modeling fuzzy set connectives. IEEE Trans. on Fuzzy Systems 2:151–161

77. Hoffmann F, Nelles O (2001) Genetic programming for model selection of TSK-fuzzy systems. Inform. Sci. 136(14):7–28
78. Hoffmann F, Pfister G (1997) Evolutionary design of a fuzzy knowledge base for a mobile robot. Int. J. Approximate Reasoning 17(4):447–469
79. Holland J (1975) Adaptation in Natural and Artificial Systems. University of Michigan Press, Ann Arbor
80. Holland J, Reitman JS (1978) Cognitive systems based on adaptive algorithms. In: [176]
81. Homaifar A, McCormick E (1995) Simultaneous Design of Membership Functions and Rule Sets for Fuzzy Controllers using Genetic Algorithms. IEEE Trans. on Fuzzy Systems 3(2):129–138
82. Hudson DL, Cohen ME, Anderson MF (1991) Use of neural network techniques in a medical expert system. Int. J. Intell. Syst. 6:213–223
83. Ishibuchi H, Murata T, Turksen IB (1997) Single-objective and two-objective genetic algorithms for selecting linguistic rules for pattern classification problems. Fuzzy Sets and Systems 89:135–150
84. Ishibuchi H, Nakashima T, Murata T (1999) Performance evaluation of fuzzy classifier systems for multidimensional pattern classification problems. IEEE Trans. System Man Cybernet. 29:601–618
85. Ishibuchi H, Nakashima T, Murata T (2001) Three-objective genetics-based machine learning for linguistic rule extraction. Inform. Sci. 136(14):109–133
86. Ishibuchi H, Nozaki K, Yamamoto N, Tanaka H (1995) Selecting Fuzzy If-Then Rules for Classification Problems Using Genetic Algorithms. IEEE Trans. on Fuzzy Systems 3(3):260–270
87. Jang JR (1993) ANFIS: Adaptive-network-based fuzzy inference system. IEEE Trans. Syst., Man, Cybern. 23(3):665–685
88. Jang JSR, Sun CT (1993) Functional equivalence between radial basis function networks and fuzzy inference systems. IEEE Trans. Neural Networks 4:156–159
89. Jang JSR, Sun CT, Mizutani E (1997) NeuroFuzzy and Soft Computing. Prentice-Hall, Englewood Cliffs, NJ
90. Janson DJ, Frenzel JF (1993) Training product unit neural networks with genetic algorithms. IEEE Expert, 8:26–33
91. Jin Y (2000) Fuzzy modeling of high-dimensional systems: complexity reduction and interpretability improvement. IEEE Trans. Fuzzy Systems 8(2):212–220
92. Juang C, Lin C (1998) An on-line self-constructing neural fuzzy inference network and its applications. IEEE Trans. Fuzzy Syst. 6:12–32
93. Kang SJ, Woo CH, Hwang HS, Woo KB (2000) Evolutionary Design of Fuzzy Rule Base for Nonlinear System Modeling and Control. IEEE Trans. on Fuzzy Systems 8(1):37–45
94. Kasabov N (1996) Foundations of Neural Networks, Fuzzy Systems and Knowledge Engineering. MIT Press, Cambridge, MA
95. Keller JK, Hunt DJ (1985) Incorporating fuzzy membership functions into the perceptron algorithm. IEEE Trans. Pattern Anal. Mach. Intell. 7:693–699
96. Keller JM, Yager RR, Tahani H (1992) Neural network implementation of fuzzy logic. Fuzzy Sets Syst. 45:1–12
97. Kim HB, Jung SH, Kim TG, Park KH (1996) Fast learning method for back-propagation neural network by evolutionary adaptation of learning rates. Neurocomput. 11(1):101–106

98. Kim J, Kasabov N (1999) HyFIS: Adaptive Neuro-Fuzzy Inference Systems and Their Application to Nonlinear Dynamical Systems. Neural Networks 12:1301–1319
99. Kinnear KE (ed) (1994) Advances in Genetic Programming. MIT Press, Cambridge, MA
100. Kosko B (1991) Neural Networks and Fuzzy Systems. Prentice-Hall, Englewood Cliffs, NJ
101. Koza JR (1992) Genetic Programming: On the Programming of Computers by Means of Natural Selection. MIT Press, Cambridge, MA
102. Koza JR (1994) Genetic Programming II: Automatic Discovery of Reusable Programs. MIT Press, Cambridge, MA
103. Koza JR, Bennett FH III, Andre D, Keane MA (1999) Genetic Programming III: Darwinian Invention and Problem Solving. Morgan Kaufmann
104. Koza JR, Keane MA, Streeter MJ, Mydlowec W, Yu J, Lanza G (2003) Genetic Programming IV: Routine Human-Competitive Machine Intelligence. Kluwer Academic Publishers
105. Krishnamraju PV, Buckley JJ, Reilly KD, Hayashi Y (1994) Genetic learning algorithms for fuzzy neural nets. In Proc. Third IEEE Int. Conf. on Fuzzy Systems (FUZZ-IEEE94), Orlando, FL, USA 1969–1974
106. Kupinski MA, Maryellen ML (1997) Feature selection and classifiers for the computerized detection of mass lesions in digital mammography. In: Proc. 1997 IEEE Int. Conf. Neural Networks., pp. 2460–2463
107. Lee S-W (1996) Off-line recognition of totally unconstrained handwritten numerals using multilayer cluster neural network. IEEE Trans. Pattern Anal. Machine Intell. 18:648–652
108. Lee H-M, Chen C-M, Chen J-M, Jou Y-L (2001) An efficient fuzzy classifier with feature selection based on fuzzy entropy. IEEE Trans. Systems Man Cybernet. B 31(3):426–432
109. Leitch DD (1995) A New Genetic Algorithm for the Evolution of Fuzzy Systems. PhD thesis, Department of Engineering, Oxford University
110. Lim MH, Rahardja S, Gwee BH (1996) A GA Paradigm for Learning Fuzzy Rules. Fuzzy Sets and Systems 82:177–186
111. Lin CT, George Lee CS (1996) Neural Fuzzy Systems A NeuroFuzzy Synergism to Intelligent Systems. Prentice-Hall, Englewood Cliffs, NJ
112. Linkens DA, Nyongesa HO (1995) Evolutionary learning in fuzzy neural control systems. In Proc. Third European Congress on Fuzzy and Intelligent Technologies (EUFIT95), Aachen, Germany, 990–995
113. Liu Y, Yao X (1998) Toward designing neural network ensembles by evolution. In: [47]
114. Marin FJ, Sandoval F (1993) Genetic synthesis of discrete-time recurrent neural network. In: Proc. Int. Workshop Artificial Neural Networks (IWANN93), pp. 179–184
115. Michalewicz Z (1992) Genetic Algorithms + Data Structure = Evolution Programs. Springer-Verlag
116. Miller GF, Todd PM, Hegde SU (1989) Designing neural networks using genetic algorithms. In: Proc. 3rd Int. Conf. Genetic Algorithms and Their Applications, pp. 379–384
117. Mitra S (1994) Fuzzy MLP based expert system for medical diagnosis. Fuzzy Sets Syst. 65:285–296

118. Mitra S, Hayashi Y (2000) Neuro-Fuzzy Rule Generation: Survey in Soft Computing Framework. IEEE Trans. Neural Networks 11(3):748–768
119. Mitra S, Pal SK (1995) Fuzzy multilayer perceptron, inferencing and rule generation. IEEE Trans. Neural Networks 6:51–63
120. Mitra S, Pal SK (1996) Fuzzy self organization, inferencing and rule generation. IEEE Trans. Syst., Man, Cybern. 26:608–620
121. Mizutani E, Takagi H, Auslander DM, Jang J-SR (2000) Evolving colour recipes. IEEE Trans. Systems Man Cybernet. 30(4):537–550
122. Nauck D, Klawonn F, Kruse R (1997) Foundations of NeuroFuzzy Systems. Wiley, Chichester, U.K.
123. Nauck D, Kruse R (1992) A Neuro-Fuzzy Controller Learning by Fuzzy Error Propagation. In: Proc. of Conf. of the North Amer. Fuzzy Inf. Proc. Soc. (NAFIPS92), pp. 388–397
124. Nauck D, Kruse R (1997) New Learning Strategies for NEFCLASS. In: Proc. of Seventh Intern. Fuzzy Systems Ass. World Congress (IFSA97), pp. 50–55
125. Nauck D, Kruse R (1999) Neuro-Fuzzy Systems for function approximation. Fuzzy Sets and Systems 101:261–271
126. Nauck D, Nauck U, Kruse R (1996) Generating Classification Rules with Neuro-Fuzzy System NEFCLASS. In: Proc. of the Biennal Conf. of the North Amer. Fuzzy Inf. Proc. Soc. (NAFIPS96)
127. Paetz J (1995) Evolutionary Optimization of Weights of a Neuro-Fuzzy Classifier and the Effects on Benchmark Data and Complex Chemical Data. In Proc. of NAFIPS 2005 - Annual Meeting of the North American Fuzzy Information Processing Society, 615–620
128. Pal SK, Ghosh A (1996) Neuro-fuzzy computing for image processing and pattern recognition. Int. J. Syst. Sci. 27:1179–1193
129. Pal SK, Mitra S (1992) Multilayer perceptron, fuzzy sets and classification. IEEE Trans. Neural Networks 3:683–697
130. Pal SK, Mitra S (1999) Neuro-fuzzy Pattern Recognition: Methods in Soft Computing. Wiley, New York
131. Parodi A, Bonelli P (1993) A new approach to fuzzy classifier systems. In: Proc. of Fifth International Conference on Genetic Algorithms (ICGA93), Morgan Kaufmann, pp. 223–230
132. Pedrycz W (1993) Fuzzy neural networks and neurocomputations. Fuzzy Sets and Systems 56:1–28
133. Pedrycz W (ed) (1996) Fuzzy Modelling: Paradigms and Practice. Kluwer Academic Press, Norwell, MA
134. Pedrycz W (ed) (1997) Fuzzy Evolutionary Computation. Kluwer Academic Publishers, Dordrecht
135. Pedrycz W (199) Computational Intelligence: An Introduction. CRC, Boca Raton, FL
136. Pedrycz W, Rocha A (1993) Knowledge-based neural networks, IEEE Trans. on Fuzzy Systems 1:254–266
137. Pedrycz W, Reformat M (1996) Genetic Optimization with Fuzzy Coding. In: [74]
138. Pena-Reyes CA, Sipper M (2001) Fuzzy CoCo: a cooperative coevolutionary approach to fuzzy modeling. IEEE Trans. Fuzzy Systems 9(5):727–737
139. Perez CA, Holzmann CA (1997) Improvements on handwritten digit recognition by genetic selection of neural network topology and by augmented training. In: Proc. 1997 IEEE Int. Conf. Systems, Man, and Cybernetics, pp. 1487–1491

140. Perneel C, Themlin JM, Renders JM, Acheroy M (1995) Optimization of Fuzzy Expert Systems Using Genetic Algorithms and Neural Networks. IEEE Trans. on Fuzzy Systems, 3(3):300–312
141. Pham DT, Karaboga D (1991) Optimum design of fuzzy logic controllers using genetic algorithms. J. System Eng. 1:114–118
142. Porto VW, Fogel DB, Fogel LJ (1995) Alternative neural network training methods. IEEE Expert 10:16–22
143. Potter MA, De Jong KA (2000) Cooperative coevolution: an architecture for evolving coadapted subcomponents. Evolutionary Comput. 8(1):1–29
144. Pujol JCF, Poli R (1998) Evolving the topology and the weights of neural networks using a dual representation. Appl. Intell. 8(1):73–84
145. Rahmoun A, Berrani S (2001) A Genetic-Based Neuro-Fuzzy Generator: NEFGEN, ACS/IEEE International Conference on Computer Systems and Applications, 18–23
146. Rojas I, Gonzalez J, Pomares H, Rojas FJ, Fernandez FJ, Prieto A (2001) Multidimensional and multideme genetic algorithms for the construction of fuzzy systems. Int. J. Approx. Reasoning 26(3):179–210
147. Ross T (1997) Fuzzy Logic with Engineering Applications. McGraw-Hill
148. Roubos H, Setnes M (2001) Compact and transparent fuzzy models through iterative complexity reduction. IEEE Trans. Fuzzy Systems 9(4):515–524
149. Rumelhart DE, McClelland JL (1986) Parallel Distributing Processing. MIT Press, Cambridge, MA
150. Russo M (1998) FuGeNeSys - A Fuzzy Genetic Neural System for Fuzzy Modeling. IEEE Trans. on Fuzzy Systems 6(3):373–388
151. Russo M, Santagati NA, Lo Pinto E (1998) Medicinal chemistry and fuzzy logic. Inform. Sci. 105(14):299–314
152. Sanchez E, Shibata T, Zadeh L (eds) (1997) Genetic Algorithms and Fuzzy Logic Systems. Soft Computing Perspectives. World Scientific, Singapore
153. Saravanan N, Fogel DB (1995) Evolving neural control systems. IEEE Expert 10:23–27
154. Sarkar M, Yegnanarayana B (1997) Evolutionary programming-based probabilistic neural networks construction technique. In: Proc. 1997 IEEE Int. Conf. Neural Networks, pp. 456–461
155. Schwefel HP (1994) On the Evolution of Evolutionary Computation. In: [195]: 116–124
156. Schwefel HP (1995) Evolution and Optimum Seeking. John Wiley & Sons, Chichester, UK
157. Seng TL, Khalid MB, Yusof R (1999) Tuning of a Neuro-Fuzzy Controller by Genetic Algorithm. IEEE Transactions on Systems, Man and Cybernetics-Part B 29(2):226-236
158. Setnes M, Babuska R, Kaymak U, van Nauta Lemke HR (1998) Similarity Measures in Fuzzy Rule Base Simplification. IEEE Trans. on Systems, Man and Cybernetics - Part B 28(3):376–386
159. Setnes M, Roubos H (2000) GA-fuzzy modeling and classification: complexity and performance, IEEE Trans. Fuzzy Systems 8(5):509–522
160. Sheble GB, Maifeld TT (1995) Refined Genetic Algorithm - Economic Dispatch Example. IEEE Trans. Power Syst. 10:117–124
161. Shim M, Seong S, Ko B, So M (1999) Application of evolutionary computation at LG electronics. In Proc. of IEEE Int. Conf. on Fuzzy Systems FUZZ-IEEE99, Seoul, South Korea, 1802–1806

162. Silva N, Macedo H, Rosa A (1998) Evolutionary fuzzy neural networks automatic design of rule based controllers of nonlinear delayed systems. In Proc. of IEEE World Congress on Computational Intelligence - Fuzzy Systems Proceedings 2:1271–1276
163. Smith SF (1980) A learning system based on genetic adaptive algorithms. PhD Thesis, University of Pittsburgh
164. Srinivas M, Patnaik LM (1991) Learning neural network weights using genetic algorithms Improving performance by search-space reduction. In: Proc. 1991 IEEE Int. Joint Conf. Neural Networks (IJCNN91), vol. 3, pp. 2331–2336
165. Subbu R, Anderson A, Bonissone PP (1998) Fuzzy logic controlled genetic algorithms versus tuned genetic algorithms: an agile manufacturing application. In: Proc. IEEE Int. Symp. on Intelligent Control (NIST)
166. Sulzberger SM, Tschicholg-Gurman NN, Vestli SJ (1998) FUN: Optimization of Fuzzy Rule Based Systems Using Neural Networks. In Proc. of IEEE Conf. on Neural Networks, pp. 312–316
167. Sun CT (1994) Rule-Base Structure Identification in an Adaptive-Network-Based Fuzzy Inference System. IEEE Trans. on Fuzzy Systems 2(1): 64–73
168. Takagi H, Suzuki N, Koda T, Kojima Y (1992) Neural networks designed on approximate reasoning architecture and their applications. IEEE Trans. Neural Networks 3:752–760
169. Tano S, Oyama T, Arnauld T (1996) Deep Combination of Fuzzy Inference and Neural Network in Fuzzy Inference. Fuzzy Sets and Systems 82(2):151–160
170. Tschichold-Gurmann N (1995) Generation and Improvement of Fuzzy Classifiers with Incremental Learning Using Fuzzy Rulenet. In: Proc. of ACM Sympos. on Appl. Comp., pp. 466–470
171. Van Le T (1995) Evolutionary fuzzy clustering. In: Proc. 2nd IEEE Conf. on Evolutionary Computation (ICEC95), Vol. 2, pp. 753–758
172. Velasco JR (1998) Genetic-based on-line learning for fuzzy process control. Int. J. Intelligent System 13(1011):891–903
173. Venturini G (1993) SIA: a supervised inductive algorithm with genetic search for learning attribute based concepts. In Proc. European Conf. on Machine Learning, Vienna, pp. 280–296
174. Vonk E, Jain LC, Johnson R (1995) Using genetic algorithms with grammar encoding to generate neural networks. In: Proc. IEEE Int. Conf. Neural Networks, pp. 1928–1931
175. Wang D, Keller JM, Carson CA, McAdoo-Edwards KK, Bailey CW (1998) Use of fuzzy-logic-inspired features to improve bacterial recognition through classifier fusion. IEEE Trans. Syst., Man, Cybern. 28:583–591
176. Waterman DA, Hayes-Roth F (eds) (1978) Pattern-Directed Inference Systems. Academic Press, New York
177. Watts MJ, Kasabov K (1998) Genetic Algorithms for the design of fuzzy neural networks. In: Proc. of the Fifth Int. Conf. on Neural Information Processing (ICONIP'R98), pp. 793–796
178. Whitley D, Starkweather T, Bogart C (1990) Genetic algorithms and neural networks: Optimizing connections and connectivity. Parallel Comput. 14(3) pp. 347–361
179. Wolpert DH, Macready WG (1997) No-Free Lunch Theorems for Optimization. IEEE Trans. Evolutionary Computation 1:67–82
180. Wong CC, Feng SM (1995) Switching-Type Fuzzy Controller Design by Genetic Algorithms. Fuzzy Sets and Systems, 74(2):175–185

181. Yao X (1999) Evolving Artificial Neural Networks. Proc. IEEE 87(9):1423–1447
182. Yao X, Liu Y (1997) A new evolutionary system for evolving artificial neural networks. IEEE Trans. Neural Networks 8:694–713
183. Yao X, Liu Y (1997) EPNet for chaotic time-series prediction. In: [186]
184. Yao X, Liu Y (1998) Making use of population information in evolutionary artificial neural networks. IEEE Trans. Syst., Man, Cyber. B 28:417–425
185. Yao X, Liu Y (1998) Toward designing artificial neural networks by evolution. Appl. Math. Computation 91(1):83–90
186. Yao X, Kim J-H, Furuhashi T (eds) (1997) Select. Papers 1st Asia-Pacific Conf. Simulated Evolution and Learning (SEAL96). Lecture Notes in Artificial Intelligence, vol. 1285, Springer-Verlag, Berlin, Germany
187. Yao X, Liu Y (1997) Fast Evolution Strategies. Control Cybern. 26(3) pp. 467–496
188. Yao X, Shi Y (1995) A preliminary study on designing artificial neural networks using co-evolution. In: Proc. IEEE Singapore Int. Conf. Intelligent Control and Instrumentation, pp. 149–154
189. Yan W, Zhu Z, Hu R (1997) Hybrid genetic/BP algorithm and its application for radar target classification. In: Proc. 1997 IEEE National Aerospace and Electronics Conf. (NAECON), pp. 981–984
190. Yu L, Zhang Y-Q(2005) Evolutionary fuzzy neural networks for hybrid financial prediction. IEEE Transactions on Systems, Man and Cybernetics - Part C 35(2):244249
191. Yuan B, Klir GJ, Swan-Stone JF (1995) Evolutionary fuzzy c-means clustering algorithm. In: Proc. 4th IEEE Int. Conf. on Fuzzy Systems (FUZZ-IEEE95), pp. 2221–2226
192. Zadeh LA (1965) Fuzzy Sets. Information and Control 8:338–353
193. Zadeh LA (1994) Fuzzy logic, neural networks, and soft computing, Commun. ACM 37:77–84
194. Zhang Y.-Q., Kandel A. (1998) Compensatory Genetic Fuzzy Neural Networks and Their Applications. Ser. Machine Perception Artificial Intelligence. Singapore: World Scientific, vol. 30
195. Zurada J, Marks R, Robinson C (eds) (1994) Computational Intelligence: Imitating Life. IEEE Press

3

Evolution of Fuzzy Controllers and Applications

Dilip Kumar Pratihar[1]* and Nirmal Baran Hui[2]

[1] Associate Professor
dkpra@mech.iitkgp.ernet.in
[2] Research Scholar
nirmal@mech.iitkgp.ernet.in
Soft Computing Lab.
Department of Mechanical Engineering
Indian Institute of Technology, Kharagpur
Kharagpur-721 302
India

Summary. The present chapter deals with the issues related to the evolution of optimal fuzzy logic controllers (FLC) by proper tuning of its knowledge base (KB), using different tools, such as least-square techniques, genetic algorithms, back-propagation (steepest descent) algorithm, ant-colony optimization, reinforcement learning, Tabu search, Taguchi method and simulated annealing. The selection of a particular tool for the evolution of the FLC, generally depends on the application. Some of the applications have also been included in this chapter.

Keywords: Fuzzy logic controller, Evolution, Least-square technique, Genetic-fuzzy system, Neural-fuzzy system, Ant-colony optimization, Reinforcement learning, Tabu search, Taguchi method, Simulated annealing

3.1 Introduction

Real-world problems are generally associated with different types of uncertainties. In the past, considerable effort has been made to model these uncertainties. Prior to 1965, it was considered that *probability theory* working based on Aristotelian two-valued logic was the sole agent available to deal with uncertainties. This particular logic uses the concept of the classical crisp set. That is a set with a fixed boundary. Prof. Zadeh developed the concept

* Corresponding author: Associate Professor, Department of Mechanical Engineering, Indian Institute of Technology, Kharagpur-721 302, India; http://www.facweb.iitkgp.ernet.in/~dkpra

D.K. Pratihar and N.B. Hui: *Evolution of Fuzzy Controllers and Applications*, Studies in Computational Intelligence (SCI) **66**, 47–69 (2007)
www.springerlink.com

of fuzzy sets, in the year 1965 [1]. Those are the sets having the vague boundaries. He argued that probability theory can handle only one out of several different types of possible uncertainties. Thus, there are uncertainties, which cannot be dealt with by using the probability theory. Taking an example, in which Mr. X requests Mr. Y, to bring some *red* apples for him from the market. There are two uncertainties at least, which relate to the following: (i) the availability of the apples, and (ii) a guarantee that the apple is red. Depending on the season, there is a probability of obtaining the apples, which varies between 0 and 1. But, the colour – *red* cannot be defined by the classical set. It is not between red (1) and not-red (0). In the fuzzy set, the colour – *red* can be defined as follows (Fig. 3.1) using the concept of membership of an element to a class. That is the function value (μ): If the colour is perfectly red PR, then it may be said red with a membership value of 1.0; if it is R, then it is considered to be red with a membership value of 0.65; if it is slightly red SR, then it is red with a membership value of 0.39. If it is not red (NR), then also it is red with a membership value of 0.0. In this way, the uncertainty related to the colour of the apples can be handled. Thus, a fuzzy set may be considered to be a more general concept than the classical set.

The concept of fuzzy set theory has been used in a number of applications, such as the Fuzzy Logic Controller (FLC), fuzzy clustering, fuzzy mathematical programming, fuzzy graph theory and other examples. Out of all such applications, FLC is the most popular application for the following reasons – (i) ease of understanding and implementations, (ii) ability to handle uncertainty etc. An exact mathematical formulation of the problem is not required for the development of an FLC. This feature makes it a natural choice for solving complex real-world problems. These are either difficult to model mathematically or the mathematical model becomes highly non-linear. It is to be noted that a fuzzy logic controller was first developed by Mamdani and Assilian, in the year 1975 [2]. The concept of fuzzy set was published in the

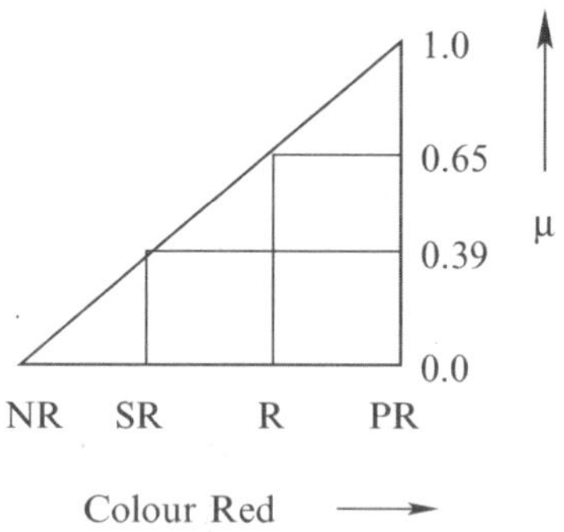

Fig. 3.1. A schematic diagram explaining the concept of membership function distribution.

year 1965. Human beings have the natural ability of determining the input-output relationships of a process. The behavior of a human being is modeled artificially, when designing a suitable FLC. The performance of an FLC depends on its knowledge base (KB), which in turn consists of both Data Base (DB) and a Rule Base (RB). The DB consists of data related to membership function distributions of the variables of the process to be controlled. Designing a proper KB of an FLC is a difficult task, which may be implemented in one of the following ways:

- Optimization of the data base only,
- Optimization of the rule base only,
- Optimization of the data base and rule base in stages,
- Optimization of the data base and rule base simultaneously.

The membership function distributions are assumed to be either Linear such as, triangular, trapezoidal or Non-Linear. The Non-Linear can be Gaussian, bell-shaped, sigmoidal in nature. To design and develop a suitable FLC for controlling a process, its variables need to be expressed in the form of some linguistic terms (such as VN: Very Near, VF: Very Far, A: Ahead for example). The relationships between the input (antecedent) and output (consequent) variables are expressed in the form of rules. For example, a rule can be expressed as indicated in Fig. 3.2:

$$\textbf{IF } I_1 \text{ is } N \textbf{ AND } I_2 \text{ is } A \textbf{ THEN } O \text{ is } AR,$$

The number of such rules will be present in the rule base. The number of linguistic terms used to represent the variables increases in order to improve the accuracy of the prediction. The computational complexity of the controller will increase with a larger number of rules. For easy implementation in either

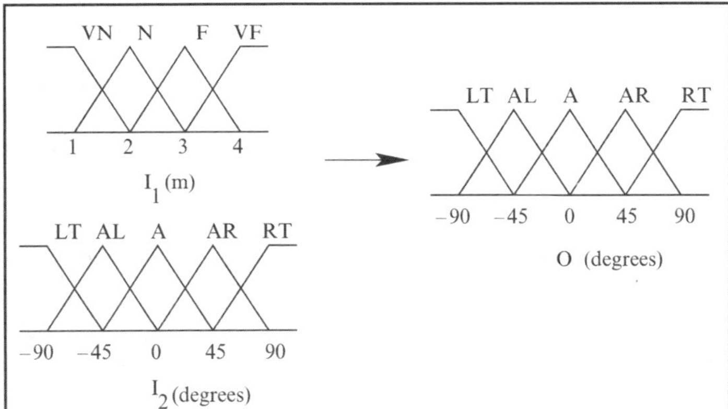

Fig. 3.2. A diagram showing some membership function distributions of input and output variables of the Fuzzy Logic Controller.

the software or the hardware, the number of rules present in the Rule Base should be as small as possible. Consequently, some investigators have tried to design and develop a hierarchical FLC, in which the number of rules will be kept to the minimum [3, 4]. It has been observed that the performance of an FLC largely depends on the rule base and optimizing the data base is a fine tuning process [5]. A fuzzy logic controller does not have an internal optimization module. An external optimizer is used to develop an optimal Knowledge Base through a proper tuning and this helps to improve the performance.

In this chapter, the focus is on the issues related to design and development of an optimal fuzzy logic controller using different optimization tools. Some of the applications of FLC are cited.

The remainder of the text is organized as follows. Two major forms of FLC are discussed in Section 2. Various methods of designing optimal FLCs are given in Section 3. A summary of this work is presented in Section 4.

3.2 Two Major Forms of Fuzzy Logic Controller

System modeling done by using the fuzzy set concept can be classified into two groups. That is *linguistic fuzzy modeling* and *precise fuzzy modeling.* Linguistic fuzzy modeling, such as *Mamdani Approach* is characterized by its high interpretability and low accuracy. The aim of precise fuzzy modeling such as *Takagi and Sugeno's Approach*, is to obtain high accuracy at the cost of interpretability. Interpretability of a fuzzy modeling is defined as a capability to express the behavior of a system in an understandable form. This is expressed in terms of compactness, completeness, consistency and transparency. The accuracy of a fuzzy model indicates how closely it can represent the system modeled. The working principles of both these approaches are briefly explained below.

3.2.1 Mamdani Approach [2]

An FLC consists of four modules namely, a fuzzy rule base, a fuzzy inference engine, fuzzification and de-fuzzification. Fig. 3.3 shows a schematic diagram explaining the working of an FLC.

(a) The condition known as the antecedent and the action called the consequent variables needed to control a process are identified and measurements are taken of all the condition variables.
(b) The measurements taken in the previous step are converted into appropriate fuzzy sets to express measurement uncertainties. This process is known as *fuzzification.*
(c) The fuzzified measurements are then used by the inference engine to evaluate the control rules stored in the fuzzy rule base and a fuzzified output is determined.

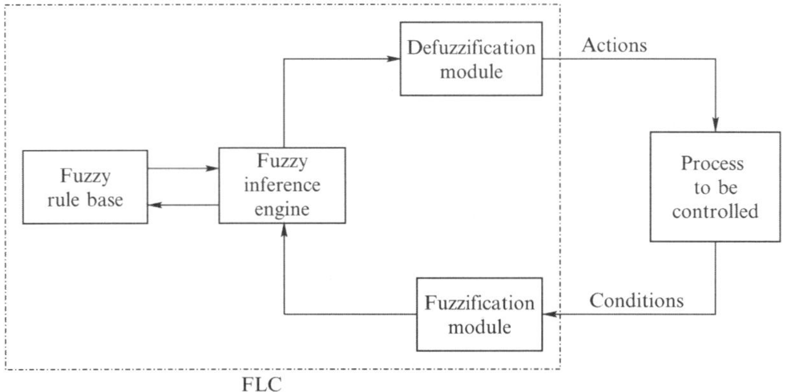

Fig. 3.3. The working cycle of an FLC.

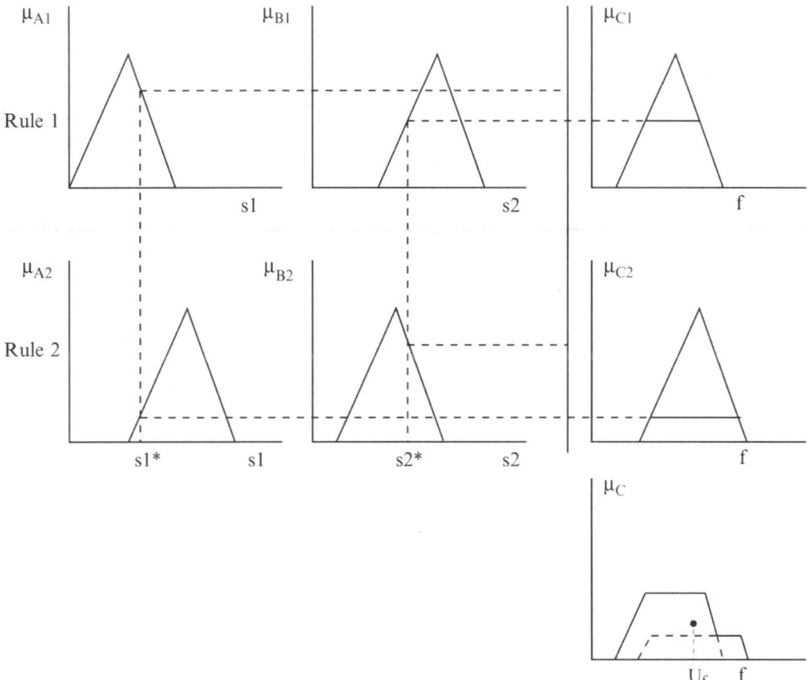

Fig. 3.4. A schematic diagram showing the working principle of an FLC.

(d) The fuzzified output is then converted into a single crisp value. This conversion is called *de-fuzzification*. The de-fuzzified values represent actions which need to be taken by the FLC in controlling the process.

The fuzzy reasoning process is illustrated in Figure 3.4. Let us assume for simplicity that only two fuzzy control rules (out of many rules present in the rule base) are being 'FIRED' as shown below. This is for a set of inputs – ($s1^*$, $s2^*$).

RULE 1: IF $s1$ is $A1$ and $s2$ is $B1$ THEN f is $C1$
RULE 2: IF $s1$ is $A2$ and $s2$ is $B2$ THEN f is $C2$.

If $s1^*$ and $s2^*$ are the inputs for fuzzy variables $s1$ and $s2$. If μ_{A1} and μ_{B1} are the membership function values for A and B, respectively, then the grade of membership of $s1^*$ in $A1$ and the grade of membership of $s2^*$ in $B1$ are represented by $\mu_{A1}(s1^*)$ and $\mu_{B1}(s2^*)$, for rule 1.

Similarly, for rule 2, where $\mu_{A2}(s1^*)$ and $\mu_{B2}(s2^*)$, are used to represent the membership function values. The firing strengths of the first and second rules are calculated as follows:

$$\alpha_1 = \min\left(\mu_{A1}(s1^*), \mu_{B1}(s2^*)\right), \tag{3.1}$$

$$\alpha_2 = \min\left(\mu_{A2}(s1^*), \mu_{B2}(s2^*)\right). \tag{3.2}$$

The membership function of the combined control action C is given by

$$\mu_C(f) = \max\left(\mu^*_{C1}(f), \mu^*_{C2}(f)\right). \tag{3.3}$$

There are several methods of defuzzification (shown in Fig. 3.5). These are explained below.

1. **Center of Sums Method:** According to this method of defuzzification (refer to Fig. 3.5(a)), the crisp output can be determined by the following.

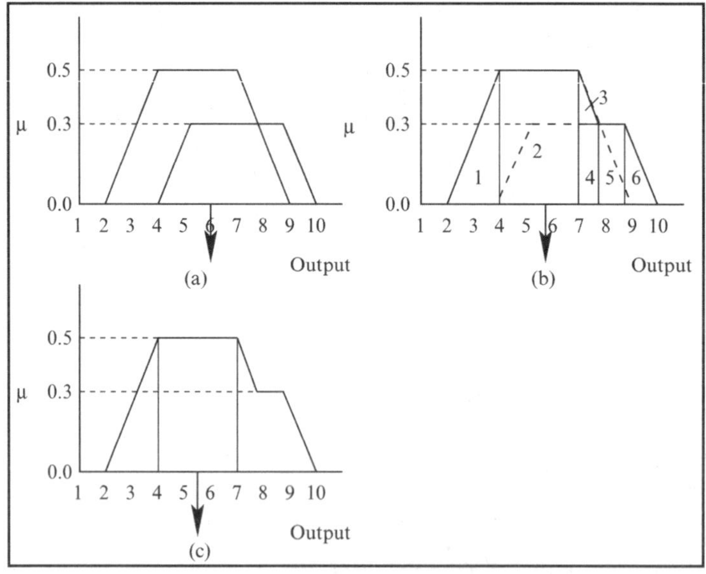

Fig. 3.5. Different methods of defuzzification.

$$U_f = \frac{\sum_{j=1}^{p} A(\alpha_j) \times f_j}{\sum_{j=1}^{p} A(\alpha_j)}, \quad (3.4)$$

Where U_f is the output of the controller. $A(\alpha_j)$ represents the firing area of the j-th rule. p is the total number of the fired rules. f_j represents the centroid of a membership function.

2. **Centroid Method:** The total area of the membership function distribution used to represent the combined control action is divided into a number of standard sub-areas. Their area and the center of area can be determined easily (refer to Fig. 3.5(b)). The crisp output of the controller can be calculated by using the expression given below.

$$U_f = \frac{\sum_{i=1}^{N} A_i f_i}{\sum_{i=1}^{N} A_i}, \quad (3.5)$$

Where N indicates the number of small areas or regions, A_i and f_i represent the area and the center of area of i-th small region.

3. **Mean of Maxima Method:** From the membership function distribution of the combined control action, the range of the output variable is located. This is where the maximum value of the membership function is reached. The mid-value of this range is considered to be the crisp output of the controller (refer to Fig. 3.5(c)).

3.2.2 Takagi and Sugeno's Approach [6]

Here, a rule consists of the fuzzy antecedent and the functional consequent parts. Thus, a rule can be represented as follows:

If x_1 is A_1^i and x_2 is A_2^i ..... and x_n is A_n^i
then $y^i = a_0^i + a_1^i x_1 + \ldots + a_n^i x_n$

where $a_0, a_1, \ldots, a_n$ are the coefficients. In this way, nonlinear system is considered as a combination of several linear systems. Control action of i-th rule can be determined for a set of inputs $(x_1, x_2, \ldots, x_n)$ as follows.

$$w^i = \mu_{A_1}^i(x_1)\mu_{A_2}^i(x_2)\ldots\mu_{A_n}^i(x_n), \quad (3.6)$$

Where $A_1, A_2, \ldots, A_n$ indicate the membership function distributions of the linguistic terms used to represent the input variables. The membership function value is given by μ. Thus, the combined control action can be determined as

$$y = \frac{\sum_{i=1}^{k} w^i y^i}{\sum_{i=1}^{k} w^i}, \quad (3.7)$$

where k is the total number of rules.

3.3 Methods of Designing Optimal Fuzzy Logic Controllers

In order to establish the input-output relationships of a process, a designer tries to design the KB of an FLC manually, based on a knowledge of the process. In most of the cases, it is difficult to gather prior information of a process. The manually-designed KB of the FLC may not be optimal. As an FLC does not have a in-built optimizer, an optimization tool is used, while tuning a KB. Several methods have been developed and some of these are discussed below.

3.3.1 Least-square Method

Attempts were made to determine an appropriate shape of the membership function distributions by using least-square methods. In this connection, see Pham and Valliappan [7], Bustince et al. [8]. The membership function distribution of a fuzzy set was assumed to follow a power function such as $\mu_A(x_i) = ax_i^b$. Here x indicates a variable represented by a fuzzy set A, $i = 1, 2, \ldots, n$, n is the number of training cases, μ_A is the membership function value of the fuzzy set A lying between 0 and 1, a (greater than zero) and b are the constants to be determined by the least-square method. Two equations were solved for this [8]:

$$nlna + \left(\sum_{i=1}^{n} lnx_i\right) b = \sum_{i=1}^{n} ln\mu_A(x_i) \tag{3.8}$$

$$\left(\sum_{i=1}^{n} lnx_i\right) lna + \left(\sum_{i=1}^{n} ln^2 x_i\right) b = \sum_{i=1}^{n} lnx_i ln\mu_A(x_i) \tag{3.9}$$

where $ax_i^b \leq 1$.

3.3.2 Genetic-Fuzzy System

Genetic algorithm (GA) [9] is a population-based search and optimization technique based on the principle of natural selection and mechanics of natural genetics, was used by several researchers, for a genetic-fuzzy system. The performance of a Fuzzy Logic Controller (FLC) is dependent on its KB. Fig. 3.6 shows the schematic diagram of the genetic-fuzzy system. Here, a GA is used to determine optimal KB of the FLC. Thus, the GA improves the performance of the FLC. During optimization of the FLC, the feedback which is a deviation in prediction is calculated. This is based on a set of training cases and it is utilized as the fitness of the GA. A GA is computationally expensive and the tuning is done off-line. Once optimized, the FLC will be able to predict the outputs for a set of inputs, within a reasonable accuracy

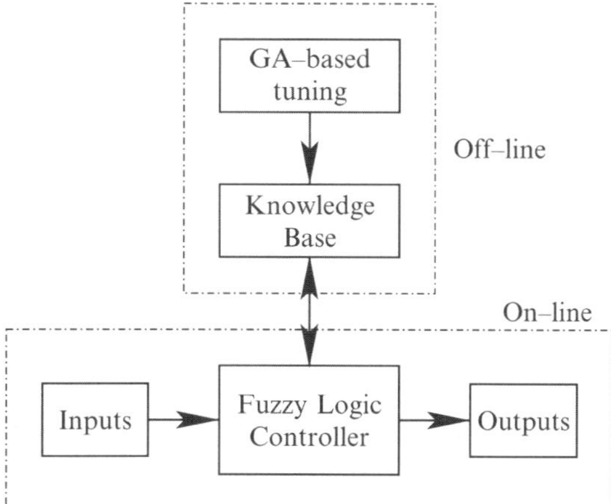

Fig. 3.6. A schematic diagram showing a genetic-fuzzy system

limit. This concept has been used to solve a number of physical problems. See Karr [10], Thrift [11], Pham and Karaboga [14]. A detailed review on this scheme is done by Cordon et al. [12].

There are three basic approaches of this scheme, the Pittsburgh [13, 14], Michigan [15] and iterative rule learning [16, 17] approaches. In *Pittsburgh approach*, the entire rule base of the FLC is represented by a GA-string. Thus, the GA-population indicates the population of candidate rule sets. The genetic operators are used to modify the rule sets and obtain the optimal rule base. In the *Michigan approach*, members of the population are individual rules. Thus, a rule set is represented by the entire population. The main drawback of these two approaches lies in the fact that for the large number of fuzzy rules, the GA requires a huge amount of computer memory. To overcome the problem, using an *iterative rule learning approach*, chromosomes code individual rule, a new rule is added to the rule set, in an iterative fashion, for every run of GA. It requires a proper encoding scheme for extracting the rules from a chromosome. In this approach, the evolved RB of the FLC may contain some redundant rules, due to the iterative nature of the GA.

A considerable amount of work has been carried out in this field of research. Some of these attempts are mentioned below. Furuhashi et al. [18] developed a variable length decoding method, known as the *Nagoya Approach*. Using this approach, as the lengths of the chromosomes are not fixed, it is difficult to implement the necessary crossover operation in GA. Again the simultaneous design of the data base and rule base requires a proper optimization procedure. This can tackle both continuous as well as integer variables. Wang and Yen [19] proposed a method, in which a GA was used to extract the rule base, and the data base of an FLC was optimized using a Kalman filtering technique. Farag et al. [20] developed a new multi-resolutional dynamic GA for this purpose. In this, the initial parameters of the data base of an FLC were determined by using Kohonen's self-organizing feature map algorithm and optimization was done by using a GA. Fuzzy rule generation and tuning using a GA was also tried by Ishibuchi et al. [21]. Recently, Abdessemed et al. [22] proposed a GA-based procedure for designing an FLC, to control the end effector's motion of a planar manipulator. Yupu et al. [23] used a GA to search for appropriate fuzzy rules. The membership function distributions were optimized by using a neural network. The FLC is becoming more popular nowadays, developing a suitable knowledge base for it, is not easy. The designer requires much time, to initially design the knowledge base (KB). It is further improved by using GA-based tuning. Thus, the designer must have a knowledge of the process to be controlled by the FLC. To overcome this requirement, a few investigators [24, 25] tried to automatically design the FLC by using a GA. Using search, the GA will develop the optimized data base and rule base for the FLC.

A GA is basically a fitness function-driven search method, therefore, it is blind for any other aspect that is not explicitly considered on fitness function. Hence, a GA might evolve some redundant rules, that have limited influence on the process to be controlled. Redundant rules are to be removed to make the rule base compact. This makes the implementation of the controller easier, particularly when it is done by hard-ware. Thus, there is a need to determine the contribution of each rule. In this context, the work of Nawa et al. [26], Ishibuchi and Nakashima [27], Ghosh and Nath [28], Hui and Pratihar [35] are important. Nawa et al. [26] measured the quality of a rule by determining its accumulated truth value. The accumulated truth value was considered to be the sum of probability of occurrences of a rule in the training data. A rule is said to be good, if its accumulated truth value is high. Ishibuchi and Nakashima [27] made an attempt to assign an importance factor to each rule. They calculated the importance factor of a rule, by considering the way it interacts with the neighbors. An evolutionary technique was utilized to find the interaction effect. Ghosh and Nath [28] investigated the effectiveness of a rule by measuring three parameters, namely *support count*, *comprehensibility* and *interestingness*. *Support count* of an item set is defined by the number of records in the data base that contains all the items of that set. *Comprehensibility* is used to justify the understandability of a rule. A rule is said to be more comprehensive, if the number of attributes associated with the antecedent part of the rule is less and *interestingness* is represented by the probability of generating a rule during the learning process. It was a theoretical approach of finding interesting rules in the rule base and is unable to predict the importance of a rule for a fixed number of attributes in both antecedent as well as in the consequent parts. The above methods considered the probability of occurrence of a rule only, for the determination of a good rule base. No attention was paid to calculate the contribution effect of a rule with respect to a specific objective. Hui and Pratihar [35] proposed a method of determining *importance factor* for each rule contained in the RB of an FLC, to check the redundancy, if any. The importance factor of a rule is calculated by considering its probability of occurrence and worth (goodness). A rule is said to be redundant and thus may be eliminated, if its importance factor comes out to be smaller than a pre-specified value and the removal of which does not lead to any non-firing situation.

The genetic-fuzzy system has been developed by the authors also, following the two different approaches discussed below.

Approach 1: GA-based tuning of the manually-constructed KB of the FLC. The KB of the FLC is designed manually and is based on the designer's experience of the problem to be solved. But, it may not be optimal in any sense. GA-based tuning is adopted, to further optimize the KB, to improve the performance. As a GA is found to be computationally expensive, the GA-based tuning is carried out, off-line. During optimization, the GA-string will carry information for both the data base as well as the rule base. The GA-search will find the optimal KB of the FLC. Once optimized, the FLC is able to determine its outputs in the optimal sense.

Approach 2: Automatic design of KB using a GA. In Approach 1, much time is spent on manual design of the KB of an FLC. It might be difficult beforehand to foresee the characteristics of the process to be controlled. Thus, designing a proper KB might be a difficult task. To overcome this, a method for automatic design of the KB is developed by using a GA. Here the task of designing a suitable KB is given to the GA. The GA through its exhaustive search will try to determine the optimal KB of the FLC.

The above concept has been used by the authors, to solve a number of physical problems. One of them is explained below.

Optimal Path and Gait Planning of a Six-legged Robot A six-legged robot will have to plan its time-optimal, collision-free path as well as the optimal gait, setting simultaneously the minimum number of ground-legs having the maximum average kinematic margin. This is while moving on a flat terrain with occasional hurdles, such as ditches and some moving obstacles. Its stability margin should always be positive to ensure static stability. This is a complicated task because the path planning and gait planning must be done simultaneously [29]. Fig. 3.7 shows the optimal path and gait for a six-legged robot. It has planned its optimal path and gait, after starting from an initial position S to reach the final position G. It faces three moving obstacles and a ditch on the way to-wards its goal. The total movement of the robot has been achieved through a number of segments called motion segments. The robot plans its optimal path and gait on-line, for each motion segment. The robot shown in Fig. 3.7, is found to reach its goal in the time-optimal sense at 79-th motion segment, after avoiding collision with the moving obstacles and generating its optimal gaits.

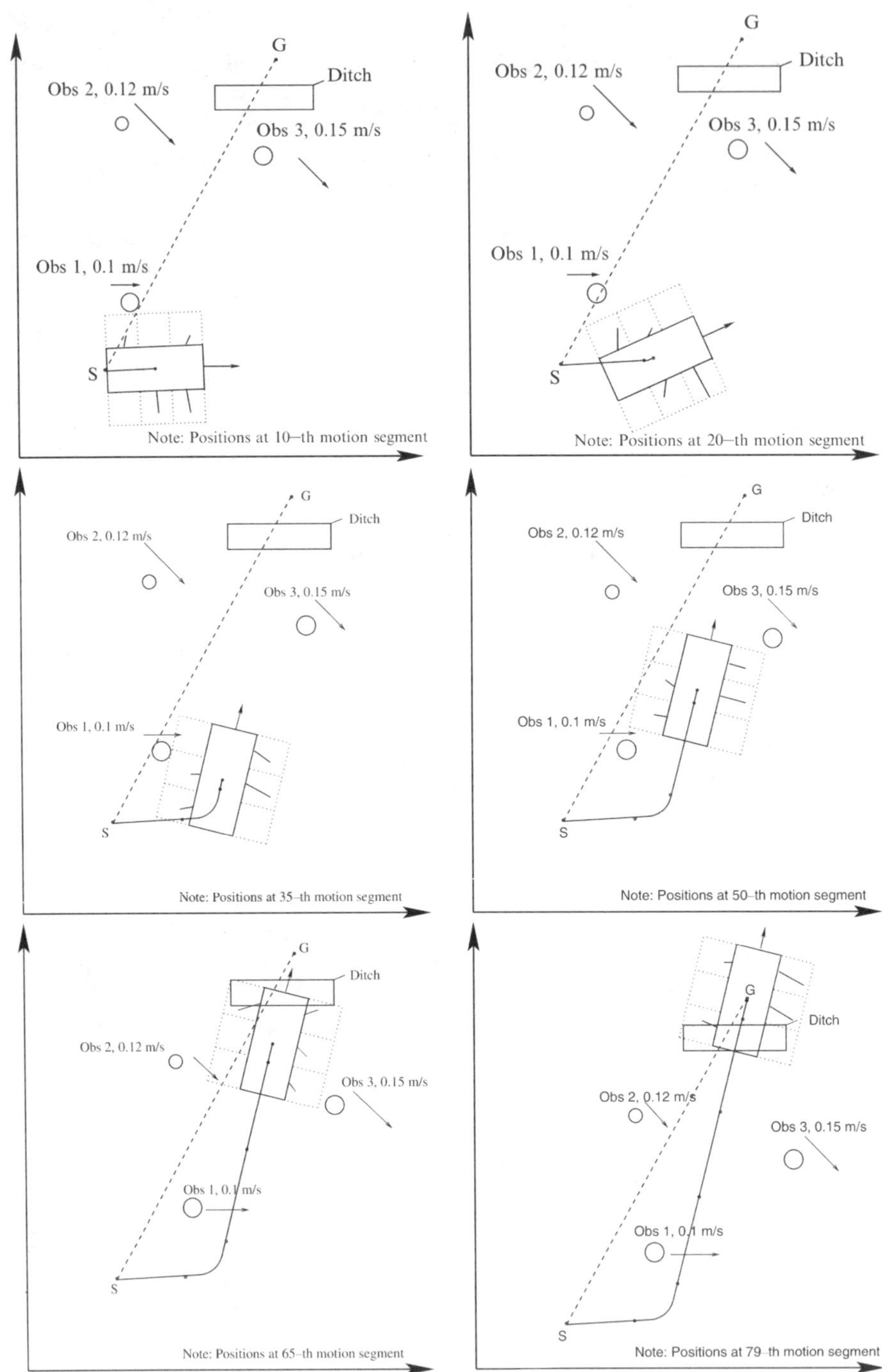

Fig. 3.7. Optimal path and gaits of a six-legged robot obtained using the genetic-fuzzy system [29]

3.3.3 Neural-Fuzzy System

The purpose of developing a neural-fuzzy system is to improve the performance of an FLC by using neural network-based learning. It had been utilized by a number of researchers to solve a variety of problems. Some of these are mentioned below. Marichal et al. [30] proposed a neuro-fuzzy approach to generate the motion of a car-like robot navigating among static obstacles. In their approach, a least mean squared algorithm was used for the learning purposes and Kohonen's self organizing feature map algorithm was considered to obtain the initial number of fuzzy rules and fuzzy membership function centers. They did not optimize the traveling time nor the approach was tested in a dynamic environment. Song and Sheen [31] suggested a pattern recognition approach based on a fuzzy-neuro network for the reactive navigation of a car-like robot. Li et al. [32] developed a neuro-fuzzy architecture for behavior-based control of a car-like robot, that navigates among static obstacles.

The present chapter includes two schemes of neural-fuzzy system developed by the authors. These are discussed below [33].

Scheme 1: Neural-fuzzy system based on Mamdani Approach. In the developed neural-fuzzy system, a fuzzy logic controller using Mamdani Approach is expressed by utilizing the structure of a Neural Network (NN) and a back-propagation algorithm is utilized to optimize the KB of the FLC. The back-propagation algorithm is a steepest descent algorithm. Fig. 3.8 shows the schematic diagram of the five layer neural-fuzzy system– Layer 1 is the input layer, fuzzification is done in Layer 2, Layer 3 indicates the AND operation. The OR operation is carried out in Layer 4, and Layer 5 is the output layer. The training cases are passed through the network and the total error is calculated. The average error is propagated in the backward direction, to determine the updated weights. The network will try to find an optimal set of weights, corresponding to which the error is minimum.

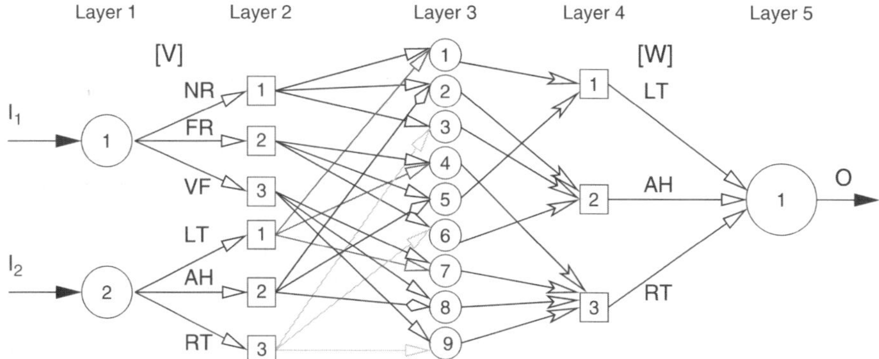

Fig. 3.8. A schematic diagram of the neural network-structured FLC

Three different approaches to Scheme 1 are developed. These are discussed in brief below.

Approach 1: NN-tuned FLC. The initial weights of the neural network representing the FLC are generated, at random. A batch mode of training is adopted. Training cases are passed through the NN (i.e., forward propagation) and average error is determined. As this error depends on the weights, it can be minimized by updating the weight values. A back-propagation algorithm is used to minimize the error.

Approach 2: Genetic-Neural-Fuzzy system. In Approach 1, the error is minimized using a steepest descent method. This may have the local minima problems. To overcome this problem, the back-propagation algorithm is replaced by a GA. As GA is a population-based search and optimization method, the chance of its solutions for getting trapped into the local minima is less. Thus, Approach 2 may be expected to perform better than Approach 1.

Approach 3: Automatic design of neural-fuzzy system. To increase the search space of the GA, a method for automatic design of neural-fuzzy system is proposed. In this approach, the outputs of different rules are evolved solely by the GA itself. The GA through its exhaustive search, determines a good rule base for the FLC. There might be some redundant rules present in the GA-designed rule base. It may happen due to the iterative nature of the GA. To identify the redundant rules, a method is proposed, in which importance of a rule is decided by considering its *frequency of occurrence* and its *worth* with respect to the objective function of the optimization problem. Based on the value of this importance factor, a decision is taken whether a particular rule will be declared as redundant.

Scheme 2: Neural-fuzzy system based on Takagi and Sugeno Approach. A neural-fuzzy system has been developed based on the Takagi and Sugeno Approach. This is known as the ANFIS (i.e., Adaptive Neuro-Fuzzy Inference Systems) [34]. An ANFIS is a multi-layered feed forward network, in which each layer performs a particular task. The layers are characterized by the fuzzy operations they perform. Fig. 3.9 shows the schematic diagram of the ANFIS structure, which consists of six layers – Layer 1 (input layer), Layer 2 (condition layer), Layer 3 (rule base layer), Layer 4 (normalization layer), Layer 5 (consequence layer) and Layer 6 (output layer). Let us assume that there are two inputs – I_1 and I_2 and one output O of the network. The first two layers perform similar tasks to those done by Layers 1 and 2 of the neuro-fuzzy system developed in Scheme 1. The functions of other layers are explained below.

Layer 3: This layer defines the rules of the fuzzy inference system. As three linguistic terms are used to represent each of the two inputs, there is a maximum of $3 \times 3 = 9$ rules present in the rule base. Each neuron in this layer represents a fuzzy rule and is termed as a rule node. The output of each neuron lying in this layer is the multiplication of their

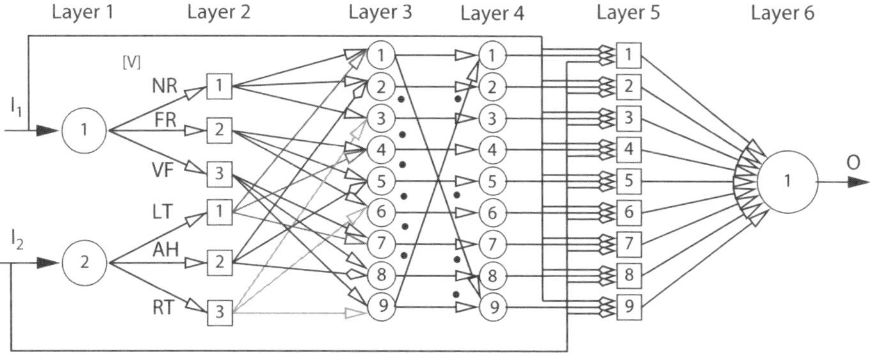

Fig. 3.9. A schematic diagram of the ANFIS architecture

respective two membership values. It is to be noted that each node output represents the firing strength of a rule.

Layer 4: This layer has the same number of nodes as the previous layer. It calculates the normalized firing strength of each node.

Layer 5: The output of a particular neuron (say, the q-th) lying on this layer is determined by

$$O_{5q} = (a_q I_1 + b_q I_2 + c_q) \tag{3.10}$$

where (a_q, b_q, c_q) represents one set of coefficients associated with the q-th node.

Layer 6: The output of the node lying on Layer 6, can be determined by summing up all incoming signals.

$$O_{61} = \sum_{q=1}^{R} O_{5q}, \tag{3.11}$$

where R indicates the total number of rules. A maximum of four rules (out of nine) will be fired, for one set of input variables. The performance of an ANFIS depends on the selection of consequence parameters and premise parameters. That is the half base-widths of the input membership function distributions. For the selection of optimal parameters, a GA might be used together with the ANFIS.

The developed neural-fuzzy systems have been used to plan collision-free, time-optimal paths of a car-like robot. This is explained below.

3.3.3.1 Collision-free, Time-optimal Path Planning for a Car-like Robot [33, 35]

A car-like mobile robot needs to find its time-optimal and collision-free path while navigating among some moving obstacles, and satisfy its kinematic (non-holonomic) constraints and dynamic constraints (such as sliding constraint, motor torque constraint, curvature constraint). A detailed discussion on these constraints is beyond the scope of this chapter. Interested readers may refer to [33], for the same. The total path of the robot is divided into a number of distance steps having varying lengths, each of which is traveled during a time step. To calculate total traveling time of a robot to reach its destination, the time steps are summed and the time required to align its main axis towards the goal is added. There can be a saving in traveling time, particularly if the robot does not change its direction in two successive distance steps. It is subtracted from the total traveling time. The aim is to minimize the traveling time after ensuring a collision-free movement of the robot. A high positive value penalty is added to the total traveling time, if the robot collides with any one of the obstacles. Fig. 3.10 shows the near-optimal, collision-free paths of a robot in the presence of 16 moving obstacles. This is as obtained by using the three approaches of Scheme 1, Scheme 2 (explained above) and a

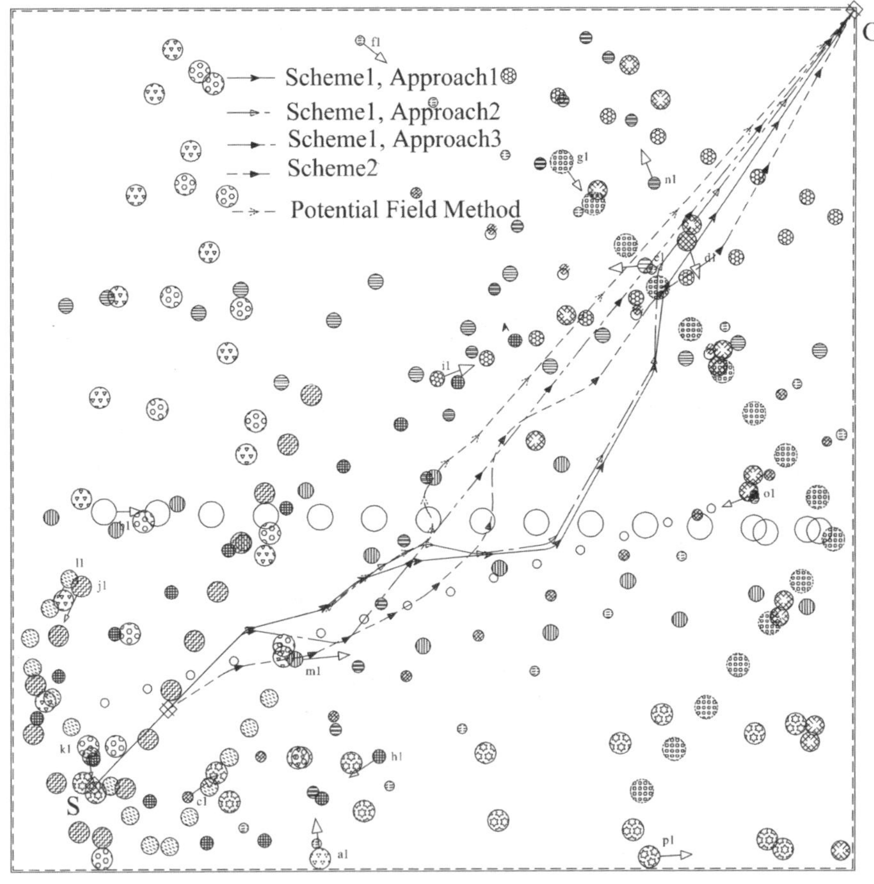

Fig. 3.10. Navigation of a robot among 16 moving obstacles

traditional motion planning scheme (potential field method) [36]. The initial position, size, velocity and direction of movement of the obstacles are created at random. The planning robot starts from the point S and reaches the goal G, by avoiding collisions with the obstacles. Soft computing-based approaches have proved their supremacy over the potential field method. It could be due to the reason that there is a chance that the solutions of the potential field method will get trapped at the local minima. On the other hand, the chance of the solutions of GA-tuned fuzzy logic controller for getting trapped into the local minima is less and it could be due to an exhaustive search carried out by the GA. Moreover, the GA is able to inject adaptability to the FLC, which has been observed from the performances of Approaches 2 and 3 of Scheme 1. Approach 3 of Scheme 1 is found to be the best of all approaches. It could be due to the fact that using this approach, a good KB of the FLC is evolved by the GA, after carrying the search in a wider space.

3.3.4 Optimization of FLC Using Ant Colony Optimization [37]

In Ant Colony Optimization (ACO) algorithm, an optimization problem is represented in the form of a graph – $G = (C, L)$. Here, C is the set of components of the problem and L indicates the possible connection or transition among the elements C. The solutions are expressed in terms of feasible paths on the graph G, after satisfying a set of constraints. Thus, the Fuzzy Rule Learning Problem (FRLP) using the ACO, is formulated as a combinatorial optimization problem. Its operational mode is composed of two stages: in the first stage, the number and antecedents of the linguistic rules are defined, and a set of consequent candidates is assigned to each rule. In the second stage, a combinatorial search is carried out to find the best consequent of each rule, according to a global error measure over the training set.

The fitness of a solution consists of two parts, namely the *functional fitness* and the *objective fitness*. The functional fitness deals with the functionality of the solutions. That is how good is the solution. The objective fitness is the measure of the quality of the solution, in terms of optimization objectives, such as area, delay, gate count, power consumption, and others.

To apply ACO algorithm to a specific problem, the following steps need to be followed:

- Represent the problem in the form of a graph or a similar easily covered structure,
- Define the way of assigning a heuristic preference to each choice that needs to be taken in each step in order to generate the solution,
- Establish an appropriate way of initializing the pheromone,
- Define the fitness function to be optimized,
- Select an ACO algorithm to determine the optimal solutions.

The Fuzzy Rule Learning Problem (FRLP) aims to obtain the rules combining the labels of the antecedents and to assign a specific consequent to each antecedent combination. This problem is interpreted as a way of assigning consequents to the rules with respect to an optimality criterion. An ant iteratively goes over each rule and chooses a consequent with a probability that depends on the pheromone trail τ_{ij} and the heuristic information η_{ij}.

3.3.5 Tuning of FLC Using Reinforcement Learning

Fuzzy rules for control can be effectively tuned by means of reinforcement learning. In this approach, the rules with their associated antecedent and consequent fuzzy sets are represented with the help of a fuzzy-neural network. For this an action selection network (ASN) is used. This network provides continuous action value and records the state of the environment and also determines the next action required. Thereafter, the actions are evaluated by means of a critic element (CE), which is a two-layer feed forward action evaluation network (AEN). It predicts the reinforcements associated with different input states and whether or not a failure has occurred. If a failure occurs, it identifies the steps leading to the failure and modifies the fuzzy sets associated with the rules. A gradient descent technique in conjunction with an average reward is used to train both the action selection network (ASN) and the action evaluation network (AEN) over a set of trials. During training, a reward is provided until a failure occurs and then a high value penalty is given.

This approach had been used by Berenji and Khedkar [38], to solve the problem of a cart-pole balancing system.

3.3.6 Optimization of FLC Using Tabu Search

Denna et al. [39] presented an approach for automatic definition of the fuzzy rules based on the Tabu Search (TS) algorithm. To determine the most appropriate rule base for solving the problem, they employed the reactive form of TS algorithm. To apply the Reactive Tabu Search (RTS) algorithm in determining the rules of a fuzzy controller, the consequent of each rule is expressed with a binary string. The learning procedure is shown in Figure 3.11. The learning begins with an initial rule base, chosen randomly at each iteration. Initial states can also be selected by following a uniform distribution over the entire state space. In such conditions, regions of interest are assigned a higher probability during the learning procedure. Performance of the rule base is then evaluated by using an error function $E(\bullet)$, over a set of typical control rules. It is important to mention that during the evaluation of $E(\bullet)$, some rules, with a smaller contribution to the system are not used. This procedure continues until a termination criterion is reached. The termination condition for each execution may be based on the following parameters:

- The number of iterations carried out,
- The Current State of the error function,
- The properties of the solution found.

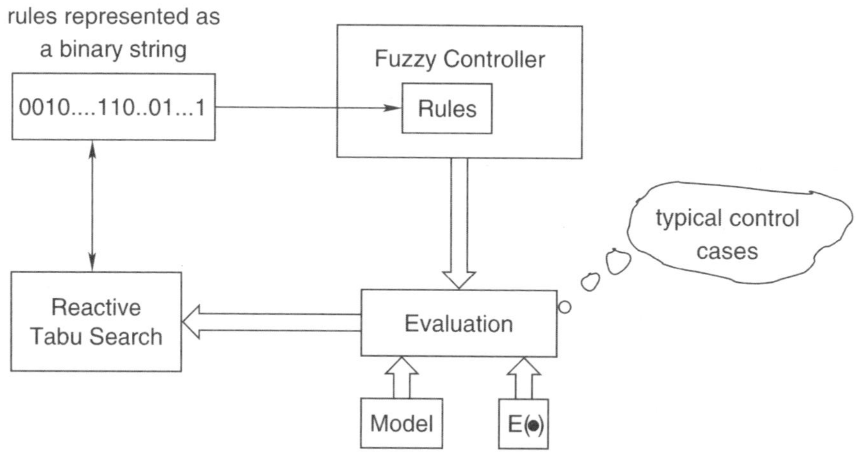

Fig. 3.11. A schematic diagram showing the learning of fuzzy rules using Tabu search [39]

Bagis [40] described a method for the determination of optimum fuzzy membership function distribution used for controlling a reservoir system of dams during floods.

3.3.7 Design of a Fuzzy Controller using the Taguchi Method

The Taguchi Method determines the parameter settings, which maximize the signal to noise (S/N) ratio in each problem by systematically performing the designed experiment. The designed experiment is composed of an inner array and an outer array. The inner array is a designed experiment using the control factors and the outer array consists of the noise factors. To design an FLC using the Taguchi method, control factors are considered as the membership parameters and different system conditions are assumed to be the noise factors. If the inner array is made up of m rows and the outer array contains n rows, then each of the m rows can obtain n performance characteristics. These n data are used to calculate the S/N ratio, for each row of the inner array. The optimal parameter settings are determined by analyzing the S/N ratio data. To check the adequacy of the model, Analysis of Mean (ANOM) and Analysis of Variance (ANOVA) are carried out. Later, a verification experiment is conducted to test the performance of the model. Kim and Rhee [41] utilized the Taguchi method, to design a suitable fuzzy logic controller, in which the following steps were used:

- Identify the performance characteristic to be observed,
- Identify important noise factors and their ranges,
- Identify the control factors and their levels,
- Construct the inner array and the outer array,

- Conduct the designed experiment,
- Analyze the data and determine optimal levels for the control factors,
- Conduct the verification experiment.

3.3.8 Fuzzy Logic Controller Tuned by Simulated Annealing

Simulated Annealing (SA) is one of the most popular non-traditional methods of optimization, in which the cooling process of a molten metal has been artificially modeled. Alfaro and Garcia [42] described a method for development of a fuzzy logic controller applied to path planning and navigation of mobile robots, by using a simulated annealing. Most of the researchers tried to optimize the membership function distributions of the FLC by utilizing the SA. In this approach,the cost function was defined as follows:

$$F = \frac{1}{N}\sum_{k}^{N}\left(y_k - \hat{y_k}\right)^2, \tag{3.12}$$

where $k = 1, 2, \ldots, N$. N is the number of learning samples, (x_k, y_k) is the kth learning sample and $\hat{y_k}$ is the output of the fuzzy system corresponding to the input vector x_k. The optimization algorithm tunes the parameters (spread and shape) of membership function distributions. This is in order to minimize the cost function.

Consider the membership functions of the input variables to be Gaussian in nature, as shown below.

$$Gaussian(x;\sigma;c) = e^{-(\frac{x-c}{\sigma})^2} \tag{3.13}$$

where c and σ indicate the Center and Width, respectively, of the membership function distribution. In SA, the following steps are to be considered in order to optimize the Gaussian membership function distribution:

1. Set an Initial Temperature T to a high value and generate initial parameters c_j^i and σ_j^i, randomly and compute the cost function (F_{old}).
2. Generate a set of new parameters c_j^i and σ_j^i and compute the new cost function (F_{new}). Obtain the change in the cost function $\delta = F_{new} - F_{old}$. If $\delta < 0$, memorize the new set of membership functions and proceed until the termination criterion is reached. Otherwise, go to Step 3.
3. If $\delta > 0$ and probability of accepting the new set of membership functions $P(\delta) = exp(-\delta/T) \leq random[0, 1]$), the center and width values are not changed. Now, go to Step 2 by reducing the temperature T to the half of its previous value.
4. Repeat Steps 2 and 3 until an acceptable solution has been found or until a specified number of iterations has been reached.

3.4 Summary

Fuzzy logic controllers have proved their worth and are popular nowadays to solve real-world complex problems. As the performance of an FLC depends on its KB, several attempts had been made to design a suitable KB. Several methods had been tried by various investigators, to solve the problem. There is a chance of further improvement and much further work is necessary.

Both the Linguistic as well as Precise Fuzzy Modeling have been used separately, to solve a variety of problems and some satisfactory results have been obtained. Linguistic fuzzy modeling ensures better interpretability, but precise fuzzy modeling aims to achieve higher accuracy. It is obvious that as interpretability of the fuzzy model increases, its accuracy will decrease and vice-versa. Thus, depending on the physical problem, a particular type of fuzzy modeling is chosen. It is challenging to obtain a proper balance between interpretability and accuracy of a fuzzy model. These two properties are inversely related and it is important to investigate as to whether a pareto-optimal front exists.

References

1. Zadeh, L.A.: Fuzzy sets. Information and Control **8** (1965) 338–353
2. Mamdani, E.H., Assilian, S.: An experiment in linguistic synthesis with a fuzzy logic controller. International Journal on Man-Machine Studies **7** (1975) 1–13
3. Wang, L.X.: Analysis and design of hierarchical fuzzy systems. IEEE Trans. on Fuzzy Systems **7** 5 (1999) 617–624
4. Lee, M.L., Chung, H.Y., Yu, F.M.: Modeling of hierarchical fuzzy systems. Fuzzy Sets and Systems **138** (2003) 343–361
5. Pratihar, D.K., Deb, K., Ghosh, A.: A genetic-fuzzy approach for mobile robot navigation among moving obstacles. International Journal of Approximate Reasoning **20** (1999) 145–172
6. Takagi, T., Sugeno, M.: Fuzzy identification of systems and its application to modeling and control. IEEE Trans. on Systems, Man and Cybernetics **SMC-15** (1985) 116–132
7. Pham, T.D., Valliappan, S.: A least square model for fuzzy rules of inference. Fuzzy Sets and Systems **64** (1994) 207–212
8. Bustince, H., Calderon, M., Mohedano, V.: Some considerations about a least square model for fuzzy rules of inference. Fuzzy Sets and Systems **97** (1998) 315–336
9. Goldberg, D.E.: Genetic algorithms in search, optimization, machine learning. Addison-Wesley, Reading, Mass, USA (1989)
10. Karr, C.: Genetic algorithms for fuzzy controllers. AI Expert (1991) 38–43
11. Thrift, P.: Fuzzy logic synthesis with genetic algorithms, Proc. of Fourth International Conference on Genetic Algorithms. (1991) 509–513
12. Cordon, O.: Gomide, F., Herrera, F., Hoffmann, F., Magdalena, L.: Ten years of genetic-fuzzy systems: current framework and new trends. Fuzzy Sets and Systems **141** (2004) 5–31

13. Hoffman F., Pfister, G.: Evolutionary design of a fuzzy knowledge base for a mobile robot. Intl. Jl. of Approximate Reasoning **17** 4 (1997) 447–469
14. Pham, D.T., Karaboga, D.: Optimum design of fuzzy logic controllers using genetic algorithms. Journal of Syst. Engg. **1** (1991) 114–118
15. Ishibuchi, H., Nakashima, T., Murata, T.: Performance evaluation of fuzzy classifier systems for multidimensional pattern classification problems. IEEE Trans. on Systems Man and Cybernetics **29** (1999) 601–618
16. Cordon, O., DeeJesus, M.J., Herrera, F., Lozano, M.: MOGUL: a methodology to obtain genetic fuzzy rule-based systems under the iterative rule learning approach. Intl. Jl. of Intelligent Systems **14** 11 (1999) 1123–1153
17. Gonzalez, A., Perez, R.: SLAVE: a genetic learning system based on an iterative approach. IEEE Trans. on Fuzzy Systems **7** 2 (2001) 176–191
18. Furuhashi, T., Miyata, Y., Nakaoka, K., Uchikawa, Y.: A new approach to genetic based machine learning and an efficient finding of fuzzy rules-proposal of Nagoya approach. Lecture notes on Artificial Intelligence **101** (1995) 178–189
19. Wang, L., Yen, J.: Extracting fuzzy rules for system modeling using a hybrid of genetic algorithms and Kalman filtering. Fuzzy Sets and Systems **101** (1999) 353–362
20. Farag, W.A., Quintana, V.H., Lambert-Torres, G.: A genetic-based neuro-fuzzy approach for modeling and control of dynamical systems. IEEE Trans. on Neural Networks **9** (1998) 576–767
21. Ishibuchi, H., Nil, M., Murata, T.: Linguistic rule extraction from neural networks and genetic algorithm-based rule selection, Proc. of IEEE Intl. Conf. on Neural Networks. Houston, TX (1997) 2390–2395
22. Abdessemed, F., Benmahammed, K., Monacelli, E.: A fuzzy-based reactive controller for a non-holonomic mobile robot. Robotics and Autonomous Systems **47** (2004) 1–22
23. Yupu, Y., Xiaoming, X., Wengyuan, Z.: Real-time stable self learning FNN controller using genetic algorithm. Fuzzy Sets and Systems **100** (1998) 173–178
24. Angelov, P.P., Buswell, R.A.: Automatic generation of fuzzy rule-based models from data by genetic algorithms. Information Science **50** (2003) 17–31
25. Nandi, A.K.: Pratihar, D.K.: Automatic design of fuzzy logic controller using a genetic algorithm-to predict power requirement and surface finish in grinding. Journal of Materials Processing Technology **148** (2004) 288–300
26. Nawa, N.E., Hashiyama, T., Furuhashi, T., Uchikawa, Y.: A Study on fuzzy rules discovery using pseudo-bacterial genetic algorithm with adaptive operator, Proc. IEEE Int. Conf. Evolutionary Computation. Indianapolis, USA, (1997) 13–16
27. Ishibuchi H., Nakashima, T.: Effect of rule weights in fuzzy rule-based classification systems. IEEE Trans. on Fuzzy Systems **9** 4 (2001) 506–515
28. Ghosh A., Nath, B.: Multi-objective rule mining using genetic algorithms. Information Sciences **163** (2004) 123–133
29. Pratihar, D.K., Deb, K., Ghosh, A.: Optimal path and gait generations simultaneously of a six-legged robot using a GA-Fuzzy approach. Robotics and Autonomous Systems **41** 1 (2002) 1–20
30. Marichal, G.N., Acosta, L., Moreno, L., Mendez, J.A., Rodrigo, J.J., Sigut, M.: Obstacle avoidance for a mobile robot: a neuro-fuzzy approach. Fuzzy Sets and Systems **124** (2001) 171–170
31. Song, K.T., Sheen, L.H.: Heuristic fuzzy-neuro network and its application to reactive navigation of a mobile robot. Fuzzy Sets and Systems **110** (2000) 331–340

32. Li, W., Ma, C., Wahl, F.M.: A Neuro-fuzzy system architecture for behavior based control of a mobile robot in unknown environments. Fuzzy Sets and Systems **87** (1997) 133–140
33. Hui, N.B., Mahendar, V., Pratihar, D.K.: Time-optimal, collision-free navigation of a car-like mobile robot using a neuro-fuzzy approach. Fuzzy Sets and Systems **157** 16 (2006) 2171–2204
34. Jang, J.S.R., Sun, C.T., Mizutani, E.: Neuro-Fuzzy and Soft Computing. Prentice-Hall of India Pvt. Ltd., New Delhi (2002)
35. Hui, N.B., Pratihar, D.K.: Automatic design of fuzzy logic controller using a genetic algorithm for collision-free, time-optimal navigation of a car-like robot. International Journal of Hybrid Intelligent Systems **5** 3 (2005) 161–187
36. Latombe, J.C.: Robot motion planning. Kluwer Academic Publishers (1991)
37. Casillas, J., Cordon, O., Herrera, F.: Learning fuzzy rules using ant colony optimization algorithms, Proc. of 2nd Intl. Workshop on Ant Algorithms. Brussels, Belgium (2000) 13–21
38. Berenji, H., Khedkar, P.: Learning and tuning fuzzy controllers through reinforcements. IEEE Transactions on Neural Networks **3** 5 (1992) 724–740
39. Denna, M., Mauri, G., Zanaboni, A.M.: Learning fuzzy rules with Tabu search–an application to control. IEEE Transactions on Fuzzy Systems **7** 2 (1999) 295–318
40. Bagis, A.: Determining fuzzy membership functions with tabu search – an application to control. Fuzzy Sets and Systems **139** (2003) 209–225
41. Kim, D., Rhee, S.: Design of a robust fuzzy controller for the arc stability of CO_2 welding process using the Taguchi method. IEEE Transactions on Systems, Man and Cybernetics–Part B **32** 2 (2002) 157–162
42. Alfaro, H.M., Garcia, S.G.: Mobile robot path planning and tracking using simulated annealing and fuzzy logic control. Expert Systems with Applications **15** (1998) 421–429

4

A Neuro-Genetic Framework for Multi-Classifier Design: An Application to Promoter Recognition in DNA Sequences

Romesh Ranawana and Vasile Palade

University of Oxford Computing Laboratory, Oxford, United Kingdom January 16, 2006
romesh.ranawana@comlab.ox.ac.uk
vasile.palade@comlab.ox.ac.uk

Summary. This chapter presents a novel methodology for the customization of neural network based multi-classifiers used for the recognition of promoter regions in genomic DNA. We present a framework that utilizes genetic algorithms (GA's) for the determination of optimal neural network parameters for better promoter recognition. The framework also presents a GA based method for the combination of the designed neural networks into the multi-classifier system.

4.1 Introduction

All sequenced genomes[1] are extremely data-rich. The information contained within these sequences includes the answer for a lot of the problems faced by humans, but, due to their complexity, it is near impossible to analyze them using traditional methodologies. Thus, to fully realize the value of any sequenced genome and to fully understand it, efficient and intelligent computational methods are needed in order to identify the biological relevant features in the sequences and to provide insight into their structure and function.

Due to the difficulty involved in defining these features in a manner understandable to a computer algorithm, designers are being forced into using machine learning techniques for their implementation. Machine learning algorithms allow us to design efficient programs for pattern recognition tasks which do not have well defined data sets. Some of the recently used methodologies include the usage of neural networks [13, 30, 31], decision trees [27], genetic programming [10], hidden markov models [2, 9, 11, 12] and fuzzy logic [17, 32].

[1] The genome of an organism is its set of chromosomes, containing all of its genes and associated DNA. Please see Section 4.2 for further details.

R. Ranawana and V. Palade: *A Neuro-Genetic Framework for Multi-Classifier Design: An Application to Promoter Recognition in DNA Sequences*, Studies in Computational Intelligence (SCI) **66**, 71–94 (2007)
www.springerlink.com

One observation that has been made with the use of these types of methodologies for gene recognition is the difference in results obtained through the use of each type of algorithm. Although each type of system provides similar results in terms of the accuracy displayed, slight differences in the correct and incorrect genes recognized have been observed. Thus, it can be concluded that each methodology focuses the system to specialize in specific types of genes. It is due to this reason that hybrid multi-classifiers have now become increasingly popular in the implementation of pattern recognition systems. These hybrid multi-classifier systems are implemented by building a series of classifiers to solve the same task, but by using a different implementation algorithm for each of them. The results of each of these individual classifiers are then combined to obtain the final result of the classifying system.

Due to their robustness and their remarkable ability in deriving meaning from complicated or imprecise data, neural networks can be used to extract patterns and detect trends that are too complex to be noticed by either humans or other computer based techniques. Thus, they have been used extensively in the solving of problems related to bioinformatics. Nevertheless, one of the central dogmas involved in the design of a neural network for the solving of any problem lies in the decision on the characteristics or parameters that define the network. In this chapter, we present a framework which can be utilized for the automatic design and implementation of a neural network based multi-classifier system for the identification of protein coding regions (i.e. genes) in strings of DNA.

4.2 Biological Background

4.2.1 Genes and Promoters

Gene expression involves the usage of the information contained within a gene to produce a protein, which in turn leads to the production of the phenotype determined by that particular gene. The production of these proteins is a multi-step process, which begins with transcription and translation and is followed by folding, post-translation modification and targeting. During transcription, the DNA is copied into RNA[2] by a polymerase called RNA polymerase (RNAP). This transcription yields mRNA, which is the first step of the protein production process.

Promoters within genomes sequences are sequences of DNA located upstream of protein coding regions. They are usually short sequences and are bounded by specific transcription factors[3]. RNAP has the ability to recognize

[2] Is a nucleic acid, and is used as intermediatory during the production of proteins from genomic information.

[3] A transcription factor (TF) is a protein which binds onto specific DNA sequences on a genome sequence.

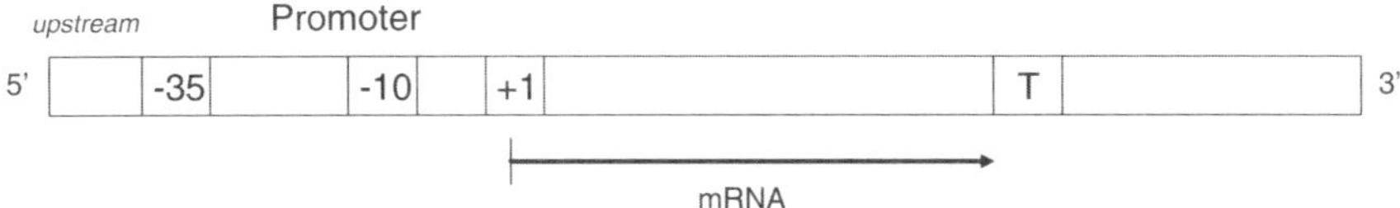

Fig. 4.1. A simple Prokaryotic Gene model

promoters of protein coding genes. Thus, RNAP uses the successful identification of promoters for the identification of protein coding regions. This paper uses the same method for the identification of genes in genomic DNA. The proposed system works on the premise that the successful identification of a promoter region leads to the successful identification of the start position of a gene.

Figure 4.1 displays a simple model of a Prokaryotic gene. The '-35' positions and '-10' positions located within the promoter. The '$+1$' position marks the beginning of the gene and the 'T' signifies the 'terminator' which marks the end of the gene. It is the DNA between these two positions which is transcribed in to mRNA, which is turn used for the production of the protein. 'In prokaryotes, the sequence of a promoter is recognized by the Sigma(σ) factor of the RNA polymerase. These sites are represented by the '-35' and '-10' symbols in Figure 4.1. E.Coli promoters are composed of two of these sites[4]. These are the locations to which E.Coli polymerase, the RNAP used by E.Coli, binds onto in order to begin the transcription of protein. These two binding sites are always located at the points known as the -35 hexamer box and the -10 hexamer box. Consensus sequences have been identified for each each of these two locations. A consensus sequence is the most probably sequence to occur at a certain position.

The spacer between the -10 hexamer box and the transcriptional has a variable length, the most probable length being 7. The spacer between the -10 site and the -35 site is also of variable length and can vary between 15 and 21 bases. An "ideal" E.Coli promoter would look something like the sequence displayed in Figure 4.2. It displays the most probable sequences that could occur at the two hexamer boxes. It is this variation that can make the recognition of these promoters difficult with traditional methodologies. Many promoter sequences have the pyrimidine[5] at position $+1$ (one nucleotide upstream of the transcription start site or the gene), and purine[6] at the transcriptional start site. In addition to the more obvious features described, a few more non-obvious features have been observed by [5] and [16].

[13] used the software available at http://www-lecb.ncifcrf.gov/toms/delia.html to display the sequence logos of 438 E.Coli promoters that were aligned according to their transcriptional start sites. Their results are displayed

[4] These sites are also known as binding sites.

[5] C or T

[6] A or G

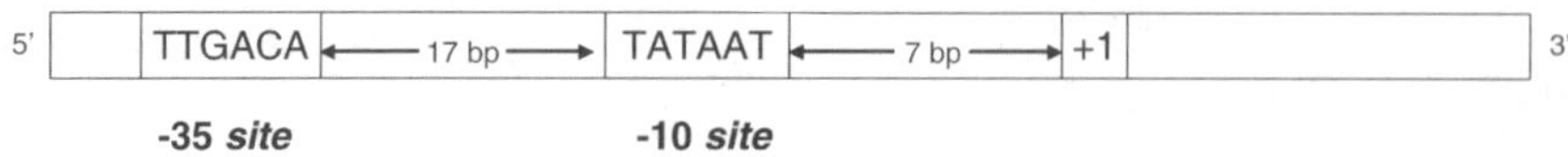

Fig. 4.2. A simple Prokaryotic Gene model

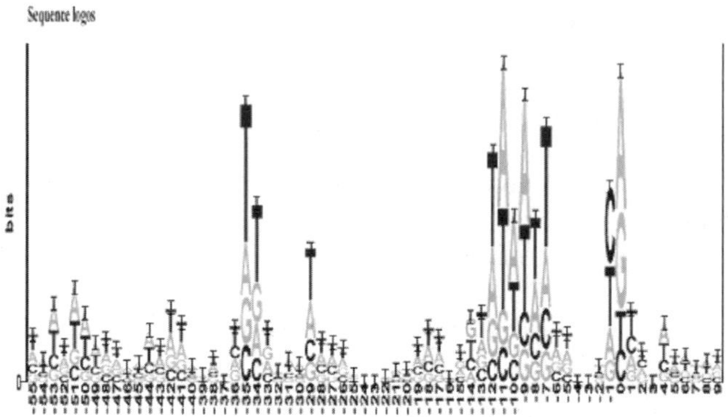

Fig. 4.3. A flow-chart of the framework utilized for the development of the system

in Figure 4.3. This software used the Shannon Entropy to measure the non-randomness of the individual bases at each position independently. We use these observations later in the paper to display the ability of our neural networks to learn these distributions automatically.

4.2.2 The Importance of Studying the E.Coli Gene

E.Coli K-12 and its close relatives have been extensively studied in genetic, molecular and biochemical studies from the dawn of modern microbiology and molecular biology. The development of a 'virtual cell' maybe far into the future, but it is a well known concept that the place to start the consideration of what will work and how to set about it lies in the development of a simple but adaptable system. In this regard, E.Coli K-12 provides an opportunity to test out approaches to modelling cellular and organismal systems. The Bioinformatics Initiative[7] lists a set of reasons giving out reasons as to why E.Coli is currently the organism of choice for such a study, both as a reference organism for many studies on prokaryotic systems and as a source of information on proteins and metabolic pathways that are shared by eukaryotics as well.

4.2.3 Methods Used for the Identification of E.Coli Promoters

Most of the methods that have been used for the detection of E.Coli promoters involve the alignment of all the given sequences at the transcriptional start site

[7] http://www.ecoli.princeton.edu/E_coli_Bioinformatics_Document.pdf

(i.e. the '+1' position) and then scanning them for the consensus sequences. They look for the primary consensus sequences at positions '−35' and '−10'. Some of the more successful methods scan for some of the weaker motifs that have been identified at positions '+1', '−22' and the '−44' position. Some examples of the use of these methods are listed out in [20] and [26]. Each of these systems programmed known knowledge about the E.Coli promoter into the system in order correctly classify unknown sequences. The main problem associated with these systems and other systems which use this technique is the uncertainty involved in the placement of the consensus sequence within the promoter. This can be caused due to the consensus sequence varying from the typical or by the consensus sequence being placed at a slightly different position. The spacer between consensus sequences is also variable.

Another method involves the use of a scoring matrix to calculate the variation of the different bases within the sequence. An example of such an implementation can be found at [4]. Neural networks have also been successfully utilized for the recognition of these E.Coli promoters [13, 14, 30]. Statistical methods like Hidden Markov Models and the EM algorithm have also been successfully utilized for this purpose.

The system presented in this paper also attempts to solve this problem, but utilizes a hybridization of a few of the methods mentioned above to achieve the objective. We built nine neural networks, each of which was individually trained and customized for better performance. These neural networks were then combined using a genetic algorithm. The results show us that the customized neural networks were able to identify the consensus sequences and identify important subsections of the DNA sequences shown to them. Manual entry of the consensus sequence information was not required.

4.2.4 The Data Set

The data set used consisted of 473 E.Coli sequences and 600 non-promoter sequences. The positive data sequence was obtained from the PromEC[8] website. Each of the obtained E.Coli sequences were 101 nucleotides in length and were aligned at the transcriptional start site. Each of the transcriptional start sites were positioned such that they each appeared at position 76 of the string. Thus, each of the positive sequences consisted of nucleotides from 101 nucleotides starting from the −75 position, 75 nucleotides upstream of the transcriptional start site, and ending at the +25 position, 25 nucleotides downstream of the transcriptional start site. The negative data set was obtained from the database compiled by the E.Coli GenomeProject[9], at the University of Wisconsin-Madison. These negative data consisted of E.Coli genes with the preceding promoter region deleted.

Using the above mentioned data set, we constructed three data sets: a training sequence, a positive test sequence and a negative test sequence.

[8] http://bioinfo.md.huji.ac.il/marg/promec/

[9] http://www.genome.wisc.edu/sequencing/k12.htm#seq

The training set consisted of 314 positive sequences and 429 negative sequences. The remaining positive and negative data were divided up into a positive test set of 159 sequences and a negative test set of 171 sequences. For the purpose of initial training and testing the originating neural networks, all three data sets were encoded using −2, −1, 1 and 2 to represent the bases A, T, C and G, respectively.

4.3 Multi-Classifier Systems

4.3.1 Classifiers

Classification, in computer science terminology, is the act of placing a given data item within a set of categories based on the properties of that particular data item. A classifier is a computer based agent who would perform this classification. For any pattern classification application research area currently under investigation, there are many possible classification algorithmic solutions which could be utilized.

Any learning problem involves the usage of a set of training examples of the form $\{(x_1, y_1), ..., (x_m, y_m)\}$ for the approximation of a function $f(x)$. The x values are usually of the form $< x_{i,1}, x_{i,2}, ..., x_{i,n} >$, and are composed of either real or discrete values. The y values are the expected outputs for the corresponding x vales, and are usually drawn from discrete sets of classes.

Thus, when presented with the training set, the learning paradigm attempts to approximate the function $f(x)$ and produces a *classifier*. The produced function is an approximation of the function $f(x)$ as the training set, more often than not, contains random noise. Thus, with the use of multiple learning paradigms, one obtains a set of classifiers $h_1, h_2, ..., h_L$, each of which attempts to approximate the function $f(x)$. Due to differences between the learning paradigms, the way that the approximation is evolved also differs. Thus, the results obtained through each classifier can also be different.

4.3.2 Why use Multiple Classifiers?

An ensemble of classifiers or a multi classifier, is a collection of classifiers whose individual decisions are combined in some way to obtain a combined decision. [8] claims that such an ensemble can only be more successful than the individual classifiers if the individual classifiers disagree with each other.

As mentioned in [33], there are two main groups of methodologies for the combinations of classifiers, namely, feature-vector-based methods[10] and syntactic-and-structural methods[11]. In addition to this distribution, each group is composed of a large variety of identifiable subgroups. For many of the

[10] E.g. Neural networks
[11] E.g. Fuzzy rule bases

classification problems encountered, it has been experimentally proven that different classifiers exhibit different degrees of success. No classifier has yet has proven to be totally perfect in the classification of any problem domain. Often, it has also been observed that different classifiers can disagree with each other on classification decisions when presented with the same training data and testing data.

Thus, Individual classifier models are recently being challenged by combined pattern recognition systems, which often show better performance. This is due to the reason that, when a system that incorporates many methodologies is used, the inadequacies of one methodology are usually nullified by the characteristics of another. That is, a multiple classifier can nullify the predictive inaccuracy obtained through the user of singular classification systems. Thus, genes not recognized by one classifier will be recognized by another and genes incorrectly recognized by one classifier will be rejected by another (Figure 4.4). The challenge in building such a system is the selection of a combination method for the results provided by each classifier in order to come up with the optimal and most popular result. Here, the performance of the entire system can be proven to be never much worse than that of the best expert [8].

[3] also states three causes as to why multiple classifiers exhibit better performance when compared to singular systems, and then goes on to state that the use of multiple classification solutions and their combinations leads to the nullification of many of the problems encountered through the use of singular classifiers.

In a situation where a number of classifiers are available, the simplest approach would be to select the best performing classifier and use it for the classification task [18]. This approach, although simple and easy to implement,

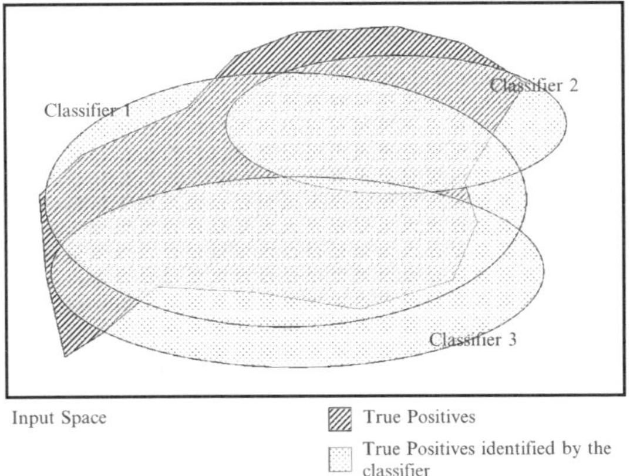

Fig. 4.4. A Classifier problem with different classifiers being used for the same data set. The shaded region represents the true positives and the ellipses represent the space classified by each classifier as being true positives

does not guarantee good performance [22]. It is highly probable that a combination of classifiers would outperform a singular classifier [25]. On the other hand, different and worse performing classifiers might only add to the complexity of the problem and provide even worse result than the worst classifier. Thus, it is a well known fact that if a multi-classifier system is to be successful, the different classifiers should have good individual performances and be sufficiently different from each other [28]. But, neither individual performances [21, 35] nor the diversity [24, 29] provides an ideal measure of how successful the combination of the classifiers will be.

As explained in [25], the core element of classifier selection is the selection criteria. The most natural choice being the combined performance, which will also the criterion for selection of the combiner. The only drawback of this methodology is the exponential complexity of testing out all possible combinations of a given set of classifiers.

It has been proven that the usage of an ensemble of neural networks for certain classification problems can improve classification performance, when compared to the use of singular neural networks. [19] and [23] provide results tested on protein secondary structure prediction. [1] also lists out an overview of applications in molecular biology.

Of the different methodologies used for the combination of multiple neural net classifiers, majority voting, neural networks, Bayesian inference and the Dempster-Shafer theories have proven the most popular [15, 33, 34]. The Dempster-Schafer method has proven to be successful, but has a considerable dependency on the function used for the alignment of probability. For the type of output produced by neural networks, posterior class-conditional probabilities can be calculated. The calculation of these probabilities becomes relatively simple, specially when the number of output classes is small.

In this Chapter, we test two methods (LAP and LOP) that can be used for the combination of results obtained through multiple neural network classifications and are compared with the use of a variation of the two methods, the LOP2 method. These two methods, the LAP and LOP methods, were introduced in (Hansen and Krogh, 1999).

4.3.3 The LAP and LOP Methods for Combining Classifiers

Introduction to the LAP and LOP Methods

[7] presents a general composition method that considers the average of results of several neural networks to obtain a combined result that would improve the overall classification performance.

The method that we used for the combination of the classifiers designed in this research was a variation of the LOP (Logarithmic Opinion Pool) [7] method, LOP2. We compare this method with the LAP (Linear Average Predictor) method, which is more commonly used. The LOP method is a general ensemble method which shows how the error or an ensemble can be written as the average error between members (called the ensemble ambiguity). [7] claims that this proves that ensembles always improve average performance.

Description of the Ensemble Composition Method used in this Research

An ensemble consists of M predictors f_i, which are combined into the combined predictor F. Let each predictor produce an output in the form of a probability vector $(f_i^1,...,f_i^N)$, where is the estimated probability that input belongs to class c_j. There is also a coefficient M associated with each ensemble of the form $(\alpha_1,...,\alpha_M)$ where $\sum_{i=1}^{M} \alpha_i = 1$.

The LAP Method

The combined predictor for the LAP of the ensemble is then defined in Equation (4.1).

$$F_{LAP}^j = \sum_{i=1}^{M} \alpha_i f_i^j \tag{4.1}$$

The LOP Method

The combined predictor for the LOP of the ensemble is then defined in Equation (4.2), where Z is a normalization factor given by Equation (4.3).

$$F_{LOP}^j = \frac{1}{Z} exp\left(\sum_{i=1}^{M} \alpha_1 log f_i^j\right) \tag{4.2}$$

$$Z = \sum_{j=1}^{N} exp\left(\sum_{i=1}^{M} \alpha_1 log f_i^j\right) \tag{4.3}$$

The LOP2 Method

The LOP algorithm was changed by removing the condition $(\alpha_1,...,\alpha_M)$ where $\sum_{i=1}^{M} \alpha_i = 1$, with α also being given the capability to take on negative values. We found that this slight alteration allowed the system to nullify small errors created by the more successful classifiers. This research attempted to utilize this methodology for the combination of the classifiers designed, and utilized genetic algorithms for the optimization of the unknown coefficients $(\alpha_1,...,\alpha_M)$.

4.4 Design of the System

4.4.1 Justification for the Development of the Framework

The development of this system was promoted through the development of the MultiNNProm System [http://web.comlab.ox.ac.uk/oucl/work/romesh.ranawana/RP2004a.pdf] and the problems encountered during its implementation. MultiNNProm provided us with very promising results in terms of the accuracy and precision displayed. The main problem encountered during the development was the large amount of time taken to properly configure the neural networks and to determine the optimal encoding methods for the given data set. All these design decisions had to be made on an ad-hoc basis, with the testing of many different configurations and encoding methods being required in order to obtain an optimal set of parameters. As each test run required the training and testing of a neural network, this process was extremely time consuming.

In order to negate these problems, we successfully designed and implemented a framework which would automatically determine the optimal parameters required for the design of the multi-classifier system. Within this framework, we utilized a genetic algorithm based method for the determination of the optimal number of layers and the optimal number of neurons on each layer. We also implemented a component of this framework to automatically determine an optimal set of encodings for the presented DNA data set. Thus, when presented with a DNA data set, the framework successfully designed and trained a set of neural networks for inclusion in a multi-classifier system. The following sub-sections describe the implementation of the framework and also lists out the results obtained.

4.4.2 Determination of Optimal Neural Networks Configurations for the given DNA Data Set

Determination of the Optimal Number of Hidden Layers

In order to obtain an optimal neural network configuration for the DNA data considered in this paper, we ran a series of simulations, each testing out the accuracy of the system for different numbers of hidden layers. We began the simulation run with a neural network that contained one hidden layer and 10 neurons on that hidden layer. The network was trained until an error rate of $1e-4$ was reached. The resulting network was then tested on a separate testing set. A similar set of training/testing runs were then conduced on separate networks, each containing one neuron more than the amount present on the previously trained/tested network. Figure 4.5 graphically illustrates the normalized error of each configuration fitted onto a third order polynomial curve. This curve identifies the fact that as the number of neurons on the first layer increases, the generalization capability of the system in turn decreases. Each of these networks demonstrated an accuracy in excess of 96% when presented with the training set.

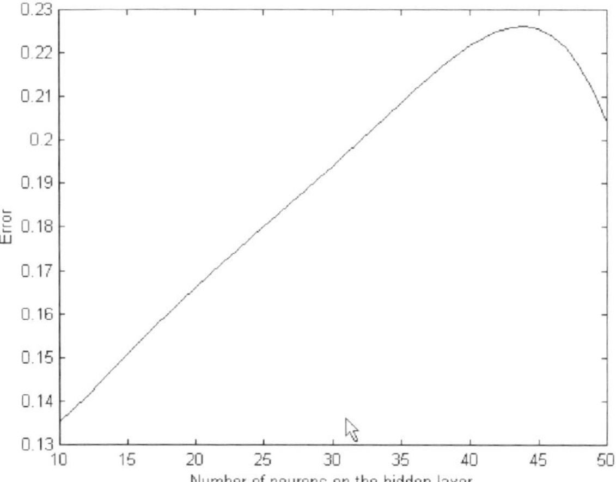

Fig. 4.5. The first order polynomial curve fit of the error of the system with respect to the number of neurons on the hidden layer of a one hidden layer neural network

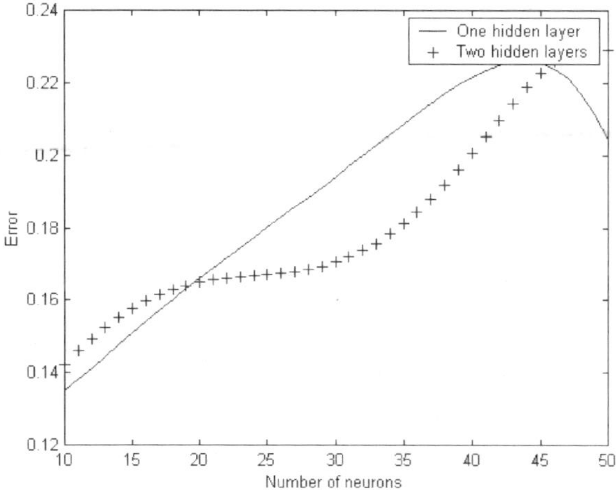

Fig. 4.6. The polynomially fit curve of the errors for the one hidden layer and two hidden layer neural networks

We then performed a similar set of simulations on networks with two hidden layers. We maintained a constant number of neurons on the first hidden layer and varied the number of units on the second hidden layer. Figure 4.6 shows the accuracy of the networks that contained 50 neurons on the first hidden layer and a varying number of neurons on the second hidden layer. It was also observed that the performance of the system with either one or two hidden layers was similar in terms of the accuracy with respect to the number of neurons on the layer before the output layer.

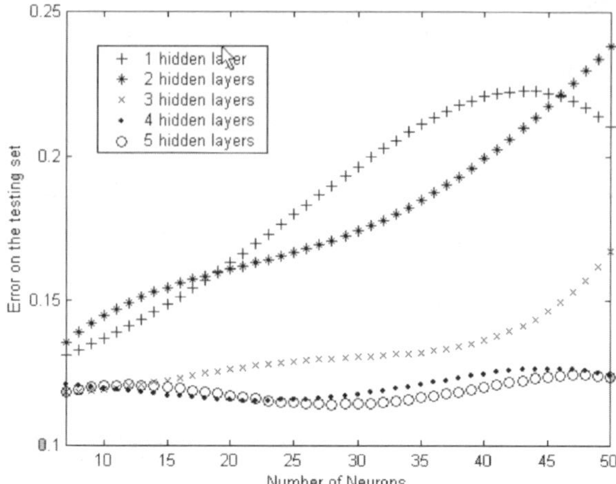

Fig. 4.7. A comparison of the errors for the systems with 1, 2, 3, 4 and 5 hidden layers

Table 4.1. Neural network configurations for the types of neural network trained and tested. Here, * indicates the variable value

Type	Configuration
One hidden Layer	*
Two hidden layer	50:*
Three hidden layer	100:50:*
Four hidden layer	100:80:50:*
Five hidden layer	125:100:80:50:*

A marked improvement in the generalization of the system was observed once the number of hidden layers exceeded 2. Figure 4.7 compares the accuracy of the systems trained and tested using 1, 2, 3, 4 and 5 with the number of neurons on the final layer being varied. Table 4.1 lists the configurations used as the base for the configuration of each of these networks. It was observed that the performance did not vary drastically once the number of layers exceeded 3. While the performance of the systems with 3 hidden layers diminished by small increments, the performances of the remaining two systems remained almost constant.

We accredited this increase in performance along with the increase in the number of hidden layers to the networks being able to identify the complexity of the given promoters more successfully. We concluded that neural networks with a higher number of hidden layers were able to recognize relationships of a higher order better than networks with one or two hidden layers. That is, they were able to correctly learn and classify second and third order relationships among the bases of the DNA string, which in turn leads to more genes being classified.

Table 4.2. The best performing neural network configurations for varying numbers of hidden layers with the error percentage observed

Number of hidden layers	Optimal Configuration	Error(%)
1	18	12.84
2	50:6	12.73
3	100:50:5	11.84
4	100:80:50:18	11.55
5	125:100:80:50:24	11.43

Listed in Table 4.2 are the minimum errors obtained for each type of neural network along with the number of neurons present on the final hidden of that optimal network. Thus, it was concluded that the optimal network with regard to accuracy, generalization and complexity was the network that contained three hidden layers, with 10 neurons on the final hidden layer. Although the network with 5 hidden layers did perform moderately better, the network with 3 hidden layers was deemed to be better due to the its reduced complexity and the lesser time required for training.

Determination of the Number of Units on Each Layer

Once the optimal number of neural network hidden layers for the dataset was determined, the next task was the determination of the number of units required on each layer in order to enhance generalization. For this purpose, we initialized a random set of vectors to represent the number of neurons on each layer. Each vector consisted of 12 integer values, where the first four values varied between 1 and 50, the second four between 1 and 20 and the remainder between 1 and 10. The sum of the first, the second and final four values were used to represent the number of neurons on the first, second and third hidden layers respectively. We divided the number on neurons in each layer within each chromosome in order to maximize the effect of mutation and crossover when used within the neural network. If each unit count was used by itself, individual chromosomes would only have included three values, and would have consequently nullified most of the advantageous effects of the genetic algorithm.

Four initial populations, each consisting of 60 vectors were initialized. Each of these populations was then exercised through a genetic algorithm in order to obtain an optimal set of vectors representing the configuration of the network. The 15 best performing vectors of each population was then extracted to create a new population, and was exercised through a genetic algorithm in order to obtain an optimal set of configurations for the neural network. This stage was required for the nullification of any local minimums that the genetic algorithm would have encountered.

The fitness function utilized for the selection of vectors was the classification error exhibited by the neural network in terms of the percentage of

Table 4.3. The best performing neural network configurations with respect to the number of units on each hidden layer and the error displayed by each system.

Label	Hidden Layer 1	Hidden Layer 2	Hidden Layer 3	Error(%)
a	103	63	22	9.82
b	75	19	17	8.36
c	56	43	54	8.36

false positives and false negatives with respect to the total amount of testing data presented[12]. Thus, the genetic algorithm attempted to minimize the total error displayed by the system with respect to the test data. Each neural network was trained until the error reached a value of $1E - 3$ or the number of training epochs reached 100. The genetic algorithm was implemented using a ***precision*** of 0.9, which meant that the 6 best performing vectors were included in their entirety within the next generation. Also, a ***crossover probability*** of 0.7 and a ***mutation probability*** of 0.1 was used along with roulette wheel selection [6] for its implementation. Each evolutionary function was iterated 60 times before the selection of the winning vectors was made.

Table 4.3 lists out the 3 best performing configurations along with the error displayed by each of them. The best performing system during the determination of the number of layers displayed an error rate of 11.43%, whereas with the introduction of the genetic algorithm and better unit number selections, the recognition rate was increased to produce a minimum error rate of 8.36%. Thus, it can be inferred that the correct configuration of the layers leads to a better rate of recognition. When the true positives and true negatives of configurations b and c were compared, it was found that the sequences correctly classified by each network differed by around 3%. This lead us to the conclusion that different network configurations can lead different networks specializing in specific classes of promoters and non-promoters.

4.4.3 Method used for the Determination of Optimal Encoding Methods

To test the hypothesis that the encoding method can affect the performance of the system, we ran a series of genetic algorithms which tested different encoding methods. We initialized ten random populations of 60 vectors, where each vector was represented by four random real values which ranged from -10 to $+10$. Here, the four values on the vector corresponded to the four DNA bases.

Each population was then exercised through a genetic algorithm which used the accuracy of the system of the system with respect to the testing set as the fitness function. Each population was iterated 60 times through the evolutionary algorithm with a ***mutation probability*** of 0.8, a ***precision*** of 0.8. The algorithm used roulette wheel selection for cross-overs selections. The

[12] The inverse of the Precision

10 best performing vectors from each population were then extracted to form a new population and was then exercised through a genetic algorithm which identical attributes. This step was performed in order to obtain an optimal set of encoding schemes by using the best chromosomes identified by the individual populations and also assists with the negating of negative effects caused through local minimums. The training graph for the best performing vectors of each generation of the final optimized population is displayed in Figure 4.8 by open circles. Here, the average error of each generation is represented by a cross. These results show us that the accuracy of the system was optimized from an error rate of around 8% to a value close to 4%. Thus, the total accuracy of the system was increased from 92% to 96% through the utilization of the genetic algorithm.

Table 4.8 lists out the three pest performing encoding methods along with the accuracy displayed by the usage of each. Displayed in Figure 4.9(a) and (b) are value comparisons of the 3 best performing vectors. Figure 4.9(a) shows displays the distribution of each vector where it can be seen that the distribution of the values for the different bases displays a very distinct pattern. The numerical values of A and C remain close to each other and vary between 0 and 1, whereas, the values of T and C remain numerically distanced from each others values. It was also observed that the values of T and C always remain

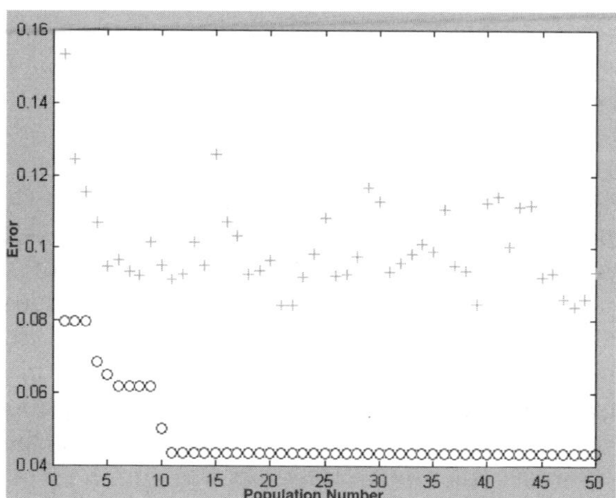

Fig. 4.8. Training graph for the encoding optimization genetic algorithm

Table 4.4. The three best performing encoding methods along with the error rate displayed by each

Label	A	T	C	G	Error(%)
e1	0.4788	1.3868	0.1692	−0.3659	0.0436
e2	0.1309	1.9883	0.6046	−0.0513	0.0489
e3	0.2702	0.0693	0.4255	1.5572	0.0409

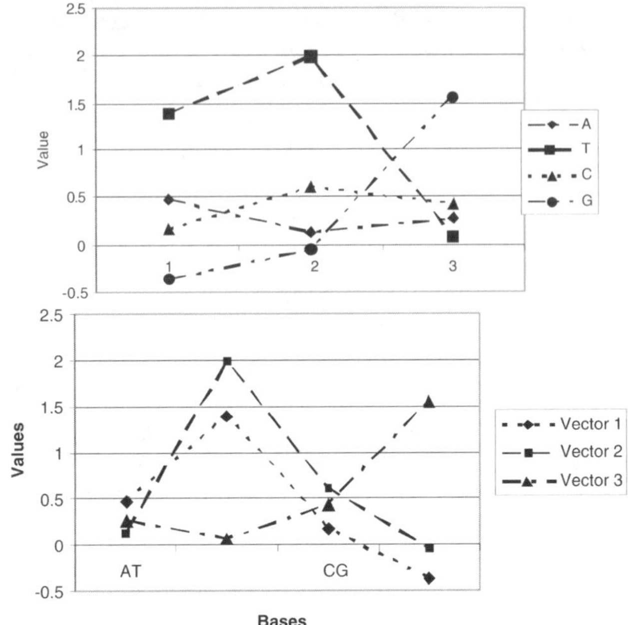

Fig. 4.9. A comparison of the best performing encoding vectors

either larger or smaller, conversely, to the values of A and C. Thus, the curve always remains either convex or concave. The line never assumes a zigzagged value. These conclusions are substantiated by Figure 4.9(b), which shows the distribution of the different bases for the three wining vectors, demonstrating the more or less constant values maintained by both A and C.

The best performing network with the optimized number of neurons on each layer displayed an error rate of 8.36%, whereas, with the introduction of an optimized set of encoding methods, the error rate was reduced to 4.36%. As with the results obtained through the optimization of the network configuration, the results obtained through the use of each encoding method produced slightly varying results with relation to the promoters classified. Thus, the underlying conclusion being that each encoding method helped the network specialize on different types of promoters present within the data set.

4.4.4 An Overview of the System

As shown in Figure 4.10, the system is a neural-network based multi-classifier system. Through the optimization process of the framework, we obtained three optimal neural network configurations; a, b and c, and three optimal sequence encoding methodologies; e1, e2 and e3. By combining each encoding method with each configuration, we developed and trained 9 neural networks, NNe1a, NNe1b, NNe1c, NNe2a, NNe2b, NNe2c, NNe3a, NNe3b and NNe3c. Each

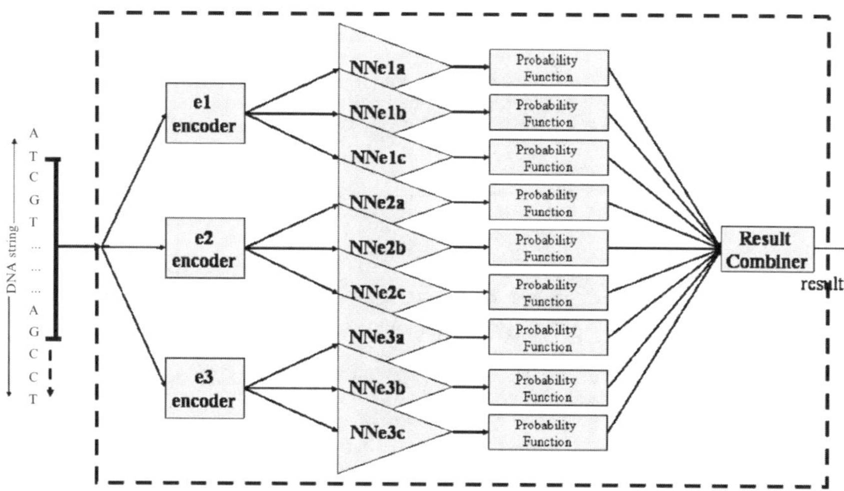

Fig. 4.10. NNGenProm Architecture

neural network was initially trained to produce an output of -1 if it predicted the sequence to be a non-promoter and an output of 1 if it predicted a promoter. The outputs of the individual neural networks were then passed onto a probability builder function which assigned probabilities as to whether the presented sequence was an E.Coli promoter or not. Finally, the outputs of the probability functions were passed onto a result combiner which combined the results and presented a final result as to whether the given sequence was an E.Coli promoter or not. The final output was of the 'yes'(1) or 'no'(0) form. The probability function used in this implementation was identical to that used during the implementation of the MultiNNPRom system.

4.4.5 The Result Combiner

The results were combined using the LOP2 method described in Section 4.3.3.2. The system was tested on three combination methods, namely, classical majority voting, LAP, and LOP2. The resulting observations showed us that the results obtained through the LOP2 method provided us with a much better recognition rate in terms of the test data, both positive and negative. The comparison of results between these three combinatorial methods is listed out in Section 4.5.

The LOP2 method was implemented as follows. We let the outputs of the nine neural networks be symbolized by O_i, where $1 \leq i \leq 9$. We also defined 9 coefficients α_i, where $1 \leq i \leq 9$. Then, the combined predictor was defined by Equations (4.4) and (4.5) where (4.6).

$$O.Positive = \frac{1}{Z} \exp\left(\sum_{i=1}^{9} \alpha_i(\log(O_i.Positive))\right) \tag{4.4}$$

$$O.Negative = \frac{1}{Z} \exp\left(\sum_{i=1}^{9} \alpha_i(\log(O_i.Negative))\right) \tag{4.5}$$

$$Z = \exp\left(\sum_{i=1}^{9} \alpha_i(\log(O_i.Positive))\right) + \exp\left(\sum_{i=1}^{9} \alpha_i(\log(O_i.Negative))\right) \tag{4.6}$$

It is one again obvious that $O.Positive + O.negative = 1$. The final conclusion on whether the given sequence was an E.Coli promoter or not was reached using equation (4.7).

$$C(O.Positive, O.Negative) = \begin{cases} Yes\ ;\ O.Positive > O.Negative \\ No\ \ ;\ Else \end{cases} \tag{4.7}$$

The methodology used to determine values for the coefficients is described in the next section. The coefficient values were determined using a method identical to that used during the design and implementation of MultiNNProm.

4.5 Results

4.5.1 Evaluation of the Performance of the Individual Classifiers

Each neural network performed perfectly on the training set and displayed a recognition rate of approximately 100% when presented with them after training. The test sequences were exercised through the 9 neural networks and the results obtained are listed out in Table 4.11. These performances are graphically in Figure 4.11. It can be observed that while some networks perform better with respect to the recognition of true negatives, others specialize in the recognition of true positives. For example, nne2b has a better

Table 4.5. The Specificity, Sensitivity and Precision of the nine trained neural networks

Network	True Negatives	True Positives	Precision(%)
nne1a	97.66	91.20	94.43
nne2a	97.66	93.71	95.69
nne3c	98.83	92.45	95.64
nne1b	98.23	88.68	93.46
nne2b	98.25	91.82	95.03
nne3b	98.83	91.20	95.01
nne1c	94.15	92.45	93.30
nne2c	98.83	92.45	95.64
nne3c	94.74	93.71	94.22

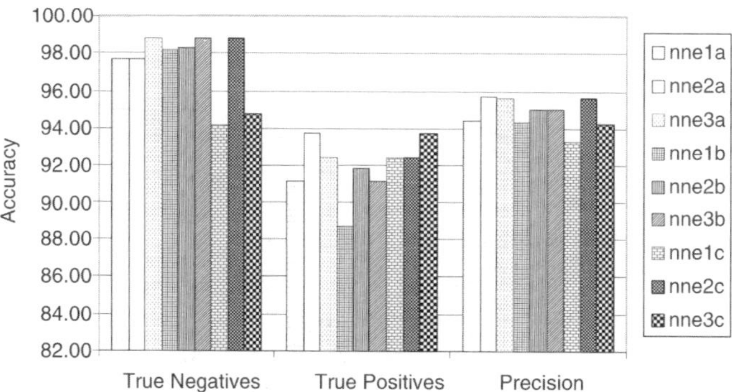

Fig. 4.11. Comparison of the true positives and true negatives recognized by each neural network

Table 4.6. The results obtained when the test data sets were exercised through the combined system

Attribute	Value
Specificity	0.9882
Sensitivity	0.9748
Precision	0.9815

recognition rate with respect to true negatives, whereas nne2c has a better recognition rate with respect to true positives. nne3a, nne3b and nne2c provide the best recognition of true negatives, whereas the best recognition rate of true positives was displayed by nne2a.

4.5.2 Performance evaluation of the Combined System

The specificity, sensitivity and precision of the neural network results combined using the LOP2 method are listed out in Table 4.6. These results are compared with the results obtained for the individual neural networks in Figure 4.12. These results indicate that the combination of these nine neural networks provides us with a far better recognition rate in terms of the precision, specificity and sensitivity. Thus, the performance of the system has been improved by a considerable margin.

It was observed that the combination of the nine neural networks using the LOP2 method did not diminish the performance by either the true positive accuracy or the true negative accuracy of the overall precision. Whereas the true negative recognition rate remains equal to the recognition rate of the best negative classifier after the application of the combinatorial function, the true positive recognition rate has been increased by a considerable fraction. Thus, as claimed by [3], the performance of the combination was not worse than the performance of the best classifier.

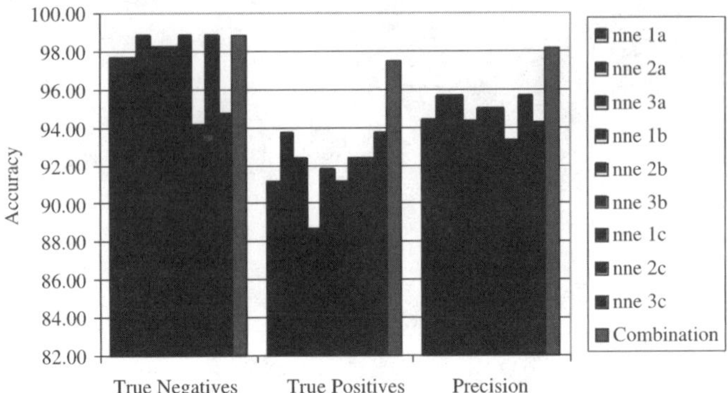

Fig. 4.12. Comparison of the specificity, sensitivity and precision of the four neural networks and the combined system

Table 4.7. Comparison of the specificity, sensitivity and precision of the research obtained using different combinatorial methods

Combinatorial method	S_p	S_n	P
Majority Voting	0.9882	0.9559	0.9720
LOP2	0.9882	0.9748	0.9815

Table 4.8. Comparison of the specificity, sensitivity and precision of MultiGenProm with previous research on the same dataset

System	S_p	S_n	P
NNGENProm	0.9882	0.9748	0.9815
Ma et al.,2001 [13]	0.9176	0.9920	0.9194
Mahadevan and Ghosh, 1994 [14]	0.9020	0.9800	0.9040

We attribute this increase in accuracy to the different specializations exhibited by each network. The combination of the networks lead to a committee of classifiers that performed better than the individual classifiers. Thus, the inaccurate predictions of one network were negated by the correct predictions of another.

The results of this combination were also compared with the results obtained through the usage of majority voting. Table 4.8 lists out this comparison. These results point our the fact that, although the combination of the 9 neural networks using either method provides us with better results than the individual neural networks, the results obtained through the LOP2 method are better.

Table 21 compares the results obtained through this rerearch with previous work done on the same problem. Thus, our system is seen as a considerable improvement compared with recent research on the recognition of E.Coli promoters.

Table 4.9. The performance of the system after each step of the framework was completed

Step	Lowest Error (%)
Determination of the number of hidden layers	11.43
Determination of the number of units on each hidden layer	8.36
Determination of the optimal encoding methods for the DNA data set	4.36
Design and implementation of the networks using optimal configurations and encodings	4.36
Combination of the neural networks	1.85

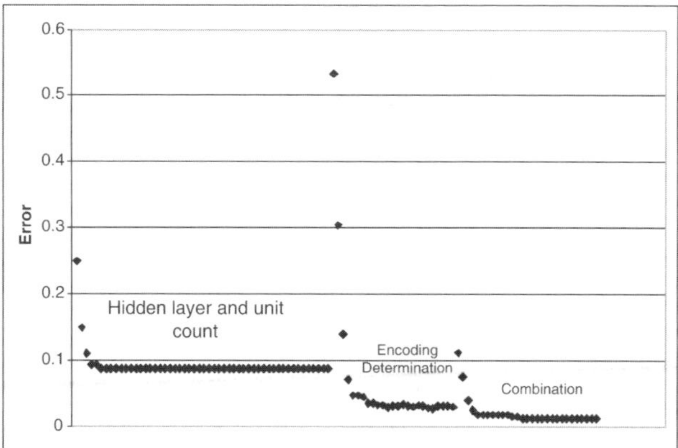

Fig. 4.13. The performance of the system through the different steps of the framework

4.5.3 Summary of the Results Obtained through the use of the Presented Framework

As tabulated in Table 4.9, the different steps presented within the framework gradually help decrease the error in the recognition rate displayed by the system. These observations are graphically illustrated in Figure 4.13. The spikes on the graph are due to random chromosomes being used for each initial population. This framework presents a methodology that assumes the determination of all the unknown parameters to be an optimization problem, and uses genetic algorithms in order to determine them.

4.6 Discussion and Future Work

In this chapter, we presented a novel approach for the recognition of E.Coli promoters in strings of DNA. The proposed system showed a substantial improvement in the recognition rate of these promoters, specifically in the recognition of true positives, i.e. rejection of non-promoters. This caused our system to display a higher specificity than all other systems developed thus far.

We conclude the reasons for this improvement to be threefold. Firstly, it was the use of larger neural networks than those that have been used thus far. This led to a far better rate of recognition and generalization. We also noted that an increase in the number of hidden layers on the neural network also led to a better recognition rate. Secondly, the use of multiple neural networks, each accepting the same set of inputs in different forms. We observed the fact that the false positives and false negatives of one network could be wiped out by true positives and true negatives in other networks. Thus, the combination of the opinions of more classifiers led us to a system that performed far better than the individual components. Our results substantiates Dietterich's (Dietterich, 1997) conclusion. Finally, the use of different encoding methods, which were customized to suite the particular data set, increased the recognition capabilities of the neural networks in terms of the data presented.

The above three reasons, coupled with the fact that each optimized parameter was used in conjunction with permutations of other optimized parameters lead to a system, when combined, provided us with far better accuracy when compared to singular systems or other multi-classifier systems using similar parameters.

As was pointed out earlier, one of the major obstacles in the development of the presented system was the time taken for the determination of optimized parameters for the development of the system. This problem was negated by the introduction of the neuro-genetic framework, which provides the designer with each of the optimized parameters necessary for the development of a system specially customized towards the given data set. The neuro-genetic framework provided us with the optimal number of layers to include within the neural networks, the number of neurons to include with each layer and also a set of optimized encoding methods specially customized for the given data set.

References

1. P. Baldi and S. Brunak. *Bioinformatics - The Machine Learning Approach*, volume 1. MIT Press, 1998.
2. E. Birney. Hidden markov models in biological sequence analysis. *IBM Journal of Research and Development*, 45(3/4), 2001.
3. T.G. Dietterich. Machine-learning research: Four current directions. *The AI Magazine*, 18(4):97–136, 1998.
4. D Frishman, A Mironov, HW Mewes, and M Gelfand. Combining diverse evidence for gene recognition in completely sequenced bacterial genomes [published erratum appears in Nucleic Acids Res 1998 Aug 15;26(16):following 3870]. *Nucl. Acids Res.*, 26(12):2941–2947, 1998.

5. D.J. Galas, M. Eggert, and M.S. Waterman. Rigorous pattern-recognition methods for dna sequences: Analysis of promoter sequences from E.Coli. *Journal of Molecular Biology*, 186(1):117–128, 1985.
6. D.E. Goldberg. *Genetic Algorithms in Search, Optimization and Machine Learning*. Addison-Wesley Longman Publishing Co., Inc., 1989.
7. J.V. Hansen and A. Krogh. A general method for combining in predictors tested on protein secondary structure prediction. In *Proceedings of Artificial Neural Networks in Medicine and Biology*, pages 259–264. Springer-Verlag, May 2000.
8. L. K. Hansen and P. Salamon. Neural network ensembles. *IEEE Trans. Pattern Anal. Mach. Intell.*, 12(10):993–1001, 1990.
9. J Henderson, S Salzberg, and K Fasman. Finding genes in dna with a hidden markov model. *Journal of Computational Biology*, 4(2):127–141, 1997.
10. John R. Koza and David Andre. Automatic discovery of protein motifs using genetic programming. In Xin Yao, editor, *Evolutionary Computation: Theory and Applications*. World Scientific, Singapore, 1996.
11. A Krogh. Two methods for improving performance of a hmm and their application for gene finding. In *Proceedings of the 5th International Conference on Intelligent Systems for Molecular Biology*, pages 179–186. AAAI Press, 1997.
12. D. Kulp, D. Haussler, M. G. Reese, and F. H. Eeckman. A generalized hidden markov model for the recognition of human genes in DNA. *ISMB-96*, pages 134–141, 1996.
13. Q. Ma, J.T.L. Wang, D. Shasha, and C.H. Wu. Dna sequence classification via an expectation maximization algorithm and neural networks: A case study. *IEEE Transactions on Systems, Man, and Cybernetics, part C: Applications and Reviews, Special Issue on Knowledge Management*, 31(4):468–475, November 2001.
14. I Mahadevan and I Ghosh. Analysis of E.Coli promoter structures using neural networks. *Nucl. Acids Res.*, 22(11):2158–2165, 1994.
15. E.J. Mandler and J. Schurmann. Combining the classification results of independent classifiers based on the dempster/schafer theory of evidence. *Pattern Recognition and Artificial Intelligence*, X:381–393, 1988.
16. G. Mengeritsky and T.F. Smith. Recognition of characteristic patterns in sets of functionally equivalent dna sequences. *Comput. Appl. Biosci.*, 3(3):223–227, 1987.
17. Lucila Ohno-Machado, Staal A. Vinterbo, and Griffin Weber. Classification of gene expression data using fuzzy logic. *Journal of Intelligent and Fuzzy Systems*, 12(1):19–24, 2002.
18. D. Partridge and W.B. Yates. Engineering multiversion neural-net systems. *Neural Comput.*, 8(4):869–893, 1996.
19. S.K. Riis and A. Krogh. Improving prediction of protein secondary structure using neural networks and multiple sequence alignments. *Journal of Computational Biology*, 3:163–183, 1996.
20. K. Robison. A comprehensive library of dna-binding site matrices for 55 proteins applied to the complete escherichia coli k-12 genome, 1998.
21. G. Rogova. Combining the results of several neural network classifiers. *Neural Netw.*, 7(5):777–781, 1994.
22. F. Roli and G. Giacinto. *Hybrid Methods in Pattern Recognition*, chapter Design of multiple classifier systems, pages 199–226. Worldwide Scientific Publishing, 2002.

23. B. Rost and C. Sander. Prediction of protein secondary structure at better than 70% accuracy. *Journal of Molecular Biology*, 232(2):584–599, July 1993.
24. D. Ruta and B. Gabrys. Analysis of the correlation between majority voting error and the diversity measures in multiple classifier systems. In *Proceedings of the 4th International Symposium on Soft Computing*, paper 1824–025, 2001.
25. D. Ruta and B. Gabrys. Classifier selection for majority voting. *Special issue of the journal of INFORMATION FUSION on Diversity in Multiple Classifier Systems*, 2004.
26. H Salgado, A Santos, U Garza-Ramos, J van Helden, E Diaz, and J Collado-Vides. RegulonDB (version 2.0): a database on transcriptional regulation in Escherichia coli. *Nucl. Acids Res.*, 27(1):59–60, 1999.
27. Steven Salzberg, Arthur L. Delcher, Kenneth H. Fasman, and John Henderson. A decision tree system for finding genes in dna. *Journal of Computational Biology*, 5(4):667–680, 1998.
28. A.J.C. Sharkey and N.E. Sharkey. Combining diverse neural nets. *Knowl. Eng. Rev.*, 12(3):231–247, 1997.
29. C.A. Shipp and L.I. Kuncheva. An investigation into how adaboost affects classifier diversity. In *Proc. IPMU 2002*, pages 203–208, 2002.
30. E.E. Snyder. Identification of protein coding regions in genomic dna. *Journal of Molecular Biology*, 248:1–18, 1995.
31. EC Uberbacher and RJ Mural. Locating Protein-Coding Regions in Human DNA Sequences by a Multiple Sensor-Neural Network Approach. *PNAS*, 88(24):11261–11265, 1991.
32. PJ. Woolf and Y Wang. A fuzzy logic approach to analyzing gene expression data. *Physiol. Genomics*, 3(1):9–15, 2000.
33. L. Xu, A. Krzyzak, and C.Y. Suen. Several methods for combining multiple classifiers and their applications in handwritten character recognition. *IEEE Trans. on System, Man and Cybernetics*, SMC-22(3):418–435, 1992.
34. L. Xu, A. Krzyzak, and C.Y. Suen. Associative switch for combining multiple classifiers. *Journal of Artificial Neural Networks*, 1(1):77–100, 1994.
35. G. Zenobi and P. Cunningham. Using diversity in preparing ensembles of classifiers based on different feature subsets to minimize generalization error. *Lecture Notes in Computer Science*, 2167:576–587, 2001.

5

Evolutionary Grooming of Traffic in WDM Optical Networks

Yong Xu[1] and Kunhong Liu[2]

[1] School of Medicine, University of Birmingham, Edgbaston, Birmingham, B15 2TT, UK. Email: y.xu.1@bham.ac.uk
[2] Intelligent Computing Lab, Hefei Institute of Intelligent Machines, Chinese Academy of Sciences, P.O. Box 1130, Hefei Anhui 230031, China. Email: lkhqz@163.com

Summary. The widespread deployment of WDM optical networks posts lots of new challenges to network designers. Traffic grooming is one of the most common and interesting problems. Efficient grooming of traffic can effectively reduce the overall cost of the network. But unfortunately, it has been shown to be NP-hard. Therefore, new heuristics must be devised to tackle them. Among those approaches, metaheuristics are probably the most promising ones. In this chapter, we present a thorough and comprehensive discussion on various metaheuristic approaches to the grooming of traffic in both static and dynamic patterns in WDM optical networks. Some future challenges and research directions are also discussed in this chapter.

Keywords: WDM Optical network, Traffic grooming, Genetic algorithm, Tabu search, Simulated annealing.

5.1 Introduction

Wavelength-division multiplexing (WDM) technology will undoubtedly dominate next generation's backbone transport networks. In WDM optical networks, hundreds of wavelength channels can be established in a single fiber to make the best use of its huge bandwidth. In such a network, a set of lightpaths on the optical layer, which is used for transmitting optical signal between its end nodes, define the virtual topology and the interconnections of them are implemented via optical cross-connects (OXCs). Each OXC can switch the optical signal coming in on a wavelength of an input fiber to the same wavelength in an output fiber. Such network architecture makes it flexible to various services and protection requirements. The deployment and design of such WDM optical networks are currently the subjects of intensive research and development efforts.

Y. Xu and K. Liu: *Evolutionary Grooming of Traffic in WDM Optical Networks*, Studies in Computational Intelligence (SCI) **66**, 95–137 (2007)
www.springerlink.com

Currently, the typical transmission rate on a wavelength is 10 Gbit/s (OC-192) and higher rate (40 Gbit/s or OC-768) is also commercially available. On the contrary, the requirement of end-users, such as Internet service providers (ISPs), universities, and industries is generally several to a few hundreds megabits, much lower than that rate. The huge gap between these two requires that a number of such low rate traffic streams be multiplexed onto high rate wavelengths to make full use of their capacity and at the same time to minimize the network's cost. However, multiplexing/demultiplexing a traffic stream to/from a wavelength requires the wavelength to drop at the two end nodes of it at which an electronic equipment must be used to process such operation. In SONET/WDM networks, for example, a SONET add/drop multiplexer (ADM) is required to electronically combine such low-rate traffic streams onto a high-rate circuit (wavelength). Furthermore, such electronic equipment is often very expensive for high-rate traffic transformations. Therefore it is beneficial if each wavelength drops only at those nodes onto which the low-rate streams are congregated in order to reduce the cost. This is realized by the addition of a wavelength ADM (WADM) at each node. With the WADM, each wavelength is allowed to either drop at a node or optically bypass it with no electronic equipment requirement at that node for that wavelength. A question that arises in such a system is how to groom low-rate traffic streams onto each wavelength so that the amount of electronic equipment needed in the network is minimized while as few wavelengths as possible are used. This problem, often referred to as the traffic-grooming (TG) problem, has been proved to be NP-complete [1] and drawn increasing interest in the literature recently [1–6].

The topic of TG has always been one of the popular problems in optical communication field since it was first proposed in 1998. It has been the main focus in the recent special issue on WDM-Based Network Architectures in IEEE Journal on Selected Areas in Communications [7]. The First and Second Workshop on TG in WDM Networks have been held recently [8]. Comprehensive review and specific review have also been done in this subject [4,9,10]. Due to the NP-complete property of this problem, research showed that evolutionary approaches are a very good option for traffic-grooming problems. In this chapter of the book we will provide a detailed discussion on how evolutionary approaches can be used to various grooming problems. Some future challenges and research directions are also discussed in this chapter. One purpose of this chapter is to provide the researchers and practitioners in evolutionary computation community a new application area and those in traffic grooming community a new tool to improve the grooming performance.

The rest of this chapter is organized as follows. Section 5.2 gives the basic concept of traffic grooming in WDM optical networks. Sections 5.3 and 5.4 give a thorough discussion on evolutionary grooming of both static and dynamic traffic patterns. An application of evolutionary approaches based traffic grooming technique will be discussed in Section 5.5. Some future challenges and research directions about traffic grooming are discussed in Section 5.6. Finally, conclusions are drawn in Section 5.7.

5.2 Traffic Grooming in WDM Optical Networks

TG is one of the most important traffic engineering techniques which is defined as the allocation of sub-wavelength traffic tributaries onto full wavelength channels in order to achieve efficient utilization of network resources, e.g., minimizing cost or blocking probability, or maximizing revenue. This is a complicated combinatorial optimization problem and has been shown to be NP-complete even in a single-hub unidirectional ring network with nonuniform traffic requests between each pair of nodes [1]. It was first introduced in 1998 [11–13] and has aroused much interest from the researchers in both academia and industrial societies due to its high academic and commercial values. It is also one of the most attractive issues in journals related to optical networks.

In WDM optical networks, each fiber transmits a number of wavelengths simultaneously in a WDM manner and at the same time every wavelength can carry several traffic streams in a Time Division Multiplexing (TDM) manner. By using the ADM, low rate traffic can be multiplexed onto high rate circuits. 4 OC-3 traffic streams, for instance, can be multiplexed onto an OC-12 circuit and 16 OC-3's onto an OC-48, etc. The number of low rate traffic streams a wavelength can accommodate is referred to as *traffic granularity.* Since high speed ADM is very expensive, the use of ADM's usually dominates the total cost of today's backbone networks. Hence TG is often used to avoid each wavelength dropping at too many nodes on the network to reduce such cost.

The benefit of TG in WDM optical networks can be shown by the simple example given below. Consider a 4-node unidirectional ring. The traffic request between each pair of nodes is 8 OC-3's and the wavelength capacity is OC-48. Thus each wavelength can carry two node pairs' amount of traffic. Since there are 6 pairs of traffic on the ring, 3 wavelengths are needed.

Fig. 5.1(a) shows the traffic assignment with no grooming, where wavelength λ_1 accommodates traffic requests between node pairs $1 \leftrightarrow 2$ and $3 \leftrightarrow 4$, λ_2 $2 \leftrightarrow 3$ and $1 \leftrightarrow 4$, λ_3 $1 \leftrightarrow 3$ and $2 \leftrightarrow 4$. In this configuration, every wavelength drops at every node so that 3 ADM's are needed at every node. 12 ADM's are needed on the ring.

Figure 5.1(b) gives another configuration with proper grooming. In this scenario, nodes 1, 2, and 4 equip with a WADM, which allows only wavelengths that carry traffic to or from that node to be added/dropped at it and bypasses all the other wavelengths. The assignment of traffic is as follows: $\lambda_1 : 1 \leftrightarrow 2$ and $1 \leftrightarrow 3$; $\lambda_2 : 2 \leftrightarrow 3$ and $2 \leftrightarrow 4$; $\lambda_3 : 1 \leftrightarrow 4$ and $3 \leftrightarrow 4$. There are also 3 wavelengths on the ring but since each wavelength drops at only three nodes, 9 ADM's are needed this time, 3 ADM's are saved compared to the scenario in Fig. 5.1(a).

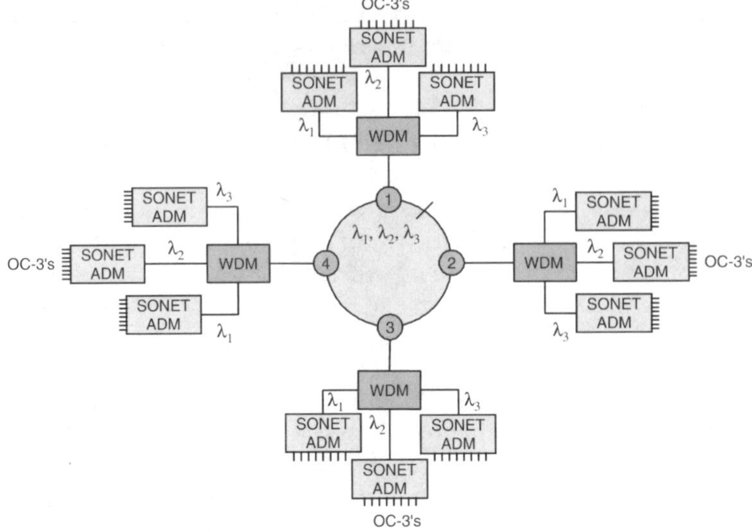

(a) A possible grooming scenario with 12 ADM's

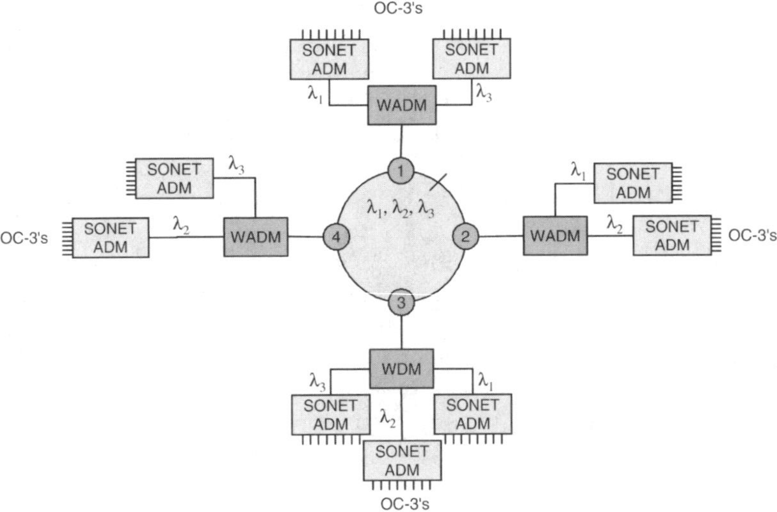

(b) Another possible grooming scenario with only 9 ADM's

Fig. 5.1. A grooming example on a 4-node unidirectional ring with two different scenarios

In these two grooming scenarios, the traffic is carried by the same number of wavelengths and every wavelength works at their full capacity, but the number of ADM's required is different. This shows that after appropriate grooming, not only can the burden of the electric devices be alleviated but the amount of them can be reduced. In large-scale networks, the reduction of the amount of AMDs due to good grooming will bring about a remarkable saving to the cost of constructing them. Therefore, traffic grooming is one of the problems that must be taken into serious account in the design of today's high-speed telecommunication networks.

According to whether traffic changes or not, there are basically two grooming categories: static grooming and dynamic grooming. The former is to groom a known and unchanged traffic pattern in order to use the minimum number of ADM's to support it. The latter is to groom a set of changing traffic patterns, either known or unknown, in order to use the minimum number of ADM's to support this changing traffic. It was shown that traffic grooming is NP-hard even for a unidirectional ring with only one egress node and nonuniform traffic. The grooming of arbitrary traffic on general network topology is even difficult. Therefore, evolutionary approach is a good candidate for it. The following sections will discuss the evolutionary approaches to the grooming of static and dynamic traffic.

5.3 Evolutionary Grooming of Static Traffic

This section presents some important algorithms for grooming of static traffic in WDM networks. Grooming of static traffic has been one of the most interested topics in the literature. Although it is more practical to groom dynamic traffic, the grooming of static traffic can not only serve as an approximate to the grooming of dynamic traffic but also provide a basic methodology to it. All the major evolutionary techniques, i.e., genetic algorithm (GA), tabu search (TS), and simulated annealing (SA), have been applied to this kind of grooming. Up to now, the main efforts are devoted to ring and mesh topologies, and we will discuss each of these approaches to this kind of grooming in the following subsections.

5.3.1 Grooming of Static Traffic in WDM Ring Networks

It is reasonable to groom traffic in ring networks due to its widespread use in today's infrastructural networks. Although many researchers have already worked in this field right from the beginning of the TG problem being proposed, and great achievements have been reached [1–3, 5, 6, 10–23], there are still some regions remained to be explored. Today, many researchers still try to develop new heuristics for solving TG problem in WDM ring networks efficiently [24, 25]. In this part, we explain the applications of evolutionary-based approaches for the static traffic grooming in WDM ring networks. Although there are many different approaches, we mainly focuse on two typical approaches: circle construction based two-step model and directly traffic matrix based operation.

5.3.1.1 Circle Construction Based Two-Step Model

The circle construction based model was first proposed in [2]. In this model, all the traffic connections are first divided into the basic traffic rate, and then as few circles as possible are constructed to include all the requested connections, where each circle consists of multiple nonoverlapping (i.e., link disjoint) connections. The circle construction heuristic is shown in Fig. 5.2. In this algorithm, it searches the traffic matrix from the beginning to the end, and then selects the traffic requests in the order of the longest to the shortest stride to construct circles. The connection will be added into the circle under the condition that at least one end node of it already existing in the circle, otherwise, it will be put into a so-called *GapMaker* list which is initially empty. When this process ends, a greedy algorithm is employed to check the GapMaker list to construct as few circles as possible with minimum number of end nodes. For example, given a 5-node ring network with a traffic matrix shown in Fig. 5.3(a). Let i and j denote the source and the destination node of a traffic request and r_{ij} represent the traffic request from node i to node j, a possible circle may be constructed by the traffic stream 0-4 and 4-0. But as r_{ij} is not necessarily equal to r_{ji}, only five circles can be constructed for r_{04} and r_{40}, and two basic traffic requests belonging to r_{40} will be put into the GapMaker list. After the circles have been constructed, another heuristic will be used to groom circles onto wavelengths with overlapping as many end nodes belonging to different circles as possible so as to reduce the number of ADM's. If the granularity of a wavelength is 8, eight circles can be groomed onto a wavelength.

This model was widely applied to the grooming of static traffic in ring networks and two SA based and one GA based approaches were proposed based on it in [3, 15, 20] respectively. For the purpose of comparison, only the GA approach is shown below.

The GA approach proposed in [15] aimed to minimize the total number of ADM's. The operation of the GA can be illustrated as follows. A binary coding method is used and each bit ${x_i}^w$ represents an ADM needed on wavelength w at node i if it equals "1" or indicates no ADM necessary if it is "0". The fitness function can be presented as minimizing:
$\sum_{w=1}^{W} \sum_{i=1}^{N} x_i^w$ (where W and N are the total number of wavelengths and nodes in the ring respectively).

The traffic request is first divided into basic elements (the lowest speed rate), then the circle construction heuristics is applied to construct C basic circles based on the traffic request. Finally a GA is employed to grooming circles onto wavelengths.

A seed obtained from the greedy circle construction grooming algorithm shown in Fig. 5.2 is inserted into the initial population and the rest chromosomes are generated by random permutation of the C circles. In decoding process, two methods, next-fit (NF) and best-fit (BF), are applied to assign circles to wavelengths in accordance with the order of them in the chromosome.

```
C2=1; // C2 is the number of circles being constructed by this algorithm so far
s0 = N-1 for unidirectional rings or s0 = N/2 for bidirectional rings; // the maximum stride
for (s = s0; s >=1; s --) {   // consider connections with longer strides first
    for (i = 0; i <N; i++) {   // one node at a time
        while (hi,s > 0) {
            for (c=1; c<=C2; c++){ // try all circles if necessary
                try to fit connection (i,s) into circle c without creating an additional gap;
                if (succeed) {
                    hi,s =hi,s -1;
                    break from the inner "for" loop; // no need to try the remaining circles
                } //else fail and then try the next circle
            }
            if (fail because (i,s) would overlap with some connections in any existing circle) {
                do {
                    generate a new circle for (i,s); hi,s =hi,s-1; C2= C2+1;
                } while (hi,s>0);
            }
            if (fail because fitting (i,s) in any existing circle would generate an additional gap)
            {
                do {
                    put (i,s) into the GapMaker list (to be processed later); hi,s =hi,s-1;
                } while (hi,s>0);
            }
        }
    }
} // all hi,s have been include either in the circles or in the GapMaker list
if(the objective is to minimize the number of circles) {
    for (each connection in the GapMaker list ){
        try to fit into existing circles;
        if (fail) { // because it overlaps with existing connections
            generate a new circle for the connection; C2= C2+1;
        }
        remove the connection from the GapMaker list;
    }
}
if (the objective is to minimize the number of end nodes) {
    for (each connection in the GapMaker list) {
        try to fit into existing circles without creating an additional gap;
        if (fail) { // because it either overlaps with existing connections or it create an additional
                gap
            generate a new circle for the connection; C2= C2+1;
        }
        remove the connection from the GapMaker list;
    }
}
```

Fig. 5.2. Circle construction heuristics (source: reference [2])

NF algorithm distributes circles evenly among all wavelengths but BF tries to fully pack every wavelength except possibly the last one. The corresponding algorithms are shown in Fig. 5.4. An exponential ranking selection scheme is adopted with the probability $p_i = (r-1)^* r^{M-i}/(r^M - 1)$ for the ith individual to be selected in M total individuals, where r is set to be 0.99. Two crossovers, partially mapped crossover (PMX) and order crossover (OX), are used to produce offspring. In mutation, the elements of an individual are right-shifted by a random number of positions in a cyclic manner.

(a)

n	0	1	2	3	4
0	0	5	3	1	5
1	4	0	1	3	4
2	1	3	0	5	3
3	2	3	6	0	5
4	7	2	4	3	0

(b)

n	0	1	2	3	4
0	0	4	6	6	2
1	3	0	3	3	4
2	5	4	0	3	4
3	6	2	5	0	3
4	3	7	1	2	0

(c)

n	0	1	2	3	4
0	0	5	6	6	5
1	4	0	3	3	4
2	5	4	0	5	4
3	6	3	6	0	5
4	7	7	4	3	0

Fig. 5.3. Three sample traffic matrices in a WDM ring network with 5 nodes (source: reference [19])

```
Next-fit greedy circle grooming algorithm
/* Mw   is the number of circles to be groomed onto wavelength w */
if( trying to distribute circles evenly onto wavelengths)
{    Cg=0; // the number of circles groomed so far
     for (w=W; w>=1; w --)
     {  W'=[(C-Cg/G)];
        Mw =[(C-Cg/W')];
        Cg =Cg +Mw   }  }
if(grooming circles onto the first possible wavelength checked in a fix order)
{    for (w=1; w <W; w ++)
        Mw =G;
        Mw =C mod G;   }
w=1;
for( c=C; c >=1; c --)
{ groom circle onto wavelength w until there are Mw circles on w;
   w++;
}
```

```
Best-fit greedy circle grooming algorithm
/* Mw   is the number of circles to be groomed onto wavelength w */
if( trying to distribute circles evenly onto wavelengths)
{    Cg=0; // the number of circles groomed so far
     for (w=W; w >=1; w--)
     {  W'=[(C-Cg/G)];
        Mw =[(C-Cg/W')];
        Cg =Cg +Mw   }  }
if(grooming circles onto the first possible wavelength checked in a fix order)
{    for (w=1; w<W; w ++)
        Mw =G;
     Mw =C mod G;   }
for (c=C; c =1; c --)
{  find wavelength w that can accommodate circle c with the addition of fewest nodes;
   if(there are multiple wavelengths found)
        choose one with the fewest circles;
   Mw = Mw-1;
   Groom circle c onto wavelength w;
}
```

Fig. 5.4. NF and BF algorithms (source: reference [15])

5.3.1.2 Direct Traffic Matrix Based Grooming Method

The GA based circle construction model may not be very efficient for large scale networks. Therefore, another GA-based approach directly operating on traffic matrix and grooming the traffic in a comprehensive way was proposed in [16]. The framework of the GA is described below.

First, the traffic matrix is converted into a vector **V**, and a random permutation of the traffic items is used as a chromosome, $\mathbf{V} = (x_0, \ldots, x_k, \ldots, x_{n(n-1)} - 1)$ if there are n nodes in a ring. The position of each item is used to determine the decoding order. The connection between the component's subscript k of the vector **V** and those of the traffic matrix can be formulated as:

$$\begin{cases} k = (2n - i - 1) \times i/2 + j - i, & 0 \leq i < n - 1,\ i + 1 \leq j < n \\ \quad \text{or} \quad 0 \leq k < n(n-1)/2 & \\ k = \dfrac{n\,(n-1) + i\,(i-1)}{2} + j, & 1 \leq i < n, \quad 0 \leq j < i \\ \quad \text{or} \quad n(n-1)/2 \leq k < n(n-1) & \end{cases} \tag{5.1}$$

The algorithm in Fig. 5.5 can be used to calculate i, j from k:

The basic ideas behind the decoding algorithm proposed in this approach can be described as follows. If a wavelength drops at π nodes, it can at most accommodate $\pi(\pi - 1)$ traffic requests. If all the $\pi(\pi - 1)$ traffic requests are assigned to the wavelength, the numbers of ADM's and wavelengths used on the ring must be minimized. Based on this observation, a first-fit heuristic incorporated with a greedy local improvement was proposed to decode each chromosome, as is shown in Fig. 5.6. The algorithm assigns the first encountered traffic items onto each wavelength until one link load on the wavelength exceeds its capacity. Then the local improvement algorithm is called to assign such a traffic item to it with no additional node added. The objective of this algorithm is to minimize the number of ADM's. Therefore, the fewer ADM's a chromosome uses, the fitter it is. If two chromosomes require the same number of ADM's, the one with smaller number of wavelengths is assigned higher fitness value. The order-mapped crossover (OMX) proposed in [26] is used to produce offspring and the inversion of a section of some randomly selected genes is used as mutation in this algorithm.

```
if(k<n*(n-1)/2)
{
   for(i=0; i<n; i++)
   { k=k-n+1; if (k<=0) break; }
   j=k+n;
}
else
{
   k=k-n*(n-1)/2-1;
   for(i=1; i<n; i++)
   {k=k-i; if (k<=0) break; }
   j=k+i-1;
}
```

Fig. 5.5. The algorithm for calculating i, j from k (source: reference [16])

```
begin with the first wavelength;
for (k=1; k <=n(n-1); k++) {
    if (the current wavelength can accommodate traffic [b_k])
        assign traffic [b_k] to the current wavelength;
    else{
        for (l=k+1; l <=n(n-1); l++) {
            if (the current wavelength can accommodate traffic [b_l]
                && no new node is added to it)
            {
                assign traffic [b_l] to the current wavelength;
                exchange gene b_k with b_l;
                break;
            }
        }
        begin with a new wavelength;
    }
}
```

Fig. 5.6. The decoding algorithm with local greedy improvement (source: reference [16])

Table 5.1. A Grooming result for the traffic matrix

Wavelength	Traffic Request	Add/Drop Nodes	Link Load
W1	$\mathbf{r}_{01}, \mathbf{r}_{10}, \mathbf{r}_{13}, \mathbf{r}_{31}$	0,1,3	8,7,7,7,7
W2	$\mathbf{r}_{02}, \mathbf{r}_{20}, \mathbf{r}_{04}, \mathbf{r}_{40}$	0,2,4	8,8,6,6,8
W3	$\mathbf{r}_{03}, \mathbf{r}_{30}, \mathbf{r}_{43}, \mathbf{r}_{34}, \mathbf{r}_{41}$	0,1,3,4	6,4,4,7,7
W4	$\mathbf{r}_{12}, \mathbf{r}_{21}, \mathbf{r}_{24}, \mathbf{r}_{42}$	1,2,4	7,5,6,6,7
W5	$\mathbf{r}_{23}, \mathbf{r}_{32}$	2,3	6,6,5,6,6
W6	$\mathbf{r}_{14}$	1,4	0,4,4,4,0

A possible grooming result based on this decoding method is shown in Table 5.1 for the traffic matrix shown in Fig. 5.3(a). In this example, it is assumed that the traffic granularity is 8, and this assumption is always adapted when discussing the traffic matrices shown in Fig. 5.3.

Fig. 5.7 gives the results of the number of ADM's and wavelengths vs. the number of nodes based on the unidirectional ring with 8–16 nodes and $G = 16$. In Fig. 5.7(a) and (b), two curves named as 'one node added' and 'no nodes added' are shown, which refer to the cases that when the greedy improvement algorithm is called to assign the additional traffic items to the spare links on existing wavelengths, whether one additional ADM is allowed to be added or not. When an additional ADM is not allowed each time, further saving on ADM's can be achieved at the cost of larger number of wavelengths. Fig. 5.7(c) shows the percentage savings on the numbers of both ADM's and wavelengths over their upper bounds and the percentage of increments of them over the theoretical lower bounds respectively.

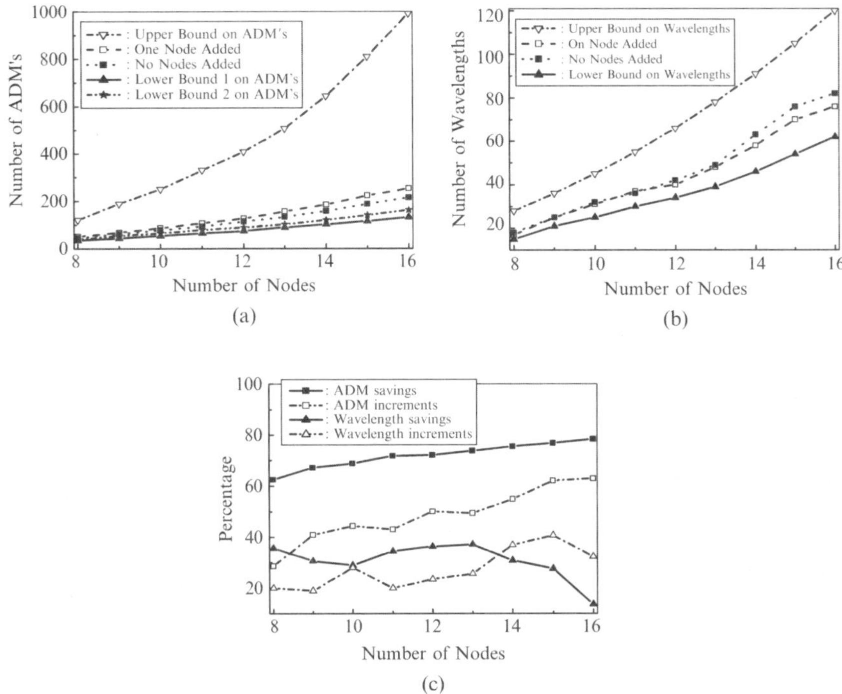

Fig. 5.7. The results of the GA (source: reference [16])

Compared with the GA approach shown in subsection 5.3.1.1, the advantage of this algorithm may be shown in the following two aspects. Firstly, the number of wavelengths can be adjusted dynamically with the assignment of the traffic requests. So a better scheme of traffic assignment can lead to a reduced number of both ADM's and wavelengths. Secondly, different from the first GA approach, fitness function in this algorithm can be used to minimize both numbers of ADM's and wavelengths. Furthermore, with only one step to execute, better results can be assured. But due to the absence of traffic splitting techniques, there would exist some spare link loads, which is a waste of resource.

5.3.1.3 Other Evolutionary Approach

Apart from the above two algorithms, there are still some other evolutionary approaches to the static grooming problem.

Two GA approaches were proposed in [14, 17]. One approach uses the chromosome to denote the state of lightpaths and a sub-chromosome L to indicate the usage of wavelengths in each lightpath. These two chromosomes were arranged in a hierarchical manner to represent the assignment of wavelengths on the ring. In another approach, the chromosomes stand for the assignment of wavelengths for the input traffic matrix. By adding a penalty term to the objective function, the illegal chromosomes were ruled out so that the chromosomes were not checked for the validity under the constraint of the wavelength capacity. Furthermore, two SA approaches can be found in [3,20]. They are not described in detail here and the interested readers can refer to the corresponding references in [3,14,17,20].

5.3.2 Grooming of Static Traffic in WDM Mesh Networks

Most of the early efforts were devoted to the traffic-grooming problems in ring networks due to its simplicity and widespread use. With the evolution of optical networks from ring to irregular topologies, the grooming of traffic in mesh networks is becoming a prominent research area as mesh topology is more general and practical, especially in IP/WDM networks where packages are usually routed in such networks.

The traffic-grooming problem in mesh networks can be divided into two sub-problems: the routing problem and the grooming problem, and each of the sub-problems is NP-hard. So it is hard to design a good evolutionary algorithm for them as well. As a result, even when the traffic pattern is static, the traffic-grooming problem in WDM mesh networks is still a great challenge. Although grooming of traffic in mesh networks has been investigated recently with many solutions based on other heuristic algorithms [27–30], there are only two papers focusing on the use of evolutionary approach [31,32], and both of them were based on GA.

However, these two approaches are too complicated to use. For example, in [31], a binary encoding scheme called the position-based bit representation (PBR) was proposed to find a proper route in mesh networks. In this scheme, each bit represents an edge of the given mesh networks, and a bit is 1 if the corresponding edge is used for setting up a routing path, and 0 if the edge is absent. So each chromosome could represent a routing path. Then the authors designed a heuristic crossover and mutation operator for avoiding producing illegal offspring and maintaining the path diversity by using a similarity metric. It is obvious that the routing solution is too complex for large-scale networks with a large number of connections (edges), because the routing search space would be very large with the increase of the possible routes. Furthermore, splitting (the corresponding concept will be explained in the next section), which is important for the efficient design of the mesh networks, is not allowed according to the coding scheme. While in [32], the algorithm avoids the problem of routing by searching the shortest path with the Dijkstra algorithm. The traffic stream was split so that a same size of the stream was processed in each route. Of course, more efficient splitting method could be used to further improve the grooming results and this will be discussed in the following subsections.

5.3.3 Conclusion

For all the algorithms shown above, they are of strong points and weak points. It is important to note that although the static grooming problem has been investigated for many years and is relatively simple compared to other grooming problems, there are still no efficient evolutionary approaches proposed based on large-scale real world network models. What is more, due to the complexity of its topology, there is still much work left for the grooming of traffic in mesh networks using evolutionary algorithms. This would be an interesting future work.

Table 5.2 summarizes various evolutionary approaches in the grooming of static traffic, from which we know the strong and weak points of each approach.

Table 5.2. Comparison of Evolutionary approaches in the Grooming of Static Traffic

Topology	Algorithm and Method	Strong Points	Weak Points	Reference
Ring Networks	GA, hierarchical chromosome	The state of lightpaths and the usage of wavelengths being marked separately	Only being applied to a very small problem	[13]
	GA, circle construction model	With clear meaning and quite straightforward	Unable to obtain the optimal results	[14]
	GA, directly operating on traffic matrix	Having the potential to obtain the optimal results	Did not consider traffic split	[15]
	GA, wavelength assignment based chromosome	With clear meaning and quite straightforward	Without an efficient method to avoid illegal individuals	[17]
	SA, circle construction model	With clear meaning and quite straightforward	Unable to obtain the optimal results	[3]
	SA, circle construction model with traffic-cutting	With clear meaning and quite straightforward	Unable to obtain the optimal results	[19]
Mesh Networks	GA, combined with clustering heuristics	With clear meaning and quite straightforward	Unable to search all routes in large-scale networks	[31]
	GA, coding the whole grooming scheme	Solving the routing problem with the Dijkstra algorithm	Unable to split the traffic efficiently	[32]

5.4 Evolutionary Grooming of Dynamic Traffic in Ring Networks

When traffic changes dynamically over time, we find that traffic requests in the traffic matrix $\boldsymbol{R}^m$ are not always at their maxima. That is, at one time, some of the traffic requests are at their higher level while others are at their lower level, at another time, the situation would be reversed. Therefore, the grooming of multiple traffic patterns can make good use of such change and enables the network designer to consider the traffic matrix in a comprehensive way in order to save more ADM's (as well as wavelengths) compared to the grooming of the maximum traffic.

Compared to static traffic grooming, it is more practical and general to groom dynamic traffic patterns since static traffic is merely a special case or a simplification of dynamic traffic. In the grooming of dynamic traffic, we often use a set of traffic patterns to represent the change of it. There are mainly two categories of grooming [6]: strictly nonblocking grooming and rearrangeably nonblocking grooming. If the same traffic request in the set of traffic patterns can be assigned to the same wavelengths without rearranging the assignments among them when traffic changes, it is strictly nonblocking grooming. By this way, all requests in any new traffic pattern can be established without being interrupted. On the contrary, rearrangeably nonblocking grooming is established by assigning each traffic request to the existing wavelengths with possibly rearranging some of the requests among them but the set of nodes each wavelength drops at is kept unchanged. Due to the possible rearrangement of the requests among different wavelengths, some of the existing traffic may be interrupted when new traffic is established. Nevertheless, this grooming category is usually more efficient in saving ADM's and wavelengths compared to the grooming of strictly nonblocking, since more ADM's can be shared by rearranging some requests in different traffic patterns.

The grooming of dynamic traffic is much more difficult than the grooming of static traffic. As for evolutionary approaches, only GA and TS have been applied to this kind of grooming by now. Furthermore, the problem of grooming dynamic traffic based on evolutionary approaches in other topologies WDM networks still remains unexplored, and the reason will be given in the Section 5.6. Below we will describe in detail these two evolutionary approaches to these two dynamic grooming categories in WDM ring networks.

5.4.1 Genetic Algorithms in Dynamic Traffic Grooming

Like static grooming, there are mainly two different approaches to realizing dynamic traffic grooming based on GA: circle construction based two-step approach and operation directly on traffic matrix. A typical GA approach based on the first model was proposed in [22] to grooming dynamic traffic in unidirectional ring network in a rearrangeably nonblocking way. Similar to the algorithm of grooming of static traffic in the ring, this approach first treats the dynamic traffic patterns at each moment as a static one and then follows the heuristic in [2] to construct circles. As this algorithm is a bit complicated and only designed for t-allowable traffic patterns and the grooming process has to be completed in three steps, it will not be discussed in detail here. Interested readers are recommended to refer to Ref. [22]. The typical GA for strictly and rearrangeably nonblocking grooming of traffic based on the second model will be discussed in the following.

5.4.1.1 Strictly Nonblocking Traffic Grooming

A GA approach was proposed in [6] to carry out strictly nonblocking grooming of dynamic traffic in ring networks which operates directly on the traffic matrix, and further improvement is given in [18] by combining splitting into this method. In the following we mainly describe the algorithm with no splitting.

Given a ring network, the dynamic traffic requests are represented by a set of M $n \times n$ traffic matrices $\boldsymbol{R} = \{\boldsymbol{R}^m\}$ $(m = 0, 1, \ldots, M-1)$, each of which represents the traffic requests at time $t = m$. Each element ${r_{ij}}^m$ of the traffic pattern $\boldsymbol{R}^m$ represents the traffic request originating from node i and terminating at node j at time t. The traffic matrix $\boldsymbol{R}^m$ is first converted into an $n(n-1)$-dimensional vector

$$X^m = (x_0^m, \ldots .x_k^m, \ldots .., x_{n(n-1)-1}^m), m = 0, 1, \ldots .., M-1. \tag{5.2}$$

The relationship between the component's subscript k of the vector $\boldsymbol{X}^m$ and those of the traffic matrix is the same as was given in Eq. (5.1). At each time only one traffic pattern in this set is activated and all the M traffic patterns will be constructed one by one as traffic changes. The chromosome representation is the same as was described in 5.3.1.2. But this time, one chromosome represents the same traffic element originating from the same node and terminating at another same node in the set of traffic vectors $\boldsymbol{X}^m$.

In decoding, a first fit approach incorporated with a greedy local improvement algorithm is proposed to assign traffic items in the chromosome to the current wavelength. Different from the decoding algorithm shown in Fig. 5.6, the algorithm takes all the M traffic matrices into consideration at the same time. That is, all the traffic requests or segments represented by the same gene in the chromosome are assigned to the same wavelength and the chromosome is decoded with each of the M traffic patterns one by one. In the decoding process, the algorithm examines the first encountered traffic item b_k in the chromosome. The traffic b_k is assigned to the current wavelength only if all the traffic requests represented by it in the set of M traffic patterns can be assigned to it. If the traffic can not be assigned to the current wavelength, the greedy local improvement algorithm is called to examine the remaining genes one by one until a new traffic b_l is found which can be assigned to the wavelength without additional ADM, or all the genes have been examined with no

```
begin with the first wavelength;
for (k=1; k <= n(n-1); k++) {
        for (m=1; m <=M; m++)
                if (the current wavelength cannot accommodate traffic [m][b_k])   break;
                if (m >M)
                        for (m=1; m <=M; m++)
                                assign traffic [m][b_k] to the current wavelength;
                else{
                        for (l=k+1; l <=n(n-1); l++) {
                                for (m=1; m <=M; m++)
                                        if (the current wavelength cannot accommodate traffic [m][b_l] ||
                                                one or two new nodes have been added to it)   break;
                                if (m > M){
                                        for (m = 1; m < = M; m++)
                                                assign traffic [m][ b_l] to the current wavelength;
                                        exchange gene b_k with b_l;
                                        break;
                                }
                        }
                        begin with a new wavelength;
                }
}
```

Fig. 5.8. Algorithm for decoding a chromosome in strictly nonblocking manner (source: reference [6])

such traffic found. In the first case, the new traffic b_l is assigned to the current wavelength, the two genes (b_k and b_l) are exchanged and then the algorithm proceeds with the next wavelength. In the second, the algorithm just stops accommodating the current wavelength and then begins with the next wavelength. This process proceeds until all the genes in the chromosome have been assigned to appropriate wavelengths. Fig. 5.8 realizes this algorithm. In this GA approach the OMX crossover operator and the random inversion mutation operator are applied.

To understand the coding algorithm, take the sample traffic matrices shown in Fig. 5.3 as an example. Let Fig. 5.3(a) and (b) denote the traffic matrices at two different times in a ring network. In addition, to illustrate the advantage of the dynamic traffic grooming, these two traffic patterns are treated as a static one, then it is necessary to groom the max-traffic matrix, which is shown as Fig. 5.3(c), to support the traffic request at these two moments. Table 5.3 shows a possible result for grooming the max-matrix in a static way and Table 5.4 gives the grooming results by a strictly nonblocking way. We can see that with a static method, 19 ADM's and 8 wavelengths are required but with a strictly nonblocking grooming method, only 18 ADM's and 7 wavelengths are required. With the increase of the number of nodes in the network, more ADM's and wavelengths can be saved by using strictly nonblocking grooming method and this is a great benefit for network designer. We will see from the further grooming results that it consumes some more

Table 5.3. Grooming results for the max-matrix in a static way

Wavelength	Traffic Request	Add/Drop Nodes	Virtual Link Load	Link Load1	Link Load2
W1	r_{01}, r_{10}	0,1	5,4,4,4,4	5,4,4,4,4	4,3,3,3,3
W2	r_{02}, r_{20}	0,2	6,6,5,5,5	3,3,1,1,1	6,6,5,5,5
W3	r_{03}, r_{30}	0,3	6,6,6,6,6	1,1,1,2,2	6,6,6,6,6
W4	r_{04}, r_{40}	0,4	5,5,5,5,7	5,5,5,5,7	2,2,2,2,3
W5	r_{12}, r_{21}, r_{13}, r_{31}	1,2,3	7,6,7,7,7	6,4,6,6,6	6,6,7,6,6
W6	r_{14}, r_{41}, r_{24}	1,2,4	7,4,8,8,7	2,4.7,7,2	7,4,8,8,7
W7	r_{23}, r_{32}	2,3	6,6,5,6,6	6,6,5,6,6	5,5,3,5,5
W8	r_{34}, r_{43}, r_{42}	2,3,4	7,7,3,5,7	7,7,3,5,7	3,3,2,3,3

Table 5.4. Grooming results for the two traffic matrices in a strictly nonblocking way

Wavelength	Traffic Request	Add/Drop Nodes	Link Load1	Link Load2
W1	r_{01}, r_{10}, r_{13}, r_{31}	0,1,3	8,7,7,7,7	6,6,6,5,5
W2	r_{02}, r_{20}, r_{04}, r_{40}	0,2,4	8,8,6,6,8	8,8,7,7,8
W3	r_{03}, r_{30}, r_{43}	0,3,4	4,4,4,2,5	8,8,8,6,8
W4	r_{12}, r_{21}, r_{24}, r_{42}	1,2,4	7,5,5,5,7	5,4,8,8,5
W5	r_{23}, r_{32}	2,3	6,6,5,6,6	5,5,3,5,5
W6	r_{14}, r_{41}	1,4	2,4,4,4,2	7,4,4,4,7
W7	r_{34}	3,4	0,0,0,5,0	0,0,0,3,0

ADM's and wavelengths compared to the splitting method. And it was found that if a pure GA with no local improvement was used, its performance is worse than the heuristic alone.

The benefit of such grooming method is that it can make full use of the dynamics of the changing traffic to make some compensation among the traffic requests between different pairs of nodes. When traffic changes dynamically, the traffic requests at their lower level can compensate for those at their higher level. In this way, more ADM's as well as wavelengths can be shared compared to the grooming of maximum traffic and as a result, fewer ADM's and wavelengths will be used.

One problem with this approach is that some "gaps" would possibly exist on each wavelength during grooming. By fitting as many gaps as possible with additional traffic, more ADM's and wavelengths can be saved. The splitting method is one of the possible ways to be combined into the GA to reduce the possible gaps. When the splitting method is applied, the framework of the GA need not be significantly adjusted except for the decoding process.

Splitting means to divide a traffic request into smaller parts or shorter segments so that they can be fitted into the gaps of wavelengths. There are generally two splitting methods: traffic-cutting and traffic-dividing [18, 19]. The first means to cut a traffic request into shorter segments so that each of them can be fitted into the gaps of the wavelengths. For instance, traffic request r_{04} can be cut into two segments: r_{02} and r_{24}, in which each segment still possesses the same traffic rate. The second is to divide a traffic request into smaller parts operating at lower rates so that the lower-rated parts can be assigned to the remained capacity of the wavelengths. For instance, if a traffic request operates at the rate of OC-24 but the current wavelength only has a capacity of OC-6 left, the part OC-6 of the traffic can be assigned to it and the rest will be assigned to other wavelength(s). A possible solution to traffic given in Figs. 5.3(a) and (b) with splitting method is shown in Tables 5.5 and 5.6. Compared to the solution without splitting in Table 5.4 we find that 1 ADM and 1 wavelength are saved with this method. It should be noted that to achieve the strictly nonblocking grooming of traffic, the same cut segment or divided parts of a traffic request at different moment must be assigned to the same wavelength. Another point is that, with *traffic-cutting* method, the original single-hopped traffic becomes multi-hopped and some intermediate nodes are needed to bridge it. A traffic flow should not be cut into too many segments to avoid making the network's virtual topology too complex. Although with the *traffic-dividing* method the traffic will remain single-hopped, the network's virtual topology will become more complex too. Hence, a traffic flow should not be divided into too many parts either.

Table 5.5. Grooming results with Traffic-Cutting method for the two traffic matrices in a strictly nonblocking way

Wavelength	Traffic Request	Add/Drop Nodes	Link Load1	Link Load2
W1	r_{01}, r_{10}, r_{13}, r_{31}	0,1,3	8,7,7,7,7	6,6,6,5,5
W2	r_{02}, r_{20}, r_{04}^{03}, r_{40}, r_{34}	0,2,3,4	8,8,6,6,8	8,8,7,7,8
W3	r_{03}, r_{30}, r_{43}, r_{04}^{34}	0,3,4	4,4,4,7,5	8,8,8,8,8
W4	r_{12}, r_{21}, r_{24}, r_{42}	1,2,4	7,5,5,5,7	5,4,8,8,5
W5	r_{23}, r_{32}	2,3	6,6,5,6,6	5,5,3,5,5
W6	r_{14}, r_{41}	1,4	2,4,4,4,2	7,4,4,4,7

Table 5.6. Grooming results with Traffic-Dividing method for the two traffic matrices in a strictly nonblocking way

Wavelength	Traffic Request 1	Traffic Request 2	Add/Drop Nodes	Link Load1	Link Load2
W1	r_{01}, r_{10}, r_{13}, r_{31}	r_{01}, r_{10}, r_{13}, r_{31}	0,1,3	8,7,7,7,7	6,6,6,5,5
W2	r_{02}, r_{20}, r_{04}, r_{40}	r_{02}, r_{20}, r_{04}, r_{40}	0,2,4	8,8,6,6,8	8,8,7,7,8
W3	r_{03}, r_{30}, r_{43}, r_{34}^{4}	r_{03}, r_{30}, r_{43}, r_{34}^{2}	0,3,4	4,4,4,6,5	8,8,8,8,8
W4	r_{12}, r_{21}, r_{24}, r_{42}	r_{12}, r_{21}, r_{24}, r_{42}	1,2,4	7,5,5,5,7	5,4,8,8,5
W5	r_{23}, r_{32}	r_{23}, r_{32}	2,3	6,6,5,6,6	5,5,3,5,5
W6	r_{14}, r_{41}, r_{34}^{1}	r_{14}, r_{41}, r_{34}^{1}	1,3,4	2,4,4,5,2	7,4,4,5,7

When applying the *traffic-cutting* method, it is better to restrict the number of cut traffic requests to about the number of the wavelengths used on the ring so that each wavelength would accommodate about two cut segments in average. At the same time, the traffic request has to be cut to fill the longest spare link on a wavelength in order to further save ADM's. Without proper control, too many traffic requests would be cut and each wavelength would contain too many short segments which would result in consuming more ADM's. For the *traffic-dividing* method, a traffic request is divided only if both of its end nodes have already been the dropping nodes of a wavelength in order not to add new ADM's during assignment and there should be enough spare capacity in that wavelength.

A more complicated splitting method can be further derived by combining these two splitting methods, i.e., by dividing a traffic request into two slices first and then cutting them into smaller segments to fit them into smaller gaps of a wavelength. This method is referred to as *Synthesized-Splitting*. To our experience, a traffic request can be divided into two or more parts, but it is better to cut it into at most two parts in order not to make the network's virtual topology too complex.

By incorporated with splitting methods, the decoding algorithm of the GA approach is a slightly different from the original one, and is shown in Fig. 5.9.

Step 1. k=0, s=0 and w=1. Set the split list empty;
Step 2. If each of the M traffics represented by the traffic item b_k in the chromosome can be assigned to wavelength w, assign each of them to it. Otherwise, go to Step 11;
Step 3. l=k+1;
Step 4. If each of the M traffics represented by the traffic item b_l can be assigned to wavelength w with no additional ADM added to it, assign each of them to it, k=k+1, exchange traffic b_k with b_l;
Step 5. l=l+1. If $l< n(n$-1), go to Step 4;
Step 6. j=0. If $j\geq s$, go to Step 10;
Step 7. If each of the M traffics represented by the traffic item b_j in the split list can be assigned to wavelength w with no additional ADM added to it, assign each of them to it. Otherwise, go to Step 9;
Step 8. Delete traffic b_j from the split list and move all the traffic items behind it backward to fill that place, s=s-1. If $j<s$, go to Step 7;
Step 9. j=j+1. If $j<s$, go to Step 7;
Step 10. k=k+1. If $k< n(n$-1), go to Step 2;
Step 11. l=k. If $l\geq n(n$-1), go to Step 17;
Step 12. i=k;
Step 13. If each of the M traffics represented by the traffic item b_i can be divided into two parts according to Conditions 3 and 4 given in Subsection 2.2, divide each of the M traffics into two parts, and assign one of them to wavelength w; the other part is left in the traffic matrix;
Step 14. i=i+1, if $i< n(n$-1), go to Step 13;
Step 15. If each of the same segments cut from the M traffics represented by the item b_l can be assigned to wavelength w according to Conditions 1 and 2 given in Subsection 2.1, cut each of the M traffics into the same two segments and assign one of them to wavelength w, put the other to the split list, k=k+1, s=s+1, and exchange traffic b_k with b_l;
Step 16. l=l+1. If $l< n(n$-1), go to Step 15;
Step 17. w=w+1, go to Step 2;
Step 18. Assign the remained traffic in the split list to wavelengths in accordance with Steps 2 to 4.

Fig. 5.9. Algorithm for decoding a chromosome with the splitting methods in strictly nonblocking manner (source: reference [18])

In this algorithm, Steps 1–11 and 17–18 are the grooming with no splitting, which are just similar to the algorithm shown in Fig. 5.8; Steps 15 and 16 are employed to accomplish the *traffic-cutting* method; Steps 12–14 accomplish the *traffic-dividing* method. If only one splitting method is applied, the corresponding steps of the other method should be ignored. And if the algorithm is executed as a whole, the *Synthesized-Splitting* algorithm is also realized.

In detail, Step 2 tries to assign the first-encountered traffic item b_k represented by the chromosome to the current wavelength. Step 4 examines the remaining items one by one to assign such traffic items to the current wavelength whose end nodes have already been the dropping nodes of it to minimize the number of ADM's. Steps 7–8 assign the traffic requests in the split list to the current wavelength with no additional nodes added. Then Step 13 and 15 try to divide and cut traffic requests represented by the chromosome into two parts and assign one of them to the current wavelength without adding nodes respectively. Step 17 starts to accommodate a new wavelength. At last, Step 18 assigns the remaining traffic requests in the split list to wavelengths without splitting. In this way, after all the traffic requests represented by the chromosome have been assigned to appropriate wavelengths, the number of ADM's and wavelengths used in this chromosome can be computed, which determine the fitness value of the chromosome.

Fig. 5.10 gives the grooming results for two traffic patterns in unidirectional ring networks. Fig. 5.10 (a) and (b) give the number of ADM's and wavelengths vs. the number of nodes in the rings respectively. The results for different number of traffic granularities are shown in Fig. 5.10 (c) and (d). From these figures, it is obvious that with splitting methods, further savings on both the number of ADM's and wavelengths can be achieved. We can see that the results obtained from the *Synthesized-Splitting* algorithm are closer to the lower bounds in nearly all the cases. With the increase of the number of nodes, the savings on ADM's and wavelengths become more obvious and the advantage of splitting method is more obvious as well. On the other hand, with the increase of traffic granularity, the different methods would lead to similar results on the number of ADM's, but with the splitting methods, there still could save some more wavelengths.

5.4.1.2 Rearrangeably Nonblocking Traffic Grooming

Rearrangeably nonblocking grooming of traffic is different from strictly nonblocking grooming. When the traffic matrices are groomed in a rearrangeably nonblocking way, the same traffic request at different time can be assigned to different wavelengths. In this way, it can be guaranteed that the same ADM's can be shared by more traffic requests to reduce the numbers of ADM's and wavelengths.

When splitting method is applied in this kind of grooming, the cut segment or divided part of a traffic request at different moment can be assigned to different wavelengths. To illustrate the difference, the possible solutions for

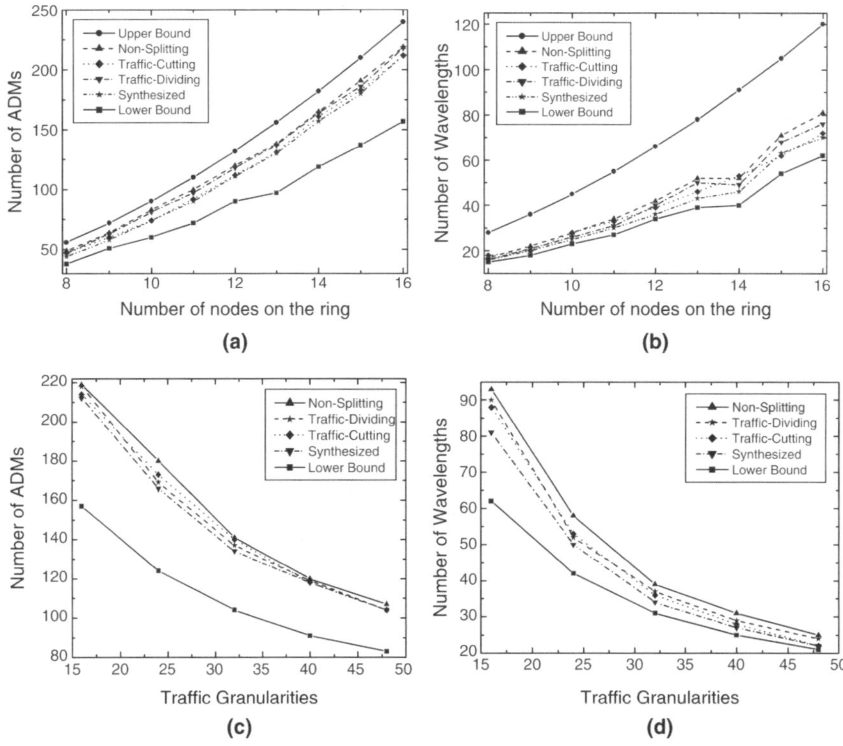

Fig. 5.10. Grooming results in strictly nonblocking manner (source: reference [18])

Table 5.7. Grooming results for the two traffic matrices in a rearrangeably nonblocking way

Wave-length	Traffic Request 1	Traffic Request 2	Add/Drop Nodes	Link Load1	Link Load2
W1	r_{01}, r_{10}, r_{03}, r_{30}, r_{13}	r_{03}, r_{30}, r_{31}	0,1,3	6,8,8,6,6	8,6,6,8,8
W2	r_{04}, r_{40}, r_{02}, r_{24}	r_{02}, r_{20}, r_{04}, r_{40}	0,2,4	8,8,8,8,7	8,8,7,7,8
W3	r_{12}, r_{21}, r_{14}, r_{41}	r_{14}, r_{41}, r_{24}, r_{42}	1,2,4	5,5,7,7,5	8,5,8,8,8
W4	r_{34}, r_{43}, r_{23}, r_{42}	r_{23}, r_{32}, r_{34}, r_{43}	2,3,4	7,7,8,5,7	7,7,5,8,7
W5	r_{20}, r_{32}	r_{01}, r_{10}, r_{12}, r_{21}	0,1,2,3	6,6,1,7,7	8,6,7,7,7
W6	r_{31}	r_{13}	1,3	3,0,0,3,3	0,3,3,0,0

grooming the two matrices shown in Fig. 5.3(a) and (b) in a rearrangeably nonblocking way are shown in Tables 5.7–5.9, without and with the splitting methods respectively. Compared to the results of the strictly nonblocking grooming shown above, it is clear that rearrangeably nonblocking grooming can achieve better performance.

Table 5.8. Grooming results with Traffic-Cutting method for the two traffic matrices in a rearrangeably nonblocking way

Wave-length	Traffic Request 1	Traffic Request 2	Add/Drop Nodes	Link Load1	Link Load2
W1	r_{01}, r_{10}, r_{03}, r_{30}, r_{13}	r_{03}, r_{30}, r_{31}	0,1,3	6,8,8,6,6	8,6,6,8,8
W2	r_{04}, r_{40}, r_{02}, r_{24}	r_{02}, r_{20}, r_{04}, r_{40}	0,2,4	8,8,8,8,7	8,8,7,7,8
W3	r_{12}, r_{21}, r_{14}, r_{41}, r_{31}^{41}	r_{14}, r_{41}, r_{24}, r_{42}, r_{13}^{12}	1,2,4	8,5,7,7,8	8,8,8,8,8
W4	r_{34}, r_{43}, r_{23}, r_{42}, r_{31}^{34}	r_{23}, r_{32}, r_{34}, r_{43}, r_{13}^{23}	2,3,4	7,7,8,8,7	7,7,8,8,7
W5	r_{20}, r_{32}	r_{01}, r_{10}, r_{12}, r_{21}	0,1,2,3	6,6,1,7,7	8,6,7,7,7

Table 5.9. Grooming results with Traffic-Dividing method for the two traffic matrices in a rearrangeably nonblocking way

Wave-length	Traffic Request 1	Traffic Request 2	Add/Drop Nodes	Link Load1	Link Load2
W1	r_{01}, r_{10}, r_{03}, r_{30}, r_{13}, r_{31}^{2}	r_{03}, r_{30}, r_{31}, r_{13}^{2}	0,1,3	8,8,8,8,8	8,8,8,8,8
W2	r_{04}, r_{40}, r_{02}, r_{24}	r_{02}, r_{20}, r_{04}, r_{40}	0,2,4	8,8,8,8,7	8,8,7,7,8
W3	r_{12}, r_{21}, r_{14}, r_{41}	r_{14}, r_{41}, r_{24}, r_{42}	1,2,4	5,5,7,7,5	8,5,8,8,8
W4	r_{34}, r_{43}, r_{23}, r_{42}	r_{23}, r_{32}, r_{34}, r_{43}	2,3,4	7,7,8,5,7	7,7,5,8,7
W5	r_{20}, r_{32}, r_{31}^{1}	r_{01}, r_{10}, r_{12}, r_{21}, r_{13}^{1}	0,1,2,3	7,6,1,8,8	8,7,8,7,7

The GA designed for the rearrangeably nonblocking traffic grooming algorithm is different from that for the strictly nonblocking and was first proposed in [19]. This GA approach is also incorporated with the splitting methods. In this algorithm, two hierarchical chromosomes, i.e., a master chromosome and a slave chromosome, are employed to represent an individual. The use of this hierarchical chromosome structure is based on the fact that in rearrangeably nonblocking grooming each wavelength would drop at the same set of nodes when traffic changes but for different traffic patterns the same traffic requests might be assigned to different wavelengths. So, the master chromosome is used to record each wavelength's dropping nodes and M slave chromosomes are used to trace the assignments of M traffic requests among wavelengths. Furthermore, the slave chromosomes are only accompanying chromosomes which are produced by a heuristic to rearrange traffic requests among wavelengths and then to delete some spare ADM's and/or wavelengths after such rearrangement. When a new master chromosome is generated by crossover or mutation, the slave chromosomes are updated accordingly.

The master chromosome is coded and decoded with the same method as that in strictly nonblocking grooming described in 5.4.1.1, and it will record each wavelength's dropping nodes. To realize the rearrangeably nonblocking grooming, it is necessary to adjust the same traffic requests among different wavelengths. The basic idea behind this adjustment is to delete some spare ADM's and/or wavelengths by rearranging the traffic requests among wave-

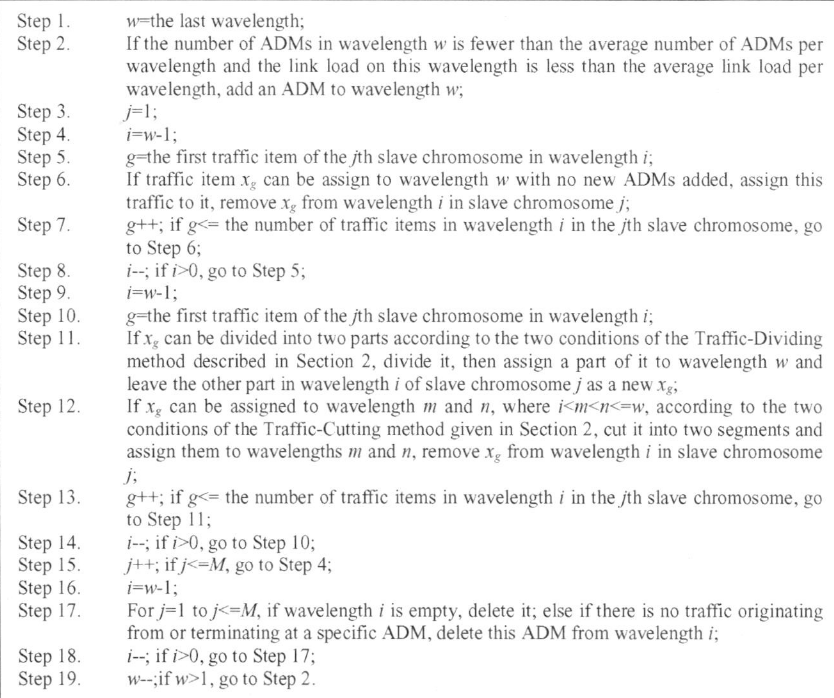

Step 1.	w=the last wavelength;
Step 2.	If the number of ADMs in wavelength w is fewer than the average number of ADMs per wavelength and the link load on this wavelength is less than the average link load per wavelength, add an ADM to wavelength w;
Step 3.	j=1;
Step 4.	i=w-1;
Step 5.	g=the first traffic item of the jth slave chromosome in wavelength i;
Step 6.	If traffic item x_g can be assign to wavelength w with no new ADMs added, assign this traffic to it, remove x_g from wavelength i in slave chromosome j;
Step 7.	g++; if $g<=$ the number of traffic items in wavelength i in the jth slave chromosome, go to Step 6;
Step 8.	i--; if $i>0$, go to Step 5;
Step 9.	i=w-1;
Step 10.	g=the first traffic item of the jth slave chromosome in wavelength i;
Step 11.	If x_g can be divided into two parts according to the two conditions of the Traffic-Dividing method described in Section 2, divide it, then assign a part of it to wavelength w and leave the other part in wavelength i of slave chromosome j as a new x_g;
Step 12.	If x_g can be assigned to wavelength m and n, where $i<m<n<=w$, according to the two conditions of the Traffic-Cutting method given in Section 2, cut it into two segments and assign them to wavelengths m and n, remove x_g from wavelength i in slave chromosome j;
Step 13.	g++; if $g<=$ the number of traffic items in wavelength i in the jth slave chromosome, go to Step 11;
Step 14.	i--; if $i>0$, go to Step 10;
Step 15.	j++; if $j<=M$, go to Step 4;
Step 16.	i=w-1;
Step 17.	For j=1 to $j<=M$, if wavelength i is empty, delete it; else if there is no traffic originating from or terminating at a specific ADM, delete this ADM from wavelength i;
Step 18.	i--; if $i>0$, go to Step 17;
Step 19.	w--;if $w>1$, go to Step 2.

Fig. 5.11. The algorithm for decoding the slave chromosomes with splitting methods (source: reference [19])

lengths with possible splitting of traffic. Each of the M slave chromosomes is the same as the master chromosome initially, but after such adjustment, the same traffic request in the M traffic patterns will be assigned to different wavelengths. The results are recorded in the M slave chromosomes which may now be different from each other. The M slave chromosomes are generated using the algorithm shown in Fig. 5.11 to represent the detailed distribution of each traffic request in the M traffic patterns in each wavelength. In this way, rearrangeably nonblocking grooming can be realized efficiently.

In this decoding algorithm for slave chromosomes, Steps 4–8 assign traffic requests to wavelengths as a whole; Steps 9–14 assign traffic requests to wavelengths with the splitting methods. The assignment is done from one slave chromosome to another. Steps 16–18 examine whether a wavelength is empty or not or whether there is any traffic originating from or terminating at a specific ADM. If not, delete the corresponding wavelength or ADM from the wavelength. In this way, spare wavelengths or ADM's will be removed from this individual. Furthermore, after such adjustment, another similar adjustment will be carried on from the first wavelength to the last so as to further reduce both the ADM's and the wavelengths.

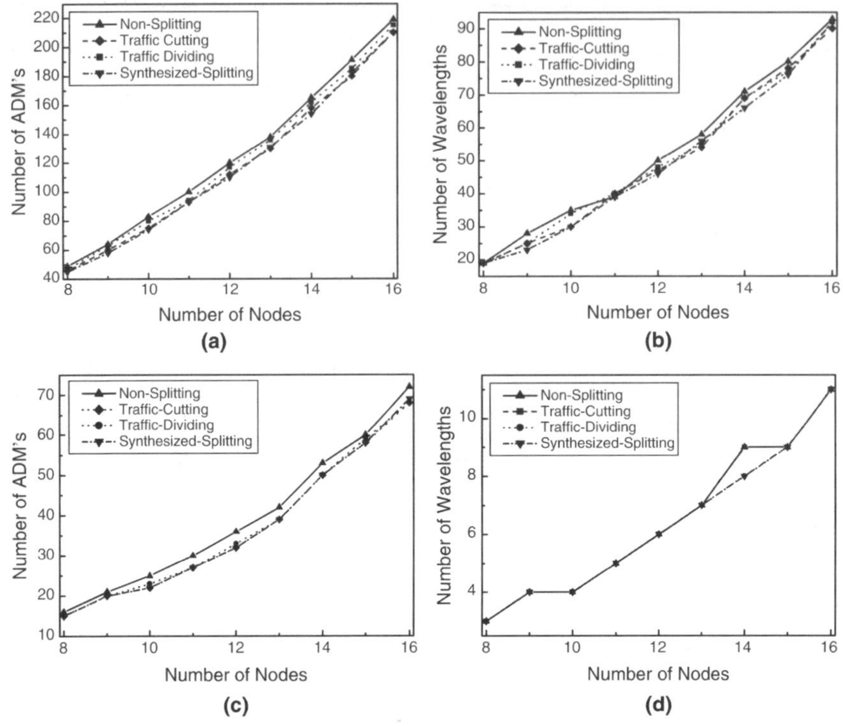

Fig. 5.12. Computer simulation results for different number of nodes when there are 2 traffic patterns in unidirectional rings (source: reference [19])

In this algorithm, it should be noted that for the rearrangeably nonblocking grooming, the condition of assigning a traffic request, the cut segment or divided part to a wavelength is different from that of strictly nonblocking grooming. The difference lies in that adding one ADM is allowed when the number of ADM's on the current wavelength is fewer than the average number of ADM's per wavelength or the link load is less than the average link load per wavelength. This is because if the condition is satisfied, this wavelength does not carry enough traffic due to its small number of dropping nodes. So in this case, by allowing adding an ADM, much more traffic requests can be assigned to it and therefore, it is likely to empty more ADM's or wavelengths after such adjustment.

The results of the number of ADM's and wavelengths vs. nodes for $g = 16$ and $g = 96$ respectively with two traffic patterns are shown in Fig. 5.12. It is clear that with the increase of traffic granularity, the difference among different splitting methods and non-splitting method becomes not so evident, but the synthesized-splitting is still the most efficient method in most cases, which is consistent with the previous observation. The comparisons between this GA approach and the one given in [22] are shown in Table 5.10. When g is small, the number of wavelengths required obtained from this GA approach will be larger than the later approach even with splitting methods, but the number of ADM's is much smaller. When the granularity is large enough, all the required wavelengths reaches their lower bounds, but the first GA approach would result in much fewer ADM's.

5.4.2 Grooming with Tabu Search Approaches

Apart from GAs, TS has also been applied to improve grooming performance in ring networks with dynamic traffic patterns. In [23], an interesting grooming problem is proposed in ring networks which is referred to as the reconfigurable grooming of dynamic traffic. In this approach, the whole process is treated as a series of additional traffic grooming. That is, based on the current wavelength assignment, when traffic changes, a grooming algorithm is applied to reconfigure the wavelength assignment according to the new traffic pattern without disrupting the old traffic assignment.

There are two kinds of reconfigurable grooming on SONET over WDM networks: best fit and full fit. Best fit means to place as much new traffic as possible using available capacity of the current configuration without increasing the number of ADM's, while full fit means to satisfy all the new traffic by adding the minimum number of ADM's. For each case, a greedy and a TS heuristic (TS-1 and TS-2) were proposed based on the two-step circle construction model proposed in [2].

For best fit case, the greedy algorithm starts from grooming the first traffic matrix $T[i, j]$ using the algorithm in [2]. In the greedy algorithm, which is shown in Fig. 5.13(a), the grooming result of the old matrix is used as the current configuration, and the new matrices are tried to be fitted into it without adding more ADM's. The information about whether the entire capacity between two nodes is occupied, whether there is an ADM at a node, and on which wavelength a circle is groomed should be recorded. When a new traffic matrix $T'[i, j]$ is given, the difference between the two matrices, $D[i,j] = T'[i,j] - T[i,j]$, is computed. If there are some connections in the old matrix that do not exist in the new one, remove them. If there are two continuous connections over two circles, merge them into a bigger one. In grooming new traffic, the traffic in the difference matrix $D[i,j]$ is assigned to the wavelengths from the shortest hop length to the longest hop length with no new ADM added.

TS-1 takes the grooming result from the greedy algorithm as its initial solution, as is shown in Fig. 5.13(b). The neighborhood is defined by swapping part of two circles of the solution in which at least one wavelength must have enough capacity. After such swapping, the remaining traffic will be assigned to the swapped wavelengths. For all the candidate solutions, the one that is not tabu and could increase the traffic amount the most is chosen as the new current solution. The search terminates after the pre-specified number of

Table 5.10. Grooming results for $g = 16$ and $g = 96$ with 16 nodes in a ring respectively (W, A represents the number of wavelengths and ADM's respectively)

Number of Traffic Patterns	$g = 16$						$g = 96$					
	Non-splitting		Synthesized-Splitting		GA Approach in [22]		Non-splitting		Synthesized-Splitting		GA Approach in [22]	
	W	A	W	A	W	A	W	A	W	A	W	A
$M = 2$	93	219	89	206	62	244	11	72	11	67	11	86
$M = 4$	98	219	91	212	/	/	12	71	11	67	/	/
$M = 8$	100	224	96	213	/	/	12	72	11	67	/	/
$M = 16$	103	224	100	216	/		12	72	11	69	/	/

(a)	(b)

1) Get grooming information for the original traffic matrix using an existing algorithm ([2]).
2) Find the difference traffic matrix.
Given the new traffic matrix, compute the difference between the old one and the new one, remove the connections between from circles over at most wavelengths.
3) Merge connections.
If there are two continuous connections over two circles, they will be merged into a bigger one on the same circle.
4) Groom new traffic.
Start from the smallest hop.
From node 0 to node N-1
From s=1to s=N /*s is the number of hops*/
while ($D[i,i+s]\neq 0$)
Try all the circles,
for each node pair ($i,i+s$) do
if capacity is available and if there is SADM on both i and $i+s$
{
Groom one unit of $D[i,i+s]$ to this circle.
set capacity occupied..
$D[i,i+s]= D[i,i+s]$-1;
}
Endwhile
End
	1) Initialization.
	Compute the initial reconfiguration solution by using the greedy algorithm. Set the initial tabu tenure value, which should be defined before the program runs.
	2) Build candidate list.
	The neighborhood of the solution x, $N(x)$, is defined as $N(x)$={x'| x' is a move by swapping part of the two circles of the solution and one part of the swap should have available capacity},
	Search all the circles. Compute $N(x)$ and put each element in $N(x)$ into the candidate list. Compute the additional new traffic amount that could been placed after such swap. Keep the maximum value of the increased traffic amount and its corresponding solution.
	3) Choose the best solution and continue.
	For all the solutions in the candidate list, choose the move that is not tabu and can increase the traffic amount the most. The values of tabu tenure of all the elements in the tabu list are decremented by one. If some element's tenure becomes zero, remove it from the tabu list. Put the current move into the tabu list.
	4) Terminate or continue with the current solution.
	If after a certain number of iterations of moves, which is specified by the user, there is no improvement, the program stops. Otherwise, go back to build the candidate list for the new movement.

Fig. 5.13. Greedy heuristic algorithm and TS-1 (source: reference [23])

iterations is reached or if there is no improvement after a certain number of steps.

A three-phase algorithm is developed for the full fit case. Firstly, a best fit algorithm (greedy or TS-1) is used to groom as much traffic as possible to wavelengths with the existing ADM's. Secondly, if there is enough capacity on a wavelength for additional connections, place an ADM at a node to groom additional traffic onto it. Finally, another TS heuristic (TS-2) is used to groom the remaining traffic onto new wavelengths. Since the structure of TS-2 is basically the same as TS-1, it is omitted here.

TS-1 can usually be applied to groom more connections than the greedy algorithm. The following example gives an illustration of these two algorithms. Assume that the old traffic matrix and the new matrix are given in Fig. 5.14 (a) and (b) respectively, then the corresponding grooming results for the greedy algorithm and the tabu search algorithm are shown in Fig. 5.15 (a) and (b) respectively. Neither of these algorithms can assign all the new traffic requests onto the existing wavelengths, and the remaining traffic matrices are shown in Fig. 5.16 (a) and (b). We can see that the TS-1 can groom two

n	0	1	2	3	4
0	0	2	6	1	0
1	7	0	6	1	3
2	2	2	0	3	3
3	5	2	7	0	2
4	4	5	7	2	0

(a)

n	0	1	2	3	4
0	0	4	6	2	5
1	7	0	8	1	4
2	3	3	0	10	11
3	5	4	7	0	2
4	6	5	7	3	0

(b)

Fig. 5.14. The old and new traffic matrices in a WDM ring network (source: reference [23])

Fig. 5.15. Greedy heuristic algorithm and TS-1 (source: reference [23])

more connections of t_{24} than the greedy algorithm, which shows the slightly superiority of the evolutionary approach over traditional heuristics.

5.4.3 Conclusion

In this section, the GA approaches to the strictly and rearrangeably nonblocking grooming of traffic with splitting methods, and the Tabu search heuristics for reconfigurable grooming of dynamic traffic have been discussed. Compared with the static grooming, dynamic grooming is much more difficult and fewer

$$\begin{array}{c|ccccc} n & 0 & 1 & 2 & 3 & 4 \\ \hline 0 & 0 & 0 & 0 & 1 & 2 \\ 1 & 0 & 0 & 0 & 0 & 1 \\ 2 & 1 & 0 & 0 & 0 & 2 \\ 3 & 0 & 2 & 0 & 0 & 0 \\ 4 & 1 & 0 & 0 & 1 & 0 \end{array}$$

(a)

$$\begin{array}{c|ccccc} n & 0 & 1 & 2 & 3 & 4 \\ \hline 0 & 0 & 0 & 0 & 1 & 2 \\ 1 & 0 & 0 & 0 & 0 & 1 \\ 2 & 1 & 0 & 0 & 0 & 0 \\ 3 & 0 & 2 & 0 & 0 & 0 \\ 4 & 1 & 0 & 0 & 1 & 0 \end{array}$$

(b)

Fig. 5.16. The remaining traffic matrices (source: reference [23])

Table 5.11. Comparison of Metaheuristics in the Grooming of Dynamic Traffic

Algorithm and Method	Strong Points	Weak Points	Reference
GA, circle construction model with a 2-step GA	No	Unable to obtain the optimal results and very time-consuming	[22]
GA, directly operating on M traffic matrices	Having the potential to obtain the optimal results	Did not consider traffic split	[6]
GA, directly operating on M traffic matrices with splitting methods	Having the potential to obtain the optimal results	Need longer time	[17, 18]
TS, circle construction model with best- and full-fit cases	With clear meaning and quite straightforward	Unable to obtain the optimal results and difficult to dynamically add new ADM's	[23]

approaches have been proposed for the ring networks. Table 5.11 summarizes various evolutionary approaches to the grooming of dynamic traffic. We can see the strong and weak points of each approach from this Table.

5.5 An Application of Evolutionary Based Traffic Grooming: Survivable Traffic Grooming

Traffic grooming is a promising technique which can be built into today's WDM optical networks for the optimal design purpose. Although newly emerged, it has already been widely combined with some other techniques, and the different applications of traffic grooming techniques are widely discussed in many fields [33–40]. As traffic grooming is NP complete in most cases by itself, when combined with other techniques to multiplex different low traffic requests onto high speed wavelengths under different network environments, efficient heuristics must be applied to search for a solution to the network configuration. Although evolutionary heuristics is an ideal candidate for traffic grooming in theory due to their powerful search ability, the applications based on evolutionary approaches have not been very well explored. A main reason is that designing a good evolutionary algorithm is a hard task due to the complexity of the real-world network. To illustrate the application of evolutionary based traffic grooming, a Tabu search algorithm for the survivable traffic grooming is discussed below.

With the widely deployment of the WDM backbone networks, the fault-recovery capability is much more critical for optical networks than before as a single failure may lead to a large volume of data and revenue loss. The network's capability of providing continuous service in the presence of failures is referred to as survivability. Generally, there are two different types of fault management schemes: protection and restoration [39, 40]. Protection refers to as a proactive procedure in which spare capacity is reserved during lightpath setup so as to achieve fast recovery when failures occur. On the contrary, restoration scheme is a reactive method, which establishes a connection after the failure occurs. It should be noted that both of these schemes require spare capacity preplanned so as to provide survivability in optical networks.

WDM survivable grooming network can be classified into two categories: dedicated-wavelength grooming networks and shared-wavelength grooming networks [41]. The former denotes that the source-destination node pairs are connected by lightpaths, where a lightpath shared by requests from a specific *s-d* pair; the latter refers that the lightpath can be shared by requests from different *s-d* pairs. The survivable traffic grooming (STG) problem addresses these two fault management schemes, protection and restoration, in these two kinds of networks. STG is a key issue to provide resource-efficient multigranularity connection services with strong survivability in WDM optical networks. The goal of STG is to provide fault recovery capability for connections with the minimum consumption of spare capacities in the network.

Network survivability can be achieved by using link-, path- or segment based fault recovery method. As these three methods are different in the ways of rerouting the fault traffic streams, different algorithms should be applied to realize different recovery method. Even though many researchers devoted to this field and many heuristics were proposed based on the two fault management schemes [42–49], STG is still a relatively unexplored field. And to the best of our knowledge, among all the evolutionary approaches, only Tabu search algorithms have been applied to STG in [50, 51], which will be discussed in the following.

In a two-layered grooming mesh network, path-protection can be applied at two levels which is referred to as protection at lightpath (PAL) and protection at connection (PAC) [42]. In references [50] and [51], the authors proposed two Tabu search algorithms for tackling the STG problem with PAL and PAC in mesh networks respectively. Since the two algorithms executed in a similar process, only the algorithm proposed in [50] is analyzed in this section.

During grooming process, the routing strategy of lightpaths and connections can be considered on the client layer and the server layer (the virtual-topology layer and the physical topology layer) separately or jointly. So there are two different models: overlay model and peer-deployment model. Based on this consideration, two different grooming algorithms were proposed, separate survivable grooming algorithm (SSGA) and integrated survivable grooming algorithm (ISGA), to solve the STG problem in the two different models respectively, and all the corresponding algorithms are shown in Fig. 5.17.

Before looking into these algorithms, there are two concepts that should be explained. The first is the shared-risk link group (SRLG) constraints, which means that (1) resource constraints: at least one wavelength is available on each of the links in a path when establishing the lightpath over the path and at the same time, at least one free transmitter and receiver at the source and destination nodes of the path is also available; (2) wavelength-continuity constraints: a lightpath must use the same wavelength on all links if the network does not have wavelength-conversion capability; (3) diverse-routing constraints: the primary and backup paths of a connection must not share a common risk in the physical topology; (4) lightpath-capacity constraint: the total bandwidth of all the connections must not be larger than the bandwidth of a lightpath. These constraints are general, so the algorithms designed must satisfy all of them. What is more, as it is proved that finding two SRLG-diverse paths between a node pair is NP-complete, it is obvious that STG problem under the SRLG constraints is also NP-complete.

The second one is the link-bundle auxiliary graph (LBAG) model proposed in [30]. Link bundling is a method to assist the management of a network with excessive links between two adjacent nodes in general multiprotocol label switching (GMPLS). With LBAG, for a network with physical topology $G_{\mathrm{p}} = (V_{\mathrm{p}}, E_{\mathrm{p}})$, the LBAG $G_{\mathrm{a}} = (V_{\mathrm{a}}, E_{\mathrm{a}})$ is constructed as a two-layer graph, the physical layer and the lightpath layer, respectively, with $|V_{\mathrm{p}}|$ nodes in each layer. For each node in the network, there are two nodes in the auxiliary graph with one in each layer. The edges in the physical layer represent the links in physical topology, named as wavelength edges (E_{w}), and there is only one wavelength edge in a link regardless of the number of wavelength or fibers in the link. The edges in the lightpath layer represent the lightpath in the virtual topology, named as lightpath edges (E_{t}). The edges between the lightpath layer and physical layers represent transceiver resource, named as transceiver edges. Figure 5.18(b) gives an LBAG example which is constructed based on the four-node network shown in Fig. 5.18(a). It should be noted that a valid path should begin and end on the lightpath layer, as shown in Fig. 5.18(c). A new lightpath should be set up when a path goes into the physical layer through a transmitter edge.

These two algorithms are both designed for the dedicated and shared path protection under the SRLG constraints.

The SSGA can be divided into two subproblems: one is to establish a virtual topology over the physical topology, which is referred to as protection-aware virtual-topology design (PAVTD) problem; the other is to assign the subwavelength connection onto the lightpaths in the virtual

Algorithm MSHT-PAVTD
Input: connection requests $\Lambda^x (x \in X)$, physical topology G_p, wavelength and transceivers, and SRLG information
Output: A virtual topology G_p
1) sum the traffic matrix set to form a single residual traffic matrix Λ^*, where
$\Lambda^x_{s,d} = \sum_{x \in X} (\Lambda^x_{s,d} \times x)$ represents the total bandwidth needed from node s to d
2) while(not all the elements of Λ^* are zero)
  select the node pair (s',d') with the maximum residual bandwidth in matrix Λ^*. Ties are broken arbitrarily.
  if(dedicated protection)
    if (TTD-RDJP(G_p, s', d', p_1, p_2))
      establish two lightpaths from s' to d' on risk-disjoint paths p_1 and p_2
    $\Lambda^x_{s,d} \leftarrow \Lambda^x_{s,d} - C$
  else if (shared protection)
    if (DIJKSTRA(G_p, s', d', p))
      establish a lightpath from s' to d' on path p
      $\Lambda^x_{s,d} \leftarrow \Lambda^x_{s,d} - C/2$

Algorithm TTD-RDJP(G_p, s, d, p, b)
Input: source node s, destination node d, topology G, usage of wavelengths and transceivers, and SRLG information
Output: risk-disjoint paths p and b, if successful, return NULL otherwise
1) update the link weights of G according to the current state of the network, and run Dijkstra's algorithm to select the shortest path as path p
2) if(p)
  delete any link in G that share at least one risk with the any link in p.
  if (shared protection)
    update the link weights of G again according to the current state of the network and the path p
  run DIJKSTRA's algorithm again on G to select the shortest path as the path b.
  if(b)
    return(p, b)
3) return NULL

Algorithm LTF-SWCSR
Input: connection requests $\Lambda^x (x \in X)$, virtual topology G_v, and SRLG information
1) sort all the connection requests in a list Q in non-increasing traffic amount order
2) while(the list Q is not empty)
  get and remove the connection request (s, d, x, y) from the head of Q.
  if(TTD-RDJP(G_v, s, d, p_1, p_2))
  if(dedicated protection)
    establish a primary connection and a backup connection from s to d on risk-disjoint path p_1 and p_2 respectively
  else if (shared protection)
    establish a connection from s to d on path p_1 and reserve bandwidth on path p_2
  else
    block the connection request (s, d, x, y)

Algorithm ISGA
Input: connection requests $\Lambda^x (x \in X)$, physical topology G_p, wavelength and transceivers, and SRLG information
Output: A virtual topology G_v and connections established over G_v
1) Initialize the LBAG G_a according to the physical topology
2) while(the list Q is not empty)
  get and remove the connection request (s, d, x, y) from the head of Q.
  if(TTD-RDJP(G_a, s, d, p, b)
    for each physical layer sub-path p_s in p or b do establish a new light path on p_s
    if(dedicated protection)
      establish a primary connection and a backup connection from the request on paths p and b respectively
    else if (shared protection)
      establish a primary connection on path p and reserve bandwidth on backup path b for the request
  else
    block the connection request.

Fig. 5.17. The corresponding algorithms for SSGA and ISGA (source: reference [50])

topology with both the primary and backup path, which is referred to as the subwavelength-connection survivable-routing (SWCSR) problem. A maximizing single-hop traffic (MSHT-PAVTD) heuristic is proposed to carry connection-oriented subwavelength channels in multiple granularities and

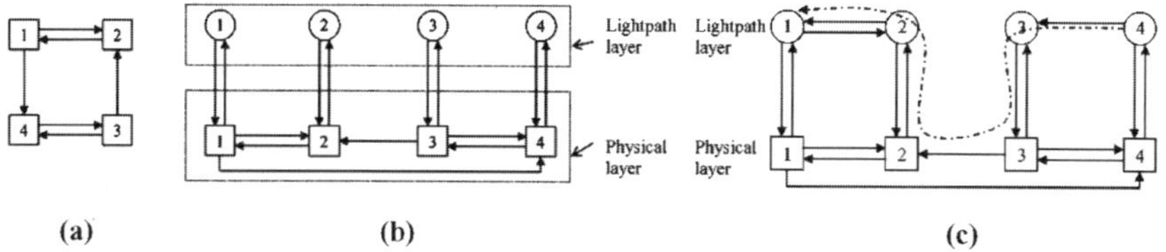

Fig. 5.18. (a) Physical topology of a four-node sample network. Each fiber link supports two wavelengths. Node 2 has full wavelength conversion capability. (b) The corresponding LBAG model. (c) A path in LBAG. (source: reference [30])

establish risk-disjoint lightpaths in dedicated protection and shared protection respectively. This algorithm tries to establish lightpath between the node pairs with the heaviest traffic load being considered first. Moreover, in MSHT-PAVTD, two times Dijkstra's risking disjoint path (TTD-RDJP) algorithm is used for finding two risk-disjoint paths by applying the Dijkstra's algorithm to find the shortest path. By deriving SRLG information for lightpaths in the virtual topology, the large-traffic-first (LTF-SWCSR) algorithm is used to solve the SWCSR problem with the aid of RDJP algorithm for finding risk-disjoint paths in virtual topology for connections.

The ISGA considers the provision of lightpaths and connections jointly, and is based on LBAG model. The objective of it is to accommodate maximum number of connections. It handles the SRLG information for lightpaths just as in SSGA, and also used TTD-RDJP algorithm to calculate a pair of risk-disjoint paths.

These two algorithms both try to assign as many sharable spare channels to a backup connection as possible, and only assign free channels to the backup connection when the capacity of the sharable spare channels is not enough. In this way, the saving of channels can be achieved. Although the results of these two algorithms are good enough already, they applied the Tabu search algorithm to further optimizing the results.

The Tabu search grooming algorithm (TSGA) is also based on the LBAG model. In TSGA, the authors defined an add or a drop operation as a move. The add operation is trying to satisfy a previously blocked connection request by finding a pair of risk-disjoint paths in the LBAG, then the objective function value increases; the drop operation tries to disconnect a satisfied connection by releasing all the bandwidth along its primary and backup paths, then the objective function value decreases. With these two defined operations, the TSGA starts with an initial solution which obtained by SSGA or ISGA, then it proceeds with an iterative process by executing a selected move. To select the best move, a move value of a connection is defined as:

$$g(c, p_1, p_2) = \begin{cases} V(c)/(WPC(p_1, p_2)) - Freq(c) & if\ add \\ -V(c)/(WPC(p_1, p_2)) - Freq(c) & if\ drop \end{cases} \quad (5.3)$$

where $g(c, p_1, p_2)$ is the move value of a connection c, p_1 and p_2 are the primary and backup paths for a satisfied connection or the path to be used for the blocked connection in the drop and add operation respectively. $V(c)$ is the revenue value of the connection c, $WPC(p_1, p_2)$ is the weighted-path cost of path p_1 and p_2. Freq(c) is the selection frequency of the connection c in the previous best move. The move with the largest value is selected at each iteration. At the same time, two diversification techniques are applied to lead the algorithm to the unexplored search spaces. The first is restart diversification, which refers that TSGA uses the initial solution from SSGA and ISGA respectively and runs twice; the second is frequency favor, which makes the less selected move more favorable than the more frequently selected one, and is incorporated into the move value function.

The authors also use an integer linear programming (ILP) approach to obtain optimal solutions to this problem. Numerical simulation results shows that the computational complexity of the ILP approach is too large even for networks of small sizes. On the other hand, ISGA performs much better than SSGA, with an average of 50% and 15% improvement in network throughput while using dedicated protection and shared protection respectively. This result implies that the integrated routing approach is superior to the overlay routing approach in terms of resource-efficiency. TSGA further improves the grooming results from ISGA by an average of about 5% at the cost of longer running time, which is required by the additional iteration optimization phase guided by the tabu search method.

5.6 Problems and Challenges

Traffic grooming is a relatively new challenge in WDM optical network design which has drawn much attention from both academia and industrial society. Efficient grooming of traffic can not only reduce the whole network's cost but also increase the signal's quality by using fewer wavelengths on each fiber. Unfortunately, due to its complexity and difficulty, high efficient heuristics must be utilized. The use of evolutionary approaches to the grooming problems enables us to achieve high quality solutions in a reasonable time. Although a significant amount of work has been done in applying evolutionary approaches to the grooming of traffic in WDM optical networks, there are still some relatively unexplored areas in this field which must be paid attention in the future.

5.6.1 Algorithms Directly Operating on Traffic Matrices

Most of the approaches for ring networks used by now are based on the circle construction method proposed in [2]. Although it is quite straightforward and reasonable for the circle construction method to be used as the basis in the evolutionary algorithms and this method also has clear meaning, there are still some problems with it. The first is that the numbers of ADM's and wavelengths cannot be minimized concurrently [1,15]. Since this method uses the minimum number of wavelengths as its searching point, the number of ADM's cannot be minimized at the same time for arbitrary nonuniform traffic patterns. The second is that this is a two-step method, or three-step method for dynamic grooming problems. Hence, even if the optimal solution can be found at each step, it is still uncertain that the optimal solution for the whole problem can also be found. Another problem with this approach is that there are always some idle ADMs with it [6]. The third is that with this method all the connections on the ring have to be divided into elementary parts and different part in the same connection will probably be assigned to different wavelength. This will inevitably increase the complex of the virtual topology and the management overhead of the network. The last is that this method can only be used in the ring networks. For more general topologies, e.g., mesh networks, it is difficult to construct circles. That is why most of the evolutionary approaches used by now dealt with only the grooming of traffic in ring networks.

Therefore, it is imperative to design evolutionary algorithms that operate directly on the traffic matrices and use only one-step approach to groom traffic onto wavelengths. Not only can these algorithms be used in ring networks but also be easily extended to other general topologies. As the exploration for grooming traffic in mesh networks is much more significant, further efforts in this aspect are obviously worthwhile. On the other hand, as it has been repeatedly shown that pure (or classical) GA or other evolutionary approaches often perform poorly in solving sophisticated combinatorial optimization problems, an efficient heuristic decoding method need also be imbedded in these algorithms to improve the grooming performance. If splitting techniques have to be included, it is ideal that traffic is split only when needed and one of the split parts must be able to be assigned to the current wavelength.

5.6.2 Grooming in Mesh and Other Topology Networks

As mentioned earlier, traffic grooming in mesh networks has been widely investigated recently. Although many researchers have already proposed efficient heuristics for grooming of dynamic traffic in mesh networks, very few of them are based on evolutionary approaches. Since traffic grooming in mesh network is much more complex than in ring network, evolutionary approaches are more than needed. Due to the "mesh" characteristic of such networks, evolutionary approaches alone, pure GAs for instance, are not powerful enough to obtain good grooming results. An ideal way to grooming traffic in such networks might be the incorporation of evolutionary approaches with other efficient local or greedy search mechanisms. The heuristic algorithms developed by now, in [27–30] for example, could be used for this purpose. The incorporation of evolutionary with other search heuristics is significant not only in traffic grooming but in other optical network designing problems, such as routing and wavelength assignment, protection, optimization of wavelength converters and optical amplifiers, etc.

At the same time, it should be noted that other irregular topologies are also of great importance in their own sense. For example, star networks are widely deployed in the interconnection of LANs or MANs, whereas most cable TV networks and PONs are based on tree topologies. So it is also of great value to discuss TG problems in these topologies. But there are very few papers discussing about TG problems in stars and trees [52–56], and none of them are based on evolutionary approaches. The TG problem in star network is NP-complete even when the traffic requests remained unchanged during the grooming process, as was proved in [52], so the grooming of arbitrary all-to-all dynamic traffic in other irregular topology networks is still a great challenge. But since the topologies of both star and tree networks are simpler than the mesh networks, it is easier to design proper evolutionary algorithms to tackle these traffic grooming problems. Therefore, the evolutionary approach to the TG problem in irregular topology networks still remains a big challenge.

5.6.3 Grooming of Dynamic Traffic

The evolutionary algorithms developed for the grooming of dynamic traffic by now are in fact offline algorithms. They usually worked on a set of known traffic matrices (patterns) and assigned them to wavelengths one by one to realize a "dynamic grooming". Obviously, this is not a real dynamic grooming in its true sense. A real dynamic grooming algorithm must be able to groom dynamically to satisfy the random connection requests on the network. This requires a fast online algorithm. Evolutionary approaches are in general not suitable for this purpose since they usually need time-consuming iterations to complete a single computation. Therefore, an alternative way to incorporate evolutionary approaches into such purpose must be explored.

A possible solution to this problem may be the combination of an offline static grooming algorithm with an online dynamic grooming one. The offline algorithm is used to perform a strictly nonblocking or rearrangeably nonblocking grooming for a set of estimated traffic patterns, which can be made by taking advantage of available information such as customer prescriptions, traffic projections, and historical data, etc. Based on the offline grooming results, a fast online algorithm can be used to dynamically assign traffic requests to the pre-designated wavelengths. If the current traffic cannot be assigned to those wavelengths, the algorithm can first assign a split part of the traffic to them then assign the remaining part to another wavelength with free capacity. In this way, the virtues of both approaches, metaheuristics for searching optimal grooming results and online heuristics for quick assigning traffic to wavelengths, can be perfectly combined together.

The effectiveness of this hybrid approach can be justified by the observations of Elwalid *et al* [64]. In the study of routing and protection in GMPLS networks they found that using the optimized paths computed offline to guide online path setups outperformed mere online algorithms under a wide range of operating conditions and was robust to inaccuracies in the estimation of traffic matrix. We also found in our previous studies that offline optimization could make full use of the dynamics of the changing traffic and make some compensation among the traffic requests between different pairs of nodes [6, 18]. That is to say, it is often that at one moment some traffic requests are at their higher level but others are at their lower level, at another the situation may be reversed. By utilizing offline optimization the traffic requests that are at their high level can be compensated by those that are at their low level, which will guarantee the changing traffic being well fitted into the pre-designated wavelengths with very few exceptions.

5.6.4 Traffic Grooming Under a Scheduled Traffic Model

In today's real-time communication system, scheduling algorithms have been widely deployed to minimize the average message delay, maximize the number of messages or increase the network's throughput [56–59]. With the aid of scheduling algorithms, time slots of the data channels can be fully filled therefore unnecessary delay can be avoided in the message transmission in the network. However, traffic grooming technique is still not widely applied to the scheduled networks. There are only just a few papers investigating the problem of grooming in scheduled traffic.

The traffic grooming in scheduled traffic model can be viewed as an extension of the traffic grooming problem. In fact, although valid in themselves, both static and dynamic traffic models are not applicable directly to deal with the applications that require capacity during specific time intervals. If it is necessary to provide some scheduled demands for bandwidth or dedicated channels at a specific time, the study on the scheduled traffic model is necessary. As the scheduled demands are dynamic and can be preplanned usually, it is possible to set up a traffic grooming network system under a scheduled traffic model. For the grooming in such a scheduled model, some more factors should be considered, such as the placement of demand time intervals, the solution of resource conflict, etc. Since both the space and the time interval should be arranged, this problem is different from other TG problems. And up to now, only Tabu search approach is proposed for further optimizing the solution achieved by other heuristics based on a sliding scheduled traffic model in [56].

For an efficient evolutionary approach to the scheduled traffic grooming, both the time and the space splitting method should be considered at the same time. So instead of traditional EA, the heuristic imbedded evolutionary algorithm may be a better candidate.

5.6.5 The GRWA Problem: Grooming, Routing and Wavelength Assignment

In wavelength routed WDM networks, in order to establish a connection between a source-destination pair, a wavelength continuous route needs to be found between the node pair. The problem of selecting routes and wavelengths to establish lightpaths is known as the Routing and Wavelength Assignment (RWA) problem. This problem has been the topic of many studies since they are very important in constructing WDM network [38, 62]. But the grooming problem is generally ignored by assuming that each traffic request takes up an entire wavelength. However, it is usually not the case in practice. It is much more practical to deal with the synthesized the RWA and grooming problem. So some researchers try to consider these three problems as a whole, which is referred as GRWA. This problem was divided into sub-problems that were solved sequentially [3, 37]. The main reason for this lies in the fact that each of these problems is already a very hard problem. This decomposition method is generally good for simplifying the problem, however, the exploration of the complete solution is not allowed with this approach and the optimal solution may be beyond the search ability. So it is necessary to consider these three problems as a whole [63].

Due to the complexity of this problem, without the aid of efficient heuristic it is difficult to find a proper solution to the GRWA problem as a whole. By now, only a tabu search algorithm has been proposed to tackle this problem [63]. With more and more complicate in today's telecommunication networks, some more efficient GRWA algorithm is very important for efficient design of such networks.

5.6.6 Traffic Grooming in Special Designed Networks

With more and more WDM optical networks deployed as the dominant transport infrastructure for national backbones, metropolitan networks, and even regional networks, it is becoming increasingly urgent to apply grooming technique in special designed network environments, such as GMPLS-Based IP/WDM networks, multicast networks, and multiple types of multiplexing networks with a hierarchical multi-commodity scenario etc. Grooming in such a complex network environment is a great challenge to both industry and academia. For example, with the development of generalized multiprotocol label switching (GMPLS) hierarchical network structure, it is ideal to have an automatic network routing and grooming algorithm at all levels. Therefore, like in MPLS, there is a great demand of efficient routing and grooming algorithms in GMPLS networks.

On the other hand, the increasing traffic request, such as video conferencing, distance learning and multi-player on-line gaming, requests the growth of the multicast connections with high quality of service (QoS). Although the grooming has been applied in multicast networks since 2003 [35], further combination of sub-rate grooming and effective service multicasting is a very important problem since it is expected that a huge portion of the future traffic in high performance optical networks may be digital multimedia multicast traffic, where user-requested contents are simultaneously transferred from a single source to multiple destinations.

At last, the inclusion of QoS requirements into traffic grooming process will be another interesting issue which will either reduce the resource usage under a certain blocking probability or lower the blocking probability under the same resource consumption. One possible way to guarantee QoS in traffic grooming process is to assign high-priority traffic flows onto separate wavelength and then assign other lower-priority traffic flows to share the same wavelength with them. In this way, we can preserve enough bandwidth for the high-priority traffic flows during dynamic provisioning of traffic to guarantee low blocking probability of them. Some other ways of realizing QoS during grooming may also be possible but researches have to be done to make the best use of wavelength's capacity and at the same time to guarantee different QoS for different traffic flows.

5.7 Conclusion

Evolutionary computing has been used in a wide variety of telecommunication problems, and it is reported that over 400 scientific papers have devoted to this filed. However, traffic grooming seems to be a relatively neglected area. In this chapter, a comprehensive review on various evolutionary approaches, especially GAs and TS, to the grooming of static and dynamic traffic in WDM optical networks has been presented. Some future challenges and research directions have also been discussed. Due to the NP-hard property of grooming problems in WDM optical networks, evolutionary approaches would be the most promising candidate to them. On the other hand, owing to its powerful search ability, the hybrid evolutionary algorithms would play a much important role in the future in solving grooming problems. As heuristics can be incorporated into evolutionary approaches, even if a heuristic can already produce good results, or even better than pure evolutionary algorithms, combining them with evolutionary algorithms would inevitably lead to better solutions. But till now, only limited work has been done in applying these evolutionary approaches to the grooming problems.

On the other hand, due to the widespread use of WDM or dense WDM technologies, more and more wavelengths will be carried by a single fiber. Hence, the design of such networks will become even more difficult. With the use of GMPLS in optical networks, a common control plane (signaling and routing) can be provided for devices that switch in different domains: packet (IP traffic), time division multiplexing (TDM), wavelength, waveband, and fiber [65,66]. This gives an important integrated multi-granularity routing and grooming possibility in such a system and 'is being exploited to move deployed systems toward a flexible, "intelligent", and more autonomous multiservice optical layer, or perhaps—as Internet traffic continues to drive network design and deployment—Internet protocol (IP)/optical networks' [66]. We believe that the use of evolutionary approaches to the grooming problems in such network systems will bring about the most cost-effective designing results for them.

References

1. A. L. Chiu and E. Modiano, Traffic grooming algorithms for reducing electronic multiplexing costs in WDM ring networks, IEEE/OSA J. Lightwave Tech., 19 (2001) 2146–2168
2. X. J. Zhang and C. M. Qiao, An effective and comprehensive approach for traffic grooming and wavelength assignment in SONET/WDM rings, IEEE/ACM Trans. Networking, 8 (2000) 608–617
3. J. Wang, W. H. Cho, V.R. Vemuri, and B. Mukherjee, Improved approaches for cost-effective traffic grooming in WDM ring networks: ILP formulations and single-hop and multihop connections, IEEE/OSA Journal Lightwave Technology, 19 (2001) 1645–1653
4. R. Dutta and G. N. Rouskas, Traffic grooming in WDM networks: past and future, IEEE Network, 16(6) (2002) 46–56
5. R. Berry and E. H. Modiano, Reducing electronic multiplexing costs in SONET/WDM rings with dynamically changing traffic, IEEE J. Selected Areas in Communications, 18 (2000) 1961–1971
6. Y. Xu, C.S. Xu, and B.X. Wu, Strictly nonblocking grooming of dynamic traffic in unidirectional SONET/WDM rings using genetic algorithm, Computer Network, 41 (2003) 227–245
7. C. M. Qiao, D. Datta, G. Ellinas, A. Gladisch, and E. Modiano, Guest editorial: Special Issue on WDM-based network architectures, IEEE J. Selected Areas in Communications, 20 (2002) 1–3
8. http://www.trafficgrooming.org/
9. E Modiano and P. J. Lin, Traffic grooming in WDM networks, IEEE Communications Magazine, 39(7) (2001) 124–129
10. Y. Xu, S. Salcedo-Sanz, and X. Yao, Metaheuristic Approaches to Traffic Grooming in WDM Optical Networks, International Journal of Computational Intelligence and Applications, 5(2) (2005) 231–250
11. O. Gerstel, P. Lin, G.H. Sasaki, Wavelength Assignment in a WDM Ring: To Minimize Cost of Embedded SONET Rings. INFOCOM (1998) 94–101
12. O. Gerstel, R. Ramaswami, G.H. Sasaki, Cost Effective Traffic Grooming in WDM Rings. IEEE INFOCOM (1998) 69–77
13. E.H. Modiano and A.L. Chiu, Traffic Grooming Algorithms for Minimizing Electronic Multiplexing Costs in Unidirectional SONET/WDM Ring Networks. Proc. of the Thirty-Second Conference on Information Science and Systems, Princeton, NJ, (1998) 653–658
14. H. W. Tzeng, J. L. Chen, and N. K. Chen, Traffic grooming in WDM networks using genetic algorithm, Proceedings of IEEE SMC '99 Conference, Tokyo, Japan, 1 (1999) 1003–1006

15. J. Xu and Q.J. Zeng, Reducing electronic multiplexing in WDM rings: an evolutionary approach, Proceedings of International Conferences on Info-tech and Info-net, Beijing, 2 (2001) 168–173
16. Y. Xu, S.C. Xu, and B.X. Wu, Traffic grooming in unidirectional WDM ring networks using genetic algorithms, Computer Communications, 25 (2002) 1185–1194
17. A. Mukhopadhyay, U. Biswas, and M.K.Naskar, A Genetic Algorithm for Traffic Grooming in Unidirectional SONET/WDM Rings. IEEE India Annual Conference, (2004) 252–255
18. K.H. Liu and Y. Xu, A New Approach to Improving the grooming performance with dynamic traffic in SONET rings, Computer Networks, 46 (2004) 181–195
19. K.H. Liu and Y. Xu, Combining GA with splitting methods to the rearrangeably nonblocking grooming of dynamic traffic in SONET ring networks, Photonic Network Communications, 10(2) (2005) 179–192
20. L. Li, S. Scott, and J. Deogun, Cost-effective approaches for circuit construction in WDM SONET rings, In Proceedings of IASTED International Conference-Wireless and Optical Communication, Banff, Canada, (2002) 333–338
21. O. Gerstel, P. Lin, and G. Sasaki, Wavelength assignment in a WDM ring to minimize cost of embedded SONET rings, Proc. IEEE INFOCOM'98, San Francisco, USA, March 1998, 94–101
22. J. Xu, Q.J. Zeng, Dynamic traffic grooming in interconnected WDM SDH/SONET rings, Proc. SPIE, Vol. 4527 (2001) 177–186
23. S. Zhang and B. Ramamurthy, Dynamic traffic grooming algorithms for reconfigurable SONET over WDM networks, IEEE J. Selected Areas in Communications, 21 (2003) 1165–1172
24. K. Roy and M. K. Naskar, A heuristic solution to SONET ADM minimization for static traffic grooming in WDM unidirectional ring networks. Photonic Network Communications, 12 (2006) 153–160
25. Y. Wang and Q.P. Gu, Efficient Algorithms for Traffic Grooming in SONET/WDM Networks, Proceedings of the 2006 International Conference on Parallel Processing, (2006) 355–364
26. Y. Xu and S.C. Xu, Heuristic operators, redundant mapping and other issues in genetic algorithms, Proceedings of CEC'01, Seoul, Korea, (2001) 1398–1405
27. Z. Ding and H. Mounir, Clustering techniques for traffic grooming in optical WDM mesh networks, Proceedings of IEEE GLOBECOM '02, Taipei, Taiwan, 3 (2002) 2711–2715
28. K. Zhu and B. Mukherjee, Traffic grooming in an optical WDM mesh network, IEEE J. Selected Areas in Communications, 20 (2002) 122–33
29. H. Wen, R. He, L. Li, and S. Wang, A plane-construction traffic grooming algorithm in WDM mesh networks. Proceedings of PDCAT'2003, IEEE, Chengdu, PR China, (2003) 263–267
30. W. Yao, B. Ramamurthy, A link bundled auxiliary graph model for constrained dynamic traffic grooming in WDM mesh networks, IEEE J. Selected Areas in Communications, 23 (2005) 1542–1555
31. C.W. Lee and E.K. Park, A genetic algorithm for traffic grooming in all-optical mesh networks, Proc. IEEE SMC'02, Hammamet, Tunisia, 7 (2002) 6–9
32. Y.G. Jiao, B.K. Zhou, H.Y. Zhang, Y.L. Guo, Grooming of Arbitrary Traffic in Optical WDM Mesh Networks Using a Genetic Algorithm, Photonic Network Communications, 10 (2005) 193–198

33. N.K. Singhal, L.H. Sahasrabuddhe, and B. Mukherjee, Optimal Multicasting of Multiple Light-Trees of Different Bandwidth Granularities in a WDM Mesh Network With Sparse Splitting Capabilities, IEEE/ACM transactions on networking, 14(5) (2006) 1104–1117
34. N. Srinivas and C. Siva Ram Murthy, Design and dimensioning of a WDM mesh network to groom dynamically varying traffic, Photonic Network Communications, 7 (2004) 19–191
35. A.E. Kamal and R.U. Mustafa, Multicast Traffic Grooming in WDM Networks, Proceedings of SPIE: the Optical Networking and Communications, (2003) 25–36
36. Y.L. Hsueh, J. Kim, C.F. Su, R. Rabbat, T. Hamada, C. Tian, and L.G. Kazovsky, Traffic Grooming on WDM Rings Using Optical Burst Transport, IEEE/OSA J. Lightwave Tech., 24 (2006) 44–53
37. J.Q. Hu and B. Leida, Traffic grooming, routing, and wavelength assignment in optical WDM mesh networks, Twenty-Third Annual Joint Conference of the IEEE Computer and Communications Societies, (2004) 495–501
38. X.W. Chu, B. Li, I. Chlamtac, Wavelength converter placement under different RWA algorithms in wavelength-routed all-optical networks, IEEE Transactions on Communications, 51(4) (2003) 607–617
39. S. Sankaranarayanan, and B. Mukherjee, Survivable WDM mesh networks—Part 1: protection, in Proc IEEE information communications (INFOCOM), (1999) 744–751
40. S. Sankaranarayanan, and B. Mukherjee, Survivable WDM mesh networks—Part 2: restoration, in Proc IEEE information communications (INFOCOM), (1999) 2023–2030
41. J. Yates, J. Lacey, and D. Everitt, Blocking in multiwavelength TDM networks, in 4th International Conference on elecommunication Systems, Modeling, and Analysis, (1996) 535–541
42. C. Ou, K. Zhu, H. Zang, L.H. Sahasrabuddhe, and B. Mukherjee, Traffic grooming for survivable WDM networks – shared protection, IEEE J. Selected Areas in Communications, 21(9) (2003) 1367–1383
43. C. Ou, K. Zhu, J. Zang, H. Zhu, B. Mukherjee, H. Zang, L.H. Sahasrabuddhe, Traffic grooming for survivable WDM networks: dedicated protection, Journal of optical networking, 3(1) (2004) 50–74
44. H. Wen, R. He, L. Li and S. Wang, Dynamic traffic-grooming algorithms in wavelength-division multiplexing mesh networks, Journal of optical networking, 2(4) (2003) 50–74
45. J. Fang and A.K. Somani, Enabling sub-wavelength level traffic grooming in survivable WDM optical network design, in proceeding of IEEE GLOBECOM'03, (2003)
46. Y. Xiong, D. Xu and C. Qiao, Achieving fast and bandwidth-efficient shared-path protection, IEEE/OSA J. Lightw. Technol., 21(2) (2003) 2683–2693
47. H. Wen, L. Li, R. He, H. Yu, S. Wang and N. Song, Dynamic grooming algorithms for survivable WDM mesh networks, Photonic Network Communications, 6(3) (2003) 253–263.
48. G. Li, B.Doverspike and C. Kalmanek, fiber span failure protection in mesh optical networks, in proce SPIE optical networking and communications, (2001) 130–141
49. C. Xin, K. Y. Ye, S. Dixit, and C. Qiao, A joint lightpath routing approach in survivable optical networks, opt. new. Mag. 3(3) (2002) 23–32

50. W. Yao, B. Ramamurthy, Survivable Traffic Grooming With Path Protection at the Connection Level in WDM Mesh Networks, IEEE J. Lightwave Technology, 23 (2005) 2846–2853.
51. W. Yao and B. Ramamurthy, Survivable traffic grooming in WDM mesh networks under SRLG constraints, in proc. IEEE int. conf. communications (ICC), (2005) 1751–1755
52. E. Angel, E. Bampis, and F. Pascual, Traffic Grooming in a Passive Star WDM Network, SIROCCO 2004, LNCS 3104, (2004) 1–12
53. R. Dutta, G.N. Rouskas, On Optimal Traffic Grooming in WDM Rings. IEEE J. Selected Areas in Communications, 20(1) (2002), 110–121
54. B. Chen, R. Dutta, G.N. Rouskas, Traffic Grooming in Star Networks, Broadnets 2004: Workshop on Traffic Grooming in WDM Networks, (2004) 25–29
55. S. Huang, R. Dutta, and G. N. Rouskas, Traffic Grooming in Path, Star, and Tree Networks: Complexity, Bounds, and Algorithms, IEEE J. Selected Areas in Communications, 24(4) (2006) 66–82
56. X.H. Huang and M. Ma, An Efficient Scheduling Algorithm for Real-Time Traffic on WDM Passive Star Optical Networks, J. Lightw. Technol., 23(11) (2005) 3683–3695.
57. R. Berry and E. Modiano, Optimal Transceiver Scheduling in WDM/TDM Networks, IEEE J. Selected Areas in Communications, 23(8) (2005), 1479–1495
58. K.S. Kim, D. Gutierrez, F.T. An and L.G. Kazovsky, Design and Performance Analysis of Scheduling Algorithms for WDM-PON Under SUCCESS-HPON Architecture, J. Lightw. Technol., 23(11) (2005), 3716–3731
59. B. Wang, T. Li, X. Luo, and Y. Fan, Traffic grooming under a sliding scheduled traffic model in WDM optical networks, in Proceedings of the 1st Workshop on Traffic Grooming in WDM Networks, San Jose, CA, (2004)
60. J. Kuri, N. Puech; M. Gagnaire, Routing and grooming of scheduled lightpath demands in a multi-granularity switching network: A mathematical model, 2005 Conference on Optical Network Design and Modelling: Towards the broadband-for-all era, (2005) 73–83
61. A. Bianco, G. Galante, D. Careglio, J.M.Finochietto, E.Leonardi, F. Neri, J.S. Pareta, S. Spadaro, Multiclass scheduling algorithms for the David metro network, IEEE J. Selected Areas in Communications, 22(8) (2004) 1483–1496
62. R. Ramaswami, K.N. Sivarajan, Routing and wavelength assignment in all-optical networks, IEEE/ACM Transactions on Networking, 3(5) (1995) 489–500
63. A.C. Houle, Addressing the GRWA problem in WDM networks with a tabu search algorithm, Canadian Conference on Electrical and Computer Engineering, (2005) 1630–1633
64. A. Elwalid, D. Mitra, I. Saniee, and I. Widjaja, Routing and protection in GMPLS networks: from shortest paths to optimized designs, IEEE/OSA J. Lightwave Tech., 21 (2003) 2828–2837
65. A. Banerjee, J. Drake, J.P. Lang, B. Turner, K. Kompella, and Y. Rekhter, Generalized multiprotocol label switching: an overview of routing and management enhancements, IEEE Commun. Mag., 39(1) (2001) 144–150
66. A. Banerjee, J. Drake, J.P. Lang, B. Turner, D. Awduche, L. Berger, K. Kompella, and Y. Rekhter, Generalized multiprotocol label switching: an overview of signaling enhancements and recovery techniques, IEEE Commun. Mag., 39(7) (2001) 144–151

6

EPSO: Evolutionary Particle Swarms

V. Miranda, *Fellow IEEE*[1,2], Hrvoje Keko[1], and Alvaro Jaramillo[1]

[1]INESC Porto, Institute of Engineering in Systems and Computers of Porto
[2]FEUP, Faculty of Engineering of the University of Porto, Portugal
vmiranda@inescporto.pt

Abstract. This chapter presents EPSO (Evolutionary Particle Swarm Optimization), as an evolutionary meta-heuristic that implements a scheme of self-adaptive recombination, borrowing the movement rule from PSO (Particle Swarm Optimization). Besides the basic model, it discusses a Stochastic Star topology for the communication among particles and presents a variant called differential EPSO or dEPSO. The chapter presents results in a didactic Unit Commitment/Generator Scheduling Power System problem and results of a competition among algorithms in an intelligent agent platform for Energy Retail Market simulation where EPSO comes out as the winner algorithm.

6.1 Introduction

This chapter presents and discusses a method that has been called EPSO – Evolutionary Particle Swarm Optimization, which has been claiming interesting properties of convergence and, especially, of robustness in convergence, when compared with classical Evolutionary Algorithms (EA) and also with Particle Swarm Optimization algorithms (PSO).

From a conceptual point of view, EPSO has an interesting feature: it allows a double interpretation on how it works, because it may be seen from two perspectives – either as a variant of PSO or as a variant of EA. The following presentation hopefully will help in understanding how the method works and how should one manipulate its characteristics to obtain better convergence characteristics in specific problems.

The name EPSO, as far as the author is aware of, has been first coined in 2002, in [1]. Subsequently, a few papers more presented some interesting results and applications [2–5]. However, more recently, some other publications have started to use the acronym EPSO for other type of algorithms, such as "extended PSO" [6], "enhanced PSO" [7] or even "emotional PSO" [8]. It is unfortunate that this may lead to some confusion to readers and researchers. We wish to make clear from the start that by EPSO we refer to a class of

V. Miranda et al.: *EPSO: Evolutionary Particle Swarms*, Studies in Computational Intelligence (SCI) **66**, 139–167 (2007)
www.springerlink.com

algorithms of the family of self adaptive EA, where the classical operators for recombination are replaced by a rule similar to the particle movement of PSO. Alternatively, we refer to a class of algorithms of the family of PSO where the weights associated to the components of the movement rule are made to evolve in a self-adaptive mechanism.

6.2 Evolutionary Algorithms

Let's recall that a general Evolutionary Algorithm may be organized in the following steps:

Procedure EA
initialize a random population P of μ elements
REPEAT
reproduction (introduce stochastic perturbations in the new population) – generate λ offspring...
...by *recombination*
...by *mutation*
evaluation – calculate the fitness of the individuals
selection – of μ survivors for the next generation, based on the fitness value
test for termination criterion (based on fitness, on number of generations, etc.)
Until test is positive
End EA

This is a general algorithm and applies to all variants of EA, whether Genetic Algorithms, Evolutionary Programming or Evolution Strategies or other of the kind.

The driving force behind any EA is the selection operator. However, it requires the action of the replication or reproduction operators – recombination and mutation – to generate new points in space to be evaluated. Some EA give relevance to the mutation operator while others relies mostly on the recombination operator.

Mutation is an operator acting on a single individual, particle or chromosome and generates a new point in the search space by modifying the coordinates of a previous point. The definition of the mutation operator may be dependent of the particular problem under analysis. In Evolution Strategies/Evolutionary Programming, where typically (not necessarily) an individual is composed of a string of real variable values, the classical mutation procedure adopts Gaussian or Lognormal random perturbations of each variable. The amplitude of mutation is governed by the variance of the distribution regulating mutations, and its square root (standard deviation) is often called *learning rate*.

In self-adaptive Evolutionary Algorithms, the learning rates are transformed into variables and added to the chromosome [9,10]. During the process, they are themselves subject to mutation and selection and eventually they acquire values that allow a near optimal progression rate towards the optimum [11]. Self-adaptation was historically used first on the operator mutation.

Recombination generates a new individual by combining features from a set of individuals in the population. For the operator recombination many forms have been proposed. Classical variants are, for example:

Uniform crossover – in this variant, the value for each variable in the newly formed individual is obtained by randomly selecting one of the μ parents to "donate" its value.

Intermediary recombination – in this variant, the value of any variable in the offspring receives a contribution from all parents. This could result either from averaging the values of all parents (global intermediary recombination) or from averaging the values from a subset of the parents only, randomly chosen (local intermediary recombination). In these processes, one may still chose to average values with equal weights or to randomly define weights for a weighted average. In the case of $\mu = 2$, one could have the value of a variable given by

$$x_k^{new} = u_k x_{k,j1} + (1 - u_k) x_{k,j2}$$

where the indices j1 and j2 denote the two parent individuals and u_k is sampled from an uniform distribution in [0,1]. This leads to new individuals in the line segment between the two parents, but the concept may extend to points in the line but external to the segment if u_k is allowed to be sampled in a larger interval.

Point crossover – in this variant, parallel to the one adopted in genetic algorithms, first one randomly defines π crossover points, common to all individuals in the set of parents, and then the offspring successively receives a part from each parent, in turns.

Good schemes of mutation and recombination must assure a convenient exploration of the complete search space, but must also allow exploitation of interesting or promising regions. Achieving a balance between these two desirable features is not at all easy. The self-adaptive EA variants, providing dynamic adaptive change in mutation rates, try to satisfy this double objective. However, although self-adaptation in crossover has been proposed, so far no scheme for a self-adaptive recombination operator has been extensively adopted. A good review is found in [12].

6.3 The Movement Rule of PSO

The classical PSO – Particle Swarm Optimization, conceived by Kennedy and Eberhart [13], relies on the parallel exploration of the search space by a set of solutions or alternatives called "particles" that are successively transformed along the process. From one iteration to the following, each particle $\mathbf{X}_i$ generates a new one according to a "movement rule", which requires that one keeps record of the best point $\mathbf{b}_i$ found by the particle in its past life and the current global best point $\mathbf{b}_g$ found by the swarm of particles in their past life.

The movement rule states that

$$\mathbf{X}_i^{new} = \mathbf{X}_i + \mathbf{V}_i^{new} \tag{6.1}$$

where $\mathbf{V}_i$ is called the particle i velocity and was classically defined by

$$\mathbf{V}_i^{new} = \mathbf{V}_i + \mathrm{Rnd}().\mathbf{Wm}_i(\mathbf{b}_i - \mathbf{X}_i) + \mathrm{Rnd}().\mathbf{Wc}_i(\mathbf{b}_G - \mathbf{X}_i) \tag{6.2}$$

where the first term of the summation represents inertia or habit (the particle keeps moving in the direction it had previously moved), the second represents memory (the particle is attracted to the best – past – point in its trajectory) and the third represents cooperation or information exchange (the particle is attracted to the best point found by all particles).

The parameters $\mathbf{W}$ are diagonal matrices with weights fixed in the beginning of the process (index m is for the memory weights and index c is for the cooperation weights). Rnd() are random numbers sampled from a uniform distribution in [0,1]. Figure 6.1 illustrates the concept.

This transformation (movement) scheme proved adequate to make the swarm converge to the zone of the optimum but failed to assure fine tuning in convergence to an accurate optimum position. Therefore, a new scheme has been proposed [11]:

$$\mathbf{V}_i^{new} = \mathrm{Dec}(t)\mathbf{Wi}_i\mathbf{V}_i + \mathrm{Rnd}().\mathbf{Wm}_i(\mathbf{b}_i - \mathbf{X}_i) + \mathrm{Rnd}().\mathbf{Wc}_i(\mathbf{b}_G - \mathbf{X}_i) \tag{6.3}$$

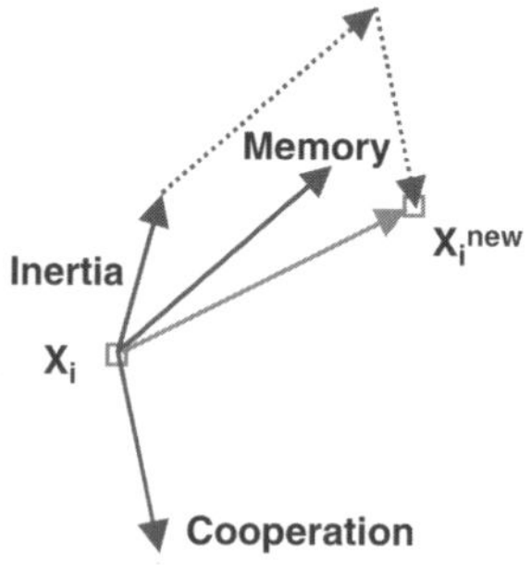

Fig. 6.1. Illustrating the movement of a particle in PSO, influenced by the three terms Inertia, Memory and Cooperation: the final transformation vector from $\mathbf{X}_i$ to $\mathbf{X}_i^{new}$ is composed of three contributions

where Dec(t) is a function decreasing with the progress of iterations, reducing progressively the importance of the inertia term, and new weights **Wi** are included.

Another version for PSO has been proposed by Maurice Clerc in [14], with the introduction of a *constriction factor* K, applied to each dimension of the particle space. In this version, the movement rule is

$$\mathbf{V}_i^{new} = \mathbf{K}\left[\mathbf{V}_i + Rnd().\mathbf{Wm}_i(\mathbf{b}_i - \mathbf{X}_i) + Rnd().\mathbf{Wc}_i(\mathbf{b}_G - \mathbf{X}_i)\right] \tag{6.4}$$

where **K** is a diagonal matrix of constriction factors. For dimension k, it is given by

$$K_k = \frac{2}{\left|2 - W_k - \sqrt{W_k^2 - 4W_k}\right|} \quad \text{with } W_k = Wm_k + Wc_k \text{ and } W_k > 4 \tag{6.5}$$

As one can see, none of these mechanisms is really adaptive and certainly not self-adaptive. Therefore, PSO is not an Evolutionary method. In fact, it does not rely on a selection operator as its driving force. It depends on *movement rule* that allows the generation of new individuals in space and this rule is such that, by itself, pushes the search towards the optimum.

6.4 EPSO as an Evolutionary Algorithm

6.4.1 Recombination in EPSO

In 2002 we have proposed an algorithm called EPSO [1] where instead of using the classical mutation and recombination operators to produce new individuals we have adopted the general scheme of the movement rule of PSO.

The general scheme of the movement rule is (whatever the variant)

$$\mathbf{V}_i^{(k+1)} = \mathbf{A}\mathbf{V}_i^{(k)} + \mathbf{B}\left(\mathbf{b}_i - \mathbf{X}_i^{(k)}\right) + \mathbf{C}\left(\mathbf{b}_G - \mathbf{X}_i^{(k)}\right) \tag{6.6}$$

If we examine this scheme, we conclude that a new particle $\mathbf{X}_i^{(k+1)}$ is formed as a combination of four other points:

- Its direct ancestor $\mathbf{X}_i^{(k)}$
- The ancestor of its ancestor $\mathbf{X}_i^{(k-1)}$
- A (possibly) distant past best ancestor $\mathbf{b}_i$
- The current global best of the swarm $\mathbf{b}_G$

We can give a different aspect to the movement rule:

$$\mathbf{X}_i^{(k+1)} = \mathbf{X}_i^{(k)} + \mathbf{A}\left(\mathbf{X}_i^{(k)} - \mathbf{X}_i^{(k)}\right) + \mathbf{B}\left(\mathbf{b}_i - \mathbf{X}_i^{(k)}\right) + \mathbf{C}\left(\mathbf{b}_G - \mathbf{X}_i^{(k)}\right) \tag{6.7}$$

$$\mathbf{X}_i^{(k+1)} = (1 + \mathbf{A} - \mathbf{B} - \mathbf{C})\mathbf{X}_i^{(k)} - \mathbf{A}\mathbf{X}_i^{(k-1)} + \mathbf{B}\mathbf{b}_i + \mathbf{C}\mathbf{b}_G \tag{6.8}$$

and we realize now that the sum of the parameters multiplying the four contributors to generate the offspring is equal to 1. It is therefore tempting to identify this expression with an intermediary recombination in EA with $\mu = 4$ and a special rule to determine who the parents are (they are not randomly selected). This means that we are considering an *enlarged population* including not only the active particles but also the direct ancestors and the set of the past best ancestors.

It is a recombination rule that has the remarkable property of pushing the population towards the optimum, as the PSO algorithms have demonstrated. Therefore, if joined together with a selection mechanism, which also pushes the population towards the optimum, one may expect that some cumulative effect may improve the performance of an optimizing algorithm.

6.4.2 Self-Adaptive Recombination

To determine the best values to use in **A**, **B** and **C**, EPSO relies on a self-adaptive mechanism. These parameters, considered as strategic, will be subject to selection and hopefully will evolve to values adapted to the type of landscape being searched.

Given a population with a set of particles, the general scheme of EPSO is the following:

- REPLICATION – each particle is replicated r times
- MUTATION – each particle has its strategic parameters mutated
- REPRODUCTION – each mutated particle generates an offspring through recombination, according to the particle movement rule, described below
- EVALUATION – each offspring has its fitness evaluated
- SELECTION – by stochastic tournament or other selection procedure, the best particles survive to form a new generation, composed of a selected descendant from every individual in the previous generation

Mutation of a parameter w into w^* is ruled by multiplicative Lognormal random numbers such as in $w_i^* = w_i[\log N(0,1)]^\tau$ or by additive Gaussian distributed random numbers such as in $w_i^* = w_i + \sigma N(0,1)$. The learning parameter (τ or σ) must be fixed externally.

The recombination rule for EPSO becomes (see Figure 6.2):

$$\mathbf{X}_i^{(k+1)} = \mathbf{X}_i^{(k)} + \mathbf{V}_i^{(k+1)} \tag{6.9}$$

$$\mathbf{V}_i^{(k+1)} = w_{i1}^* \mathbf{V}_i^{(k)} + w_{i2}^* (\mathbf{b}_i - \mathbf{X}_i) + w_{i3}^* (\mathbf{b}_G^* - \mathbf{X}_i)\,\mathbf{P} \tag{6.10}$$

where:

$\mathbf{b}_i$ – best point found by particle i in its past life up to the current generation

$\mathbf{b}_G$ – best overall point found by the swarm of particles in their past life up to the current generation

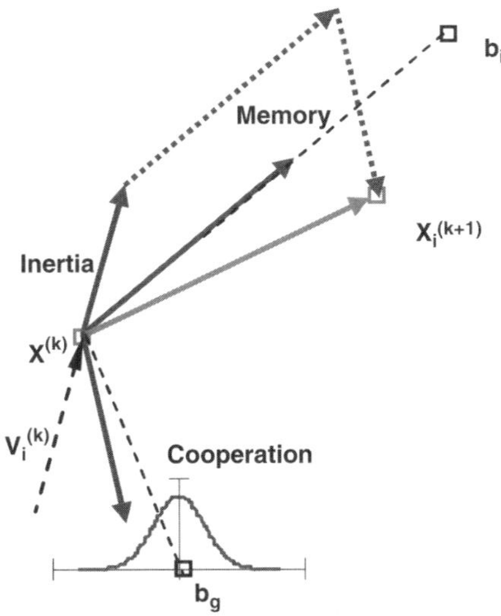

Fig. 6.2. Illustration of the EPSO movement rule. Notice that the vector associated with the *cooperation* factor does not point exactly to the global optimum b_g but to a mutated location

$\mathbf{X}_i^{(k)}$ – location of particle i at generation k
$\mathbf{V}_i^{(k)} = \mathbf{X}_i^{(k)} - \mathbf{X}_i^{(k-1)}$ – is the velocity of particle i at generation k
w_{i1} – weight conditioning the *inertia* term
w_{i2} – weight conditioning the *memory* term
w_{i3} – weight conditioning the *cooperation* or *information exchange* term
$\mathbf{P}$ – communication factor (discussed below)

The symbol * indicates that the parameter will undergo mutation. In the most effective EPSO variant, not only the weights affecting the components of movement are mutated but also the global best $\mathbf{b}_G$ is randomly disturbed to give

$$\mathbf{b}_G^* = \mathbf{b}_G + w_{i4}^* N(0,1) \tag{6.11}$$

where w_{i4} is the forth strategic parameter (target weight) associated with particle i. It controls the "size" of the neighborhood of $\mathbf{b}_G$ where it is more likely to find the real global best solution.

Selection is modeled from the Stochastic Tornament concept: among the ofspring of each particle, one compares the best one with another particle randomly sampled, and the best is selected with probability (1 – luck), where the *luck* parameter is defined in [0,1] but is usually small. If luck = 0 we have elitist selection.

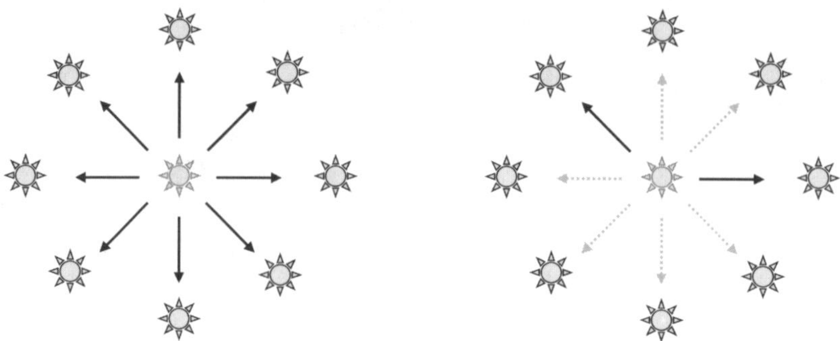

Fig. 6.3. Illustration of the star communication topology (left) and the stochastic star topology (right). In this latter case, the information of the location of the global best is only communicated to a subset of the swarm (thick arrows) and the composition of this subset varies from iteration to iteration

6.4.3 Communication Topology

The communication factor **P** induces a *stochastic star* topology for the communication among particles (Figure 6.3). It is a diagonal matrix affecting all dimensions of an individual, containing binary variables of value 1 with probability p and value 0 with probability (1-p); the p value controls the passage of information within the swarm and is 1 in classical formulations (this is the *star*).

This stochastic scheme oscillates between the star arrangement and a selfish version called *cognitive model* in [15], where no communication exists and a descendent of an individual is built only of contributions from its ancestor line.

Experimental results have suggested that a communication probability of $p = 0.20$ leads in many cases to better results than a classical deterministic star model with $p = 1$. One is lead to believe that restraining the free flow of information about the global best allows more local search by each particle and avoids premature convergence. As it is easily observed, this is yet another way of acting on the recombination operator.

The importance of not having a fixed communication topology in PSO has been suggested by works reported such as in [16] or [17]. Truly adaptive schemes for communication topology are yet to be proposed.

6.4.4 EPSO as a PSO

Because EPSO adopts the movement rule of PSO, we can always look at how it works as a swarm whose movement is ruled by weights that self-adapt in order to produce a global drift more adjusted to the landscape. Furthermore, because these weights are subject to mutation, this may give an extra chance for the swarm to escape local minima (i.e., having particles that still explore other regions of space, because they may gain enough speed).

On the other hand, EPSO also shows ability to focus and zoom in the optimum, precisely because mutations in the weights may favor the selection of the cooperation factor and reduce the importance of inertia and memory, if this strategy proves successful. This may in part explain why EPSO has shown, in many tests, robustness by consistently reaching the same optimum in a number of runs.

Lessons learned from PSO surely play a decisive role in improving EPSO algorithms. However, the point of view of EPSO as an evolutionary algorithm with self-adaptive recombination may give greater insight on how the algorithm successfully works and allow greater degree of freedom in designing better variants without being stuck with the rigid classical PSO form. One such example is described in the following section.

6.4.5 dEPSO: The Differential Evolution Version of EPSO

Differential Evolution [18], an evolutionary algorithm that uses a recombination rule that resembles some characteristics of the movement rule of PSO, has come up in the literature as an alternative method whose users claim that it displays good efficiency in reaching an optimum. There is more than a vague similarity between PSO and DE, In its variant DE2, from a particle $\mathbf{X}^{(\mathbf{k})}$ at iteration k a new individual is produced through

$$\mathbf{X}_i^{(k+1)} = \mathbf{X}_i^{(k)} + w_{i2}(\mathbf{X}_{Ai} - \mathbf{X}_{Bi}) + w_{i3}(\mathbf{b}_G - \mathbf{X}_i) \tag{6.12}$$

where $\mathbf{X}_A$ and $\mathbf{X}_B$ are two different individuals sampled from the population and distinct from the current individual $\mathbf{X}$ and $\mathbf{b}_G$ is the global best. In this recombination operator, the attraction for the particle past best is replaced by the difference between to other individuals.

The general principle of DE relies on the perception by the algorithm of the topology of the function being optimized by sensing macro-gradients through differences of points. In a way, this is exactly what a PSO algorithm also does.

We can add this flavor to EPSO by changing the movement rule precisely in the memory element, in a version we will call dEPSO. Instead of using the difference between a particle and its past best, we will use in this term the difference between two particles A and B randomly selected among the population (this can be the enlarged population including the set of particles past bests). The movement rule of dEPSO becomes

$$\mathbf{V}_i^{(k+1)} = w_{i1}^* \mathbf{V}_i^{(k)} + w_{i2}^*(\mathbf{X}_{Ai} - \mathbf{X}_{Bi}) + w_{i3}^* \left(\mathbf{b}_G^* - \mathbf{X}_i\right) \mathbf{P} \tag{6.13}$$

It is obvious that the communication topology becomes more complex, because a particle must now be aware not only of the location of the global best but also of the location of two other particles. These two particles may be chosen among the ones in the current restrict swarm or chosen among the enlarged swarm, including the set of the past best locations of all particles.

6.4.6 The Enhanced Elitist Version of EPSO

Closely observing the EPSO scheme, we become aware that that during the evolution of the swarm, selection acts only on the offspring and never on the set of parents plus offspring (in Evolution Strategies language, it is a comma strategy and not a plus strategy). To introduce a plus strategy and limited elitism, we have introduced changes in EPSO by dealing with the particle that finds the global best in a different way to the rest of the swarm: the particle at the global best is not eliminated unless one of its descendents finds a better point and replaces it – it competes with its offspring and pure elitism applies, not stochastic tournament.

Furthermore, for this particle we generate not just r descendents but a set $d > r$ – a local mini-swarm – and the mutations of the weights are done with a much narrower variance.

This variant introduces a new concept: that not all particles should be treated the same in the swarm. This reasoning opens a new path for the development of more flexible and (hopefully) more efficient variants of EPSO.

6.5 Experiments with EPSO

6.5.1 Test Functions

We now present some experiments we've designed to test the basic version of EPSO [equations (6.9) and (6.10) with P = 1] against the standard PSO 2006 by M. Clerc [19]. These experiments were based on four well-known test functions: the first and fourth functions represent difficult optima to be found, the third is the classical function that was used by ES researchers such as Beyer [11] to define the concepts of progress rate and the second exhibits a large number of deceiving local optima. All these functions describe unconstrained real-valued optimization problems.

Schaffer's function: $$\min f_1(x) = 0.5 + \frac{\left(\sin\sqrt{x^2+y^2}\right)^2 - 0.5}{(1.0+0.001(x^2+y^2))^2}$$

Alpine function: $$\max f_4(x) = \sin(x_1)\ldots\sin(x_n) \times \sqrt{x_1 \ldots x_n}$$

Sphere function: $$\min f_3(x) = \sum_{i=1}^{n} (x_i - 1)^2$$

Rosenbrock function: $$\min f_2(x) = \sum_{i=1}^{n} \left(100 \times \left(x_{i+1} - x_i^2\right)^2 + x_i - 1)^2\right)$$

The parameters used in the tests are presented in Table 6.1. The threshold used as the stopping criterion is listed in the column on the right; this threshold has been defined as a value close to the exact optimum, which is known for all problems.

6.5.2 Tests and Results

The first battery of tests defined was [1]:

1. In 500 runs randomly initiated, determine the average number of fitness evaluations needed to reach the demanding stopping threshold defined in Table 6.1.
2. Determine the average result offered by both algorithms, for a fixed number of fitness evaluations, in 500 runs randomly initiated. The limit was set at 200,000 evaluations, corresponding to 20 particles over 5,000 iterations in EPSO and 20 particles over 10,000 iterations in PSO (to have the same number of function evaluations).

The results of the tests are summarized in Tables 6.2 and 6.3.

- EPSO clearly displayed superiority over PSO in the 2-dimension Schaeffer's function. Figure 6.4 displays a typical comparison of convergence behavior between EPSO and classical PSO. Notice that the scale for the fitness (or objective) function is logarithmic.

Table 6.1. Parameters used in the test functions

Function	n	Domain	Stopping threshold
f_1 – Schaffer	2	$[-50, 50]^n$	0.0001
f_2 – Alpine	2	$[0, 100]^n$	98.9627
f_3 – Sphere	30	$[0, 100]^n$	0.01
f_4 – Rosenbrock	30	$[0, 30]^n$	1

Table 6.2. Average number of evaluations to reach the stopping threshold

Function	EPSO	PSO
$f1$ – *Schaffer 2-dimension*	4087	11165
f_2 – *Alpine 2-dimension*	16421	17007
f_3 – *sphere 30-dimension*	9789	26187
f_4 – *Rosenbrock 30-dimension*	28452	131462

Table 6.3. Average result for 200 000 evaluations in 500 runs, except for $f1$ where the average result is for 500 runs of 10 000 evaluations

Function	EPSO	PSO
$f1$ – *Schaffer 2-dimension*	1.65E-06	1.06E-03
f_2 – *Alpine 2-dimension*	98.9627	98.8484
f_3 – *sphere 30-dimension*	5.28E-09	4.40E-02
f_4 – *Rosenbrock 30-dimension*	1.91E-04	1.4243

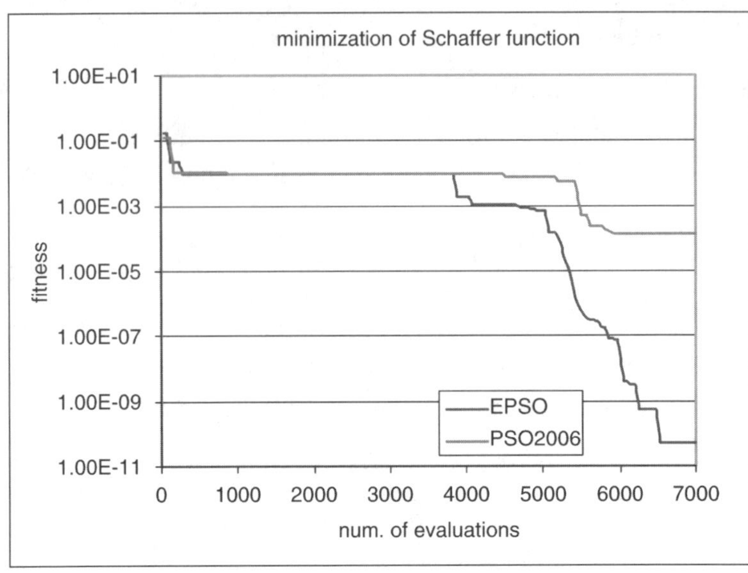

Fig. 6.4. Typical convergence of EPSO and classical PSO for Schaffer's function [1]

The Alpine function has many local optima and is therefore a good test for the ability of an algorithm to escape being trapped and instead discover the global optimum. Our first set of tests has been with the Alpine function defined in a 2-dimension space, and once more EPSO clearly displayed superiority over classical PSO. Figure 6.5 displays a typical comparison of convergence behavior between EPSO and classical PSO.

In another experiment, we used the algorithm to find the optimal maximum of the Alpine 10-dimension function in a $[0, 100]^{10}$ search space. In Figure 6.6 we can observe the average convergence in an experiment series of 20 simulations, with 40 particles. Once again we verify a superior behavior of EPSO, compared to the standard PSO. This has happened for all spaces of no matter how many dimensions.

The spherical model has been used extensively in Evolution Strategies to help in defining concepts such as "rate of progress" – as the statistical expectation of the improvement in the fitness value from any iteration to the following. Because it is an isotropic model, the improvement from iteration to iteration in an evolutionary process could be represented simply by a scalar function of the distance to the optimum. In this case, we are before a problem with a single optimum and we can test therefore the relative efficiency of EPSO against classical PSO. EPSO clearly displayed superiority over PSO in a 30-dimension sphere, and our interpretation for this fact is based on the explorative capacity conferred to EPSO by both the self-adaptive changes in the weights and on the definition of a foggy target instead of aiming the particles at the best-so-far point.

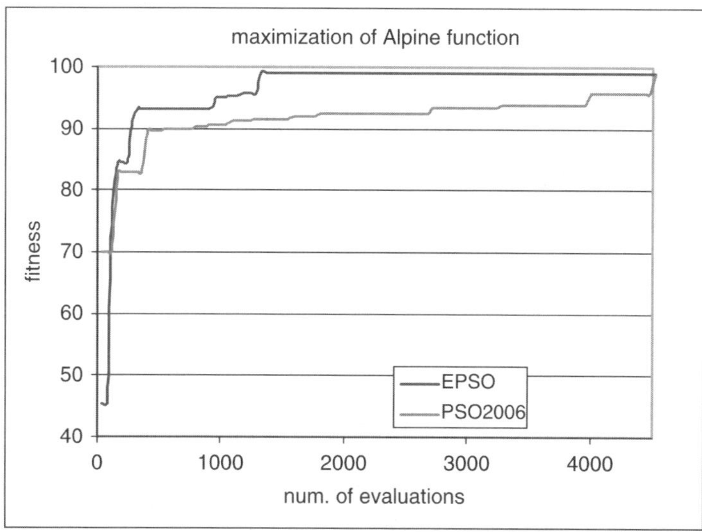

Fig. 6.5. Typical convergence of EPSO and classical PSO for the 2-dimension Alpine function [1]

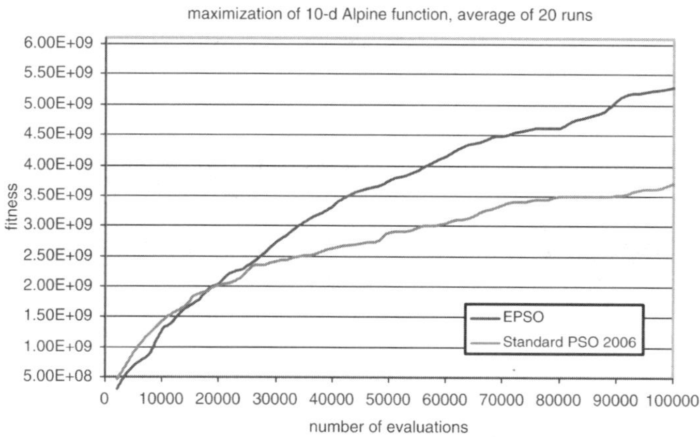

Fig. 6.6. Comparison of the average fitness evolution in EPSO and standard PSO for the 10-dimension Alpine function

Figure 6.7 displays a typical comparison of convergence behavior between EPSO and PSO. Notice again that the scale for the fitness (or objective) function is logarithmic.

We have also analyzed the effect of the variation of population size in this case. In 500 runs as above, for populations of 5, 10, 15 and 20 particles or individuals, the percentage of runs that did not satisfy a demanding threshold criterion has been registered. Figure 6.8 displays the results along

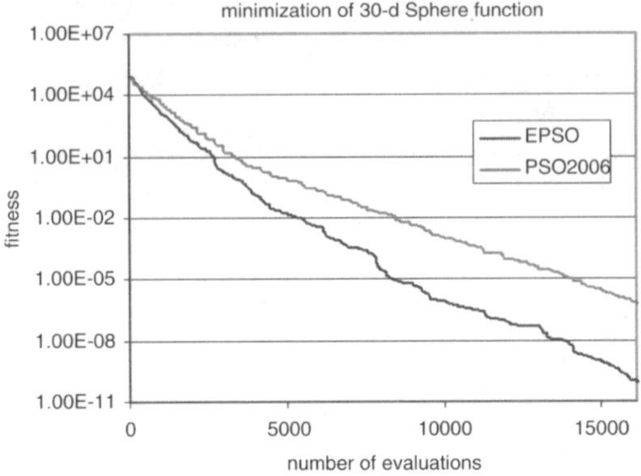

Fig. 6.7. Typical convergence of EPSO and standard PSO for the Sphere function [1]

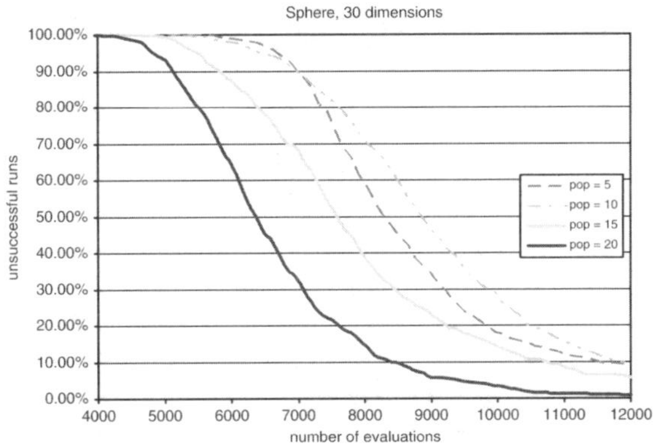

Fig. 6.8. Average error of the EPSO algorithm for different populations, measured by the percentage of runs that did not meet a given threshold close to the optimum

many iterations – it shows that enlarging the population provides a net benefit measured in the number of fitness function evaluations and suggests that in this case a population larger than 20 particles would perhaps lead to an even smaller computing effort. For a 10% failure rate, one gets below the defined threshold close to the optimum in approximately 200 EPSO iterations, which is remarkable.

The Rosenbrock function has a global optimum at point (1,...,1) with value 0 but the basin of attraction is extremely narrow and the path to reach it a deep valley, due to the twisted topology of the function in its vicinity, making it difficult to reach. It serves as a test for convergence in difficult circumstances. EPSO clearly displayed superiority over PSO in this function, defined in a 30-dimension space. Figure 6.9 displays a typical comparison of convergence behavior between EPSO and standard PSO. Notice again that the scale for the fitness (or objective) function is logarithmic. A similar comparison was

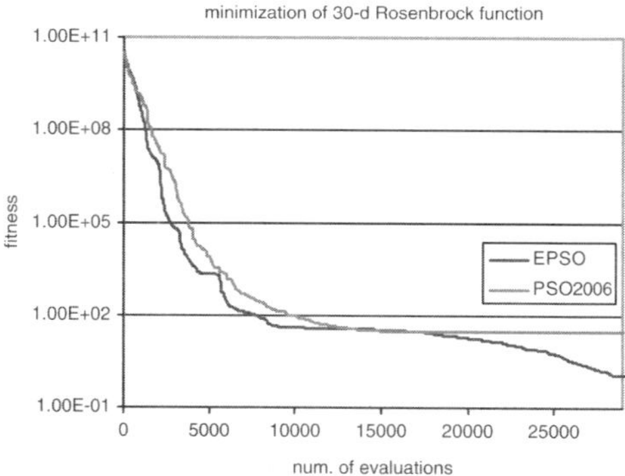

Fig. 6.9. Typical convergence of EPSO and classical PSO for Rosenbrock's function [1]

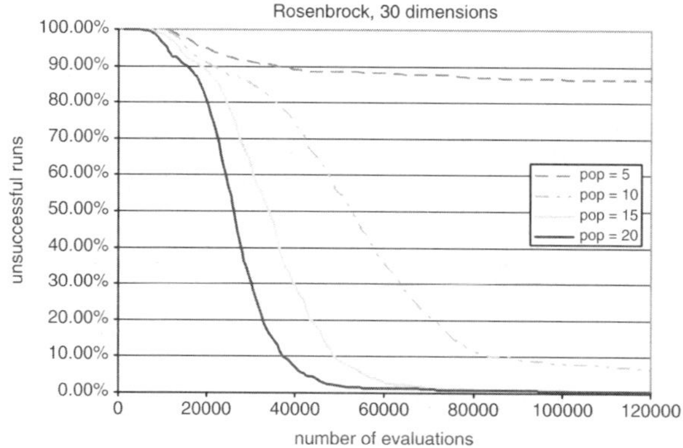

Fig. 6.10. Average error of the EPSO algorithm for different populations, measured by the percentage of runs that did not meet a given threshold close to the optimum

presented in [1] and relates to early experiments. It happens that the quality of the random number generator has shown to be important in the performance of these kinds of algorithms. Nevertheless the comparison in qualitative terms remains valid.

We presently have an efficient computer code implementing EPSO that is able to always reach the optimum of the Rosenbrock function although requiring a bit more than 40.000 fitness function evaluations for a 30 dimension space to do so.

We have also in this case analyzed the effect of the variation of population size. In 500 runs as above, for populations of 5, 10, 15 and 20 particles or individuals, the percentage of runs that did not satisfy a demanding threshold criterion has been registered.

Figure 6.10 displays the results along many iterations; one may clearly see that a population of 5 has a significantly worse behavior. The evolution in the graph and its comparison with Figure 6.8 leads us to speculate that the easier nature of the Sphere problem allows its solution by a small population, while the hard nature of the Rosenbrock function demands more exploratory power and therefore a larger population is needed.

6.5.3 Communication Probability p in the Rosenbrock Function

The effect of the communication probability p in forming a stochastic star communication scheme can be well observed in the Rosenbrock function in a space of D = 30, with a swarm of 20 particles, additive mutation with $\sigma = 0.2, r = 2$ and a maximum effort of 50,000 fitness function evaluations.

The communication probability was altered while all other parameters were left unchanged, Figure 6.11 displays the results obtained on an average of 20 runs. A number of things are immediately obvious:

- There is an optimum p value for finding the best result (closest to the optimum). The figure suggests that it may be approximately 0.75 in this case.
- There is an optimum p value that assures the best robustness to the algorithm (meaning that one has minimum variance in results for repeated runs).
- Average and robustness dependence on p are highly correlated – which is a good thing. This means that when choosing the p that leads on average to the best result, one also assures the strongest robustness in EPSO runs – which is important to reassure users that they can have confidence in the result when they perform a single run.

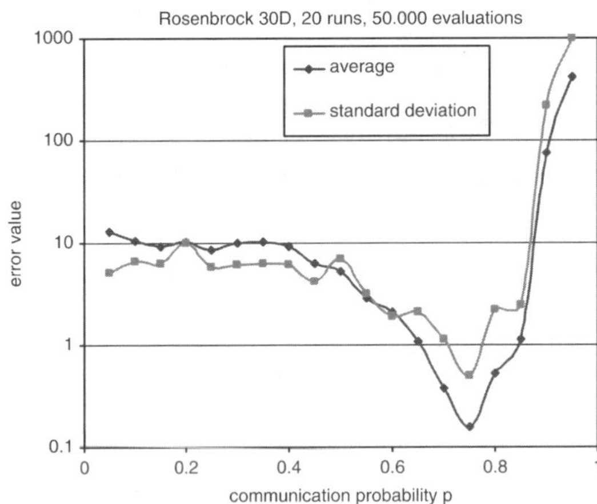

Fig. 6.11. Average error and error variance in 20 runs for the 30-dimension Rosenbrock function as a function of the communication probability p in the stochastic star

6.5.4 The Enhanced Elitist Version of EPSO

The difficulty with problems such as the Rosenbrock function is that during the evolution of the swarm, selection acts only on the offspring and never on the set of parents plus offspring (en Evolution Strategies language, it is a comma strategy and not a plus strategy). To introduce a plus strategy and limited elitism, we have introduced changes in EPSO by dealing with the particle that finds the global best in a different way to the rest of the swarm: the particle at the global best is not eliminated unless one of its descendents finds a better point and replaces it – it competes with its offspring and pure elitism applies, not stochastic tournament.

Furthermore, for this particle we generate not just r descendents but a set $d > r$ – a local mini-swarm – and the mutations of the weights are done with a much narrower variance.

Applying this elitist version to the Rosenbrock function, with the same data and $r = 2$ but $d = 4$ and $\tau = 0.005$ just for the mini-swarm, we have obtained better results than with the standard EPSO with a single main swarm. We are not presenting results because the ones available were obtained with an earlier version of EPSO and cannot be compared with the results in the previous section.

This represents an interesting improvement and suggests that exploring an elitist strategy together with the stochastic star communication topology is a way to improve the performance of EPSO in problems with narrow and difficult valleys such as is the case of the Rosenbrock function. More research is expected in this topic.

6.5.5 The Differential Evolution Version: dEPSO

The effectiveness of this model is tested in the following problem of unit commitment in power systems: given a set of generators and their generation cost curves, define which generators should be shut down and which should be in service and at which loading level, in order to minimize the overall cost (start up cost plus operation cost).

The problem is complex because of the cost functions for generators that must be considered – a mixed-integer non-linear program. Because of technical limits, the domain of a generator is not connected – there is a point (0,0) corresponding to generator shut down and then there is a gap until a point $(P_{min}, c(P_{min}))$ corresponding to the technical minimum of the machine. And the general shape of the cost functions implies that the problem has a non-convex nature – therefore, many local optima may appear.

The comparative performance of dEPSO is tested in a particular problem. The data are:

- the number of generators – ngen = 5

- the parameters of the cost function of each generator – this function is assumed to be a cubic polynomial, with 4 parameters a_i for i = 1 to 4:

$$C = a_0 + a_1 P + a_2 P^2 + a_3 P^3$$

where C is the generation cost in $/hour and P is the generator output in MW.
- the technical minimum and maximum of each generator P_{min} and P_{Max}
- the load, located at a single bus (transmission system neglected) – L = 15 MW

Cost curves and technical limits given in Table 6.4.

The objective is to minimize the sum of the costs for the five generators, noting that the domain of each variable is not continuous. The EPSO model demands an individual defined by its object parameters (5 variables) and strategic parameters (4 weights). We used Lognormal mutations with $\tau = 1$, 20 particles, r = 2 and 1000 generations. Initial values of the weights were set to 0.5 except for the target weight where the value was set to 0.001.

The cost curves are depicted in Figure 6.12 as well as the optimal solution, which is numerically presented in Table 6.5.

Table 6.4. Coefficients for the cubic cost polynomial function for each generator

Generator	a_0	a_1	a_2	a_3	Pmin	Pmax
g1	1	0,5	0,1	0,03	0 or 1	10
g2	2	0,4	0,2	0	0 or 2	10
g3	4	0,3	0,3	0	0 or 7	10
g4	6	1,5	0,15	0	0 or 2	10
g5	0	4	0	0	0 or 1	10

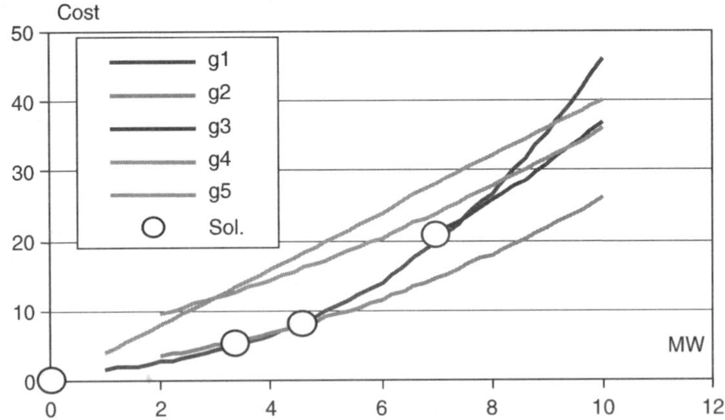

Fig. 6.12. Cost curves for the unit commitment problem and the best solution with 3 generators dispatched and 2 shut down

Table 6.5. Unit commitment and power dispatch for the optimal solution

g1	g2	g3	g4	g5	Cost
3.414	4.586	7	0	0	33.9068

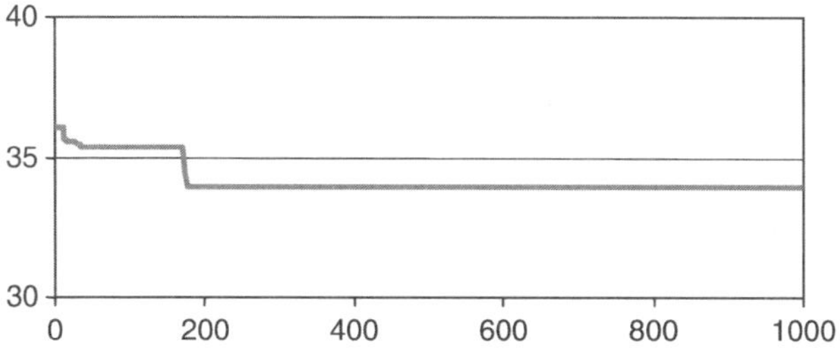

Fig. 6.13. Evolution of the best fitness during 1000 generations in one run of the EPSO algorithm

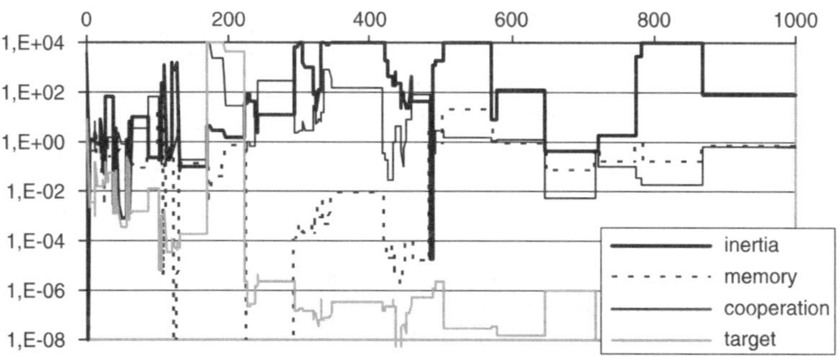

Fig. 6.14. Evolution of weights of the best particle at each iteration during the unit commitment optimization (1000 generations) using EPSO

Figure 6.13 displays the evolution of best fitness during an EPSO run and Figure 6.14 shows the weights of the global best particle kept at each iteration, for the same run. The best particle pattern shows that improvements were made till late in the process, in fact tiny adjustments after the big jump before generation 200, after which the perturbation in the target weight faded away. In the last generations, we also see that the inertia weight was dominant over the other weights.

Figure 6.15 shows the fitness evolution for a dEPSO algorithm and in Figure 6.16 shows the evolution of weights of a single specific particle in the optimization process. It suggests that the selection procedure indeed distinguished among the roles of the different components of the movement rule.

These are success cases, but how do the EPSO and dEPSO versions compare? Table 6.6 allows us to examine the best value found and the average in

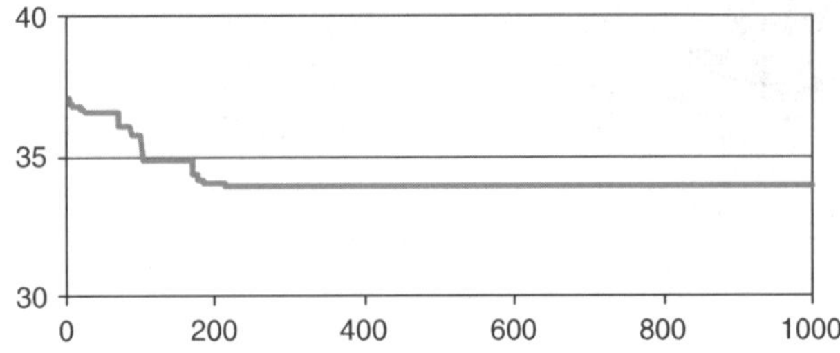

Fig. 6.15. Evolution of the best fitness during 1000 generations in one run of the dEPSO algorithm

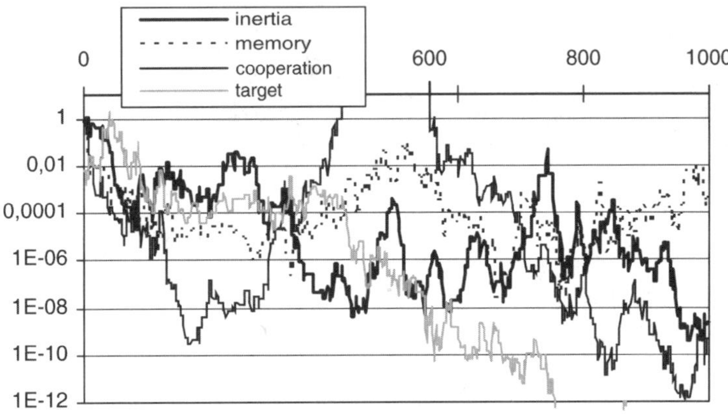

Fig. 6.16. Evolution of weights of a particle during the unit commitment optimization (1000 generations) using dEPSO

Table 6.6. Comparison of the performance of EPSO and dEPSO

	Av. 20 runs	Best found	No. times opt. found
EPSO	34,81575	33,90696	7/20
dEPSO	34,07073	33,90683	18/20

20 runs and how many times an algorithm discovers the optimum value or the optimum neighborhood, in 1000 generations, in 20 runs.

We can see that dEPSO provided a better average result and, especially, found the location of the global optimum more times than the simple EPSO. This result suggests that the dEPSO variant deserves to be looked at seriously and new research is suggested. The fact is that the problem of Unit Commitment is one very important problem in Power Systems for the economical implications it has and efficient solutions are still being sought that may compete with classical dynamic programming or branch and bound approaches, that have the known limitations of the curse of dimensionality for real systems.

6.6 Optimizing Market Competition

Comparisons of the performance of EPSO with other algorithms have been made in demanding Power System problems by other authors. In [20] the authors compared EPSO with PSO and their adaptive variant of called APSO in co-generation plant operation optimization, with the advantage of EPSO. In [21] one finds a comparison of several methods in voltage stability assessment where EPSO also emerged as the winner.

In this section we will refer to a competition among algorithms that we organized in a very demanding problem, extremely suited for meta-heuristics, given that no analytic solution may be found. It is the case of a simulation of a retail market in energy distribution, where the general objective is to discover a market strategy that will give advantage to a market actor (energy retailer) in a competitive environment.

The simulation was set over the open source intelligent agent platform JADE, which is FIPA compliant.

We will not describe this platform here but a description may be found in [22]. The simulation is composed of 19 agents running in a parallel arrangement of 5 PCs, as shown in Figure 6.17.

Each agent has an internal logic that commands its decisions, which are influenced by load forecasts, network development, reaction of consumers, prices, investment decisions, regulatory constraints, etc. Agents are assumed independent and from their interaction a complex behavior emerges that determines a dynamic market equilibrium while the simulation develops along time.

The agents represented in the platform are in Table 6.7.

One special type of agent is the retailer, which is equipped with a mechanism to predict the behavior of competitors in price setting. But the most unique feature of this agent is that, during the simulation, it performs regularly internal simulations of the market in order to try to optimize its policy for the next moves, acting on variables such as price of energy or money allocated to advertising, increasing company efficiency or investment.

This environment is dynamic because we have equipped the consumer agents with non-linear mechanisms that simulate their response to market conditions. Also, retailers depend on the action of delivery companies, because these agents make decisions about expanding their networks (having a GIS-type representation of the territory) and therefore potentially reaching new consumers.

This is an extremely complex environment to test a number of alternative meta-heuristics to optimize market strategy in the internal simulation of a retailer and to examine which agent (equipped with which algorithm) comes out as a winner (best profits) in a simulation extending for 2 years on a daily basis. We have reported some results in [23, 24] and in this paper we add comparisons including classical PSO, together with EPSO and 3 versions of Genetic Algorithms.

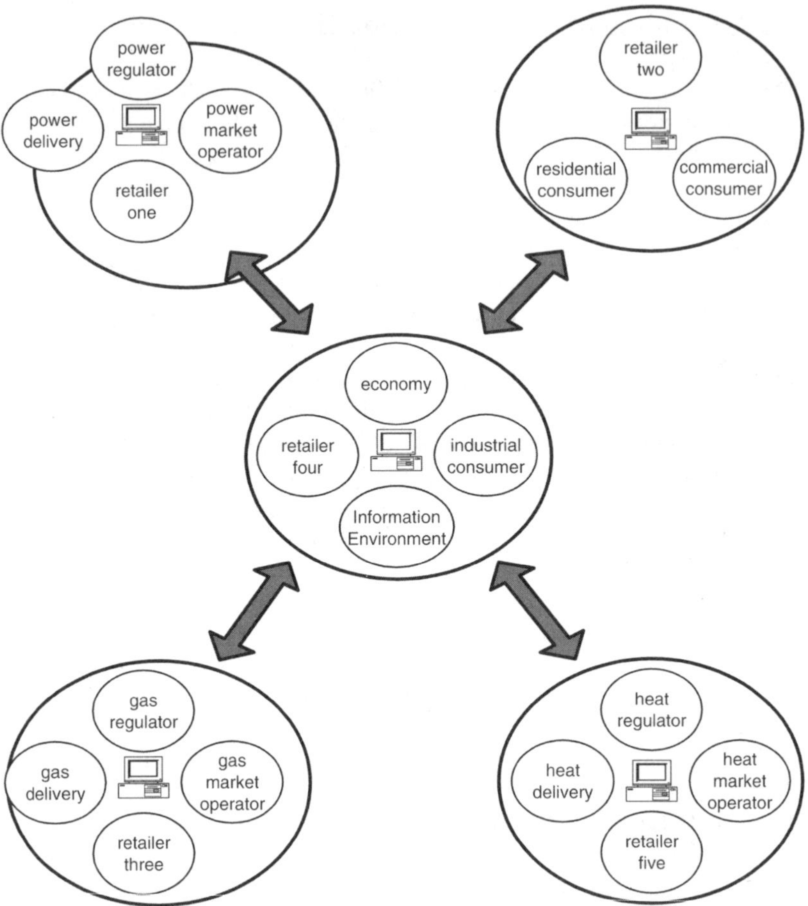

Fig. 6.17. Parallel arrangement of 19 Agents in a cluster of 5 PCs

Table 6.7. Comparison of the performance of EPSO with different values of p and local mini-swarm

AGENTS	
1 Residential cons. group	1 Commercial cons. group
1 Industrial cons. group	2 Electricity Retail Suppl.
2 Gas Retail Suppliers	1 Heat Retail Supplier
1 Power Delivery comp.	1 Gas Delivery company
1 Heat Delivery company	1 Electricity Regulator
1 Electr. Market operator	1 Gas Market operator
1 Heat Market operator	1 Gas Regulator
1 Heat Regulator	1 Economy agent
1 Information Environment	

The Genetic Algorithm versions were:

SSGA – Steady State Genetic Algorithm: a scheme where for a generation to the following we allowed the replacement of only 80% of the population, keeping the best 20%. It adds elitism to the selection, aiming at preserving in the population the most promising individuals so that they may have more chances of combining their good genes with other individuals.

DCGA – Genetic Algorithm with Deterministic Crowding: Deterministic Crowding is a selection technique where similar individuals in the population are paired before comparisons are made, and elitist selection acts on each pair formed. It aims at preserving diversity in the population.

MPGA – Multi-Population Genetic Algorithm: a scheme where we used two sub-populations, each running a SSGA, but exchanging two randomly selected individuals from one of the sub-populations to another before crossover took place. It also aims at preserving diversity.

The PSO algorithm used was the classical formulation with decreasing function in the inertia weight. The EPSO version was the basic one with Lognormal mutations.

The simulation length was of 24 months for market simulation and of 2 months for retailer internal simulations inside the evolutionary process.

The stopping criterion was the same for all algorithms: when performing the first internal simulation, the evolutionary process was stopped if there were no improvement in the fitness function after 50 consecutive generations; in all the following internal simulations, during the market simulation of 24 months, the limit was 10 generations.

In order to compensate for the influence of random events, we run every simulation 5 times. During market simulation, only one retailer was equipped with one type of EA and other retailers were not optimizing; also the investment function of Delivery agents was disconnected to avoid introducing noise in the simulation and disturb the interpretation of results.

The fitness function was the same in all cases: maximizing profit of the retailer.

In a first test, we have fixed the population for all methods as 20 individuals.

Figures 6.18 and 6.19 show the weekly profits made by a retailer agent using PSO and EPSO in five runs. One may observe that EPSO displays robustness in the results. Figure 6.20 compares the performance of the 5 algorithms and we see that EPSO was the clear winner and MPGA is second best while DCGA was the clear loser.

However, one of the possible reasons of bad performance of some algorithms could be early termination because of the application of the stopping criterion. We see clearly in Figure 6.21 that the distribution of computing effort was quite unbalanced. In an attempt to have a fair comparison, we have defined an Experiment 2 and changed the size of populations such that EPSO

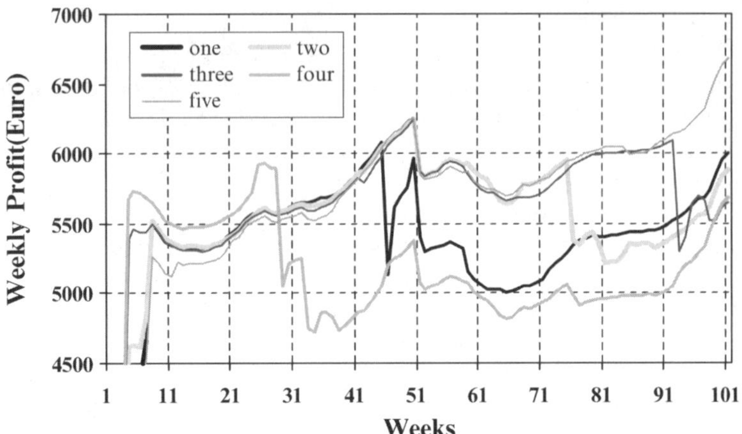

Fig. 6.18. Weekly profits made by a retailer equipped with PSO in 5 runs

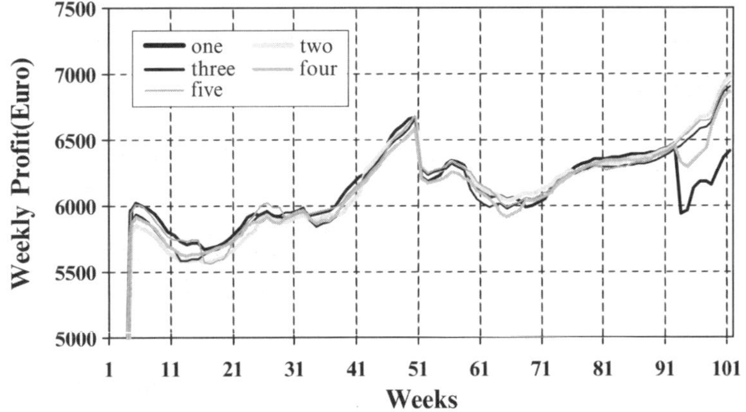

Fig. 6.19. Weekly profits made by a retailer equipped with EPSO in 5 runs

and MPGA were kept constant at 20 individuals while the population size of DCGA, SSGA, and PSO was increased to 100, 80, and 40 respectively.

Figure 6.22 shows the result as an average of five runs. Clearly, still EPSO is the winner and SSGA emerges now as the second best, while DCGA is confirmed as being the poor performer.

Figure 6.23 allows us to see that the best run of SSGA outperformed all others but that on average the profits made by a retailer with EPSO are stable on all runs and better than achieved with any other algorithm. Figure 6.24 clearly shows that now the computing effort of all five algorithms is relatively balanced.

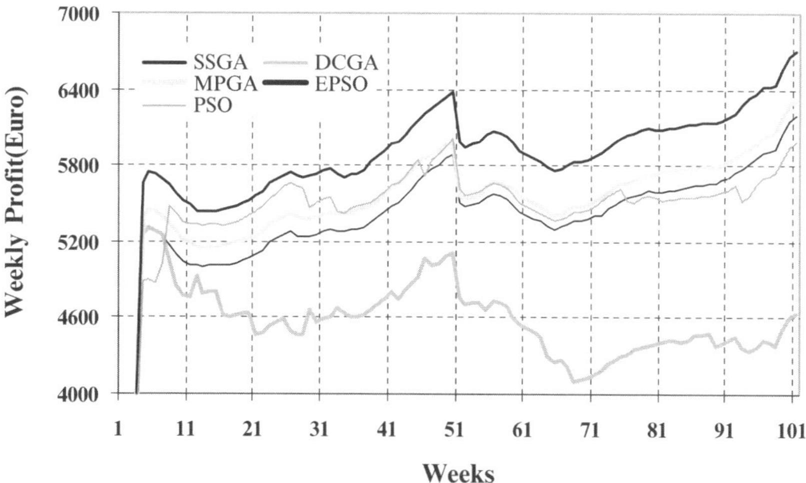

Fig. 6.20. Average of five runs for all algorithms

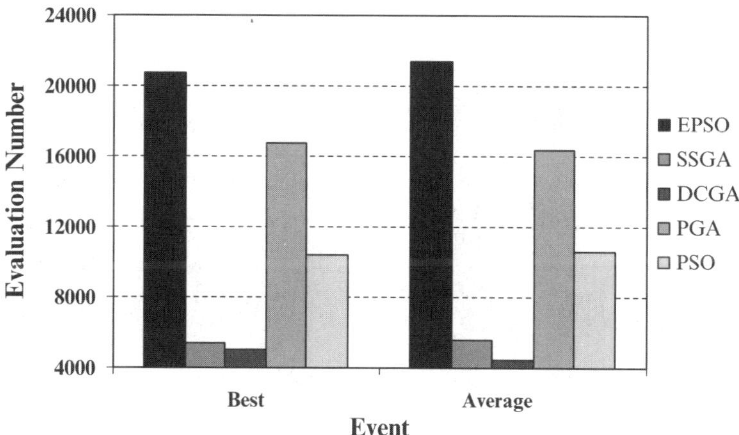

Fig. 6.21. Number of evaluations in the best run and on average of simulations done by the competing algorithms. Some algorithms experienced early termination because they could not achieve progress in a number of iterations

6.7 Conclusions

This chapter presents new ideas for improving the Evolutionary Particle Swarm Optimization algorithm and displays new evidence of the robustness of the method (in achieving consistently the same good results). The two innovations are: a random parameter affecting communication among particles, creating a topology that we call *stochastic star*, which proved valuable in cases where the algorithm needs to escape premature convergence; and a

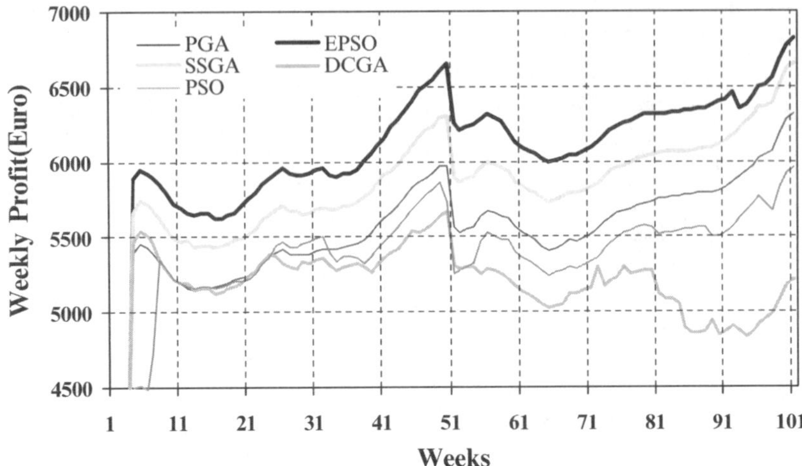

Fig. 6.22. Average of five runs with different population sizes (Exp. 2)

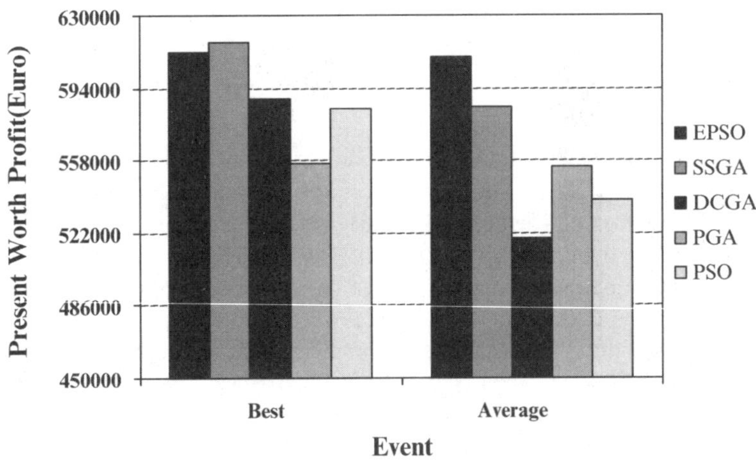

Fig. 6.23. Accumulated profits by a retailer equipped with each algorithm, for the best run and on the average of five runs of 2 year simulations (Exp. 2)

variant inspired in differential evolution that we call dEPSO, which displayed interesting behavior in the mixed integer problem of unit commitment which is a rather difficult practical problem in Power Systems.

The chapter provides also new evidence of the interest of the EPSO approach by reporting some results of a competition among distinct meta-heuristics in an extremely complex and dynamic environment: a market simulation platform composed of intelligent agents. The meta-heuristics were used to optimize the market strategy of retailer agents and EPSO emerged as the

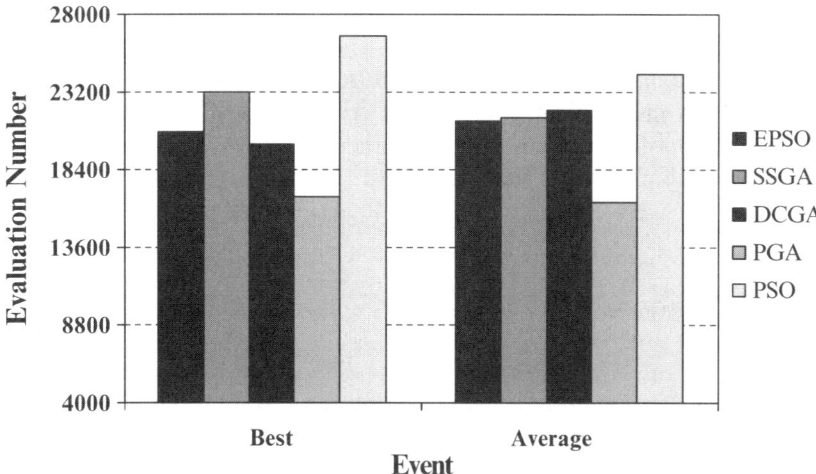

Fig. 6.24. Number of evaluations in the best run and on average of simulations done by the competing algorithms with different population sizes (Exp. 2)

algorithm that provided competitive advantage to the retailer using it to predict and select market decisions.

These results provide new incentive to further research the potential of the algorithm.

Acknowledgement

The authors wish to recognize the work of Nuno Fonseca, Naing Win Oo and Cristina Cerqueira, researchers at INESC Porto, who contributed at several stages to extensive testing of EPSO variants.

The research work has been partially funded by FCT – Foundation for Science and Technology of Portugal, under contract **POSC/EEA-ESE/60980/2004**.

References

1. V. Miranda and N. Fonseca, "EPSO – Best-of-Two-Worlds Meta-Heuristic Applied to Power System Problems", *Proceedings of WCCI/CEC – World Conference on Computational Intelligence, Conference on Evolutionary Computation*, Honolulu (Hawaii), USA, June 2002
2. V. Miranda and N. Fonseca, "Reactive Power Dispatch with EPSO - Evolutionary Particle Swarm Optimization", *Proceedings of PMAPS – International Conference on Probabilistic Methods Applied to Power Systems*, Naples, Italy, Sep 2002

3. M.S. Dias, N. Fonseca, H.M. Salgado and V. Miranda, "Optical Fibre Bragg Gratings Synthesis by Evolutionary Particle Swarm Optimization", *Proceedings of the Portuguese Physics Conference FISICA 2002*, Évora, Portugal, Sep 2002
4. V. Miranda and N. Fonseca, "EPSO - Evolutionary Particle Swarm Optimization, a New Algorithm with Applications in Power Systems ", Proceedings of IEEE/PES Transmission and Distribution Conference and Exhibition 2002: Asia Pacific, vol.2, pp.745–750, October, 2002
5. Mendonça, N. Fonseca, J.P. Lopes and V. Miranda, "Robust Tuning of Power System Stabilizers Using Evolutionary PSO", presented at ISAP 2003 – Intelligent Systems Applications to Power Systems Conference, Lemnos, Greece, September 2003
6. J.J. Xu and Z.H. Xin, "An Extended Particle Swarm Optimizer", Proceedings of the 19th IEEE International Parallel and Distributed Processing Symposium IPDPS'05 – workshop 6, Denver, Colorado, USA, April 2005
7. G.C. Chen and J.S. Yu, "Enhanced Particle Swarm Optimization and its Application in Soft Sensor" (in Chinese), Control And Decision, vol 20, no. 4, pp. 377–381, WangFa Data (ChinaInfo) and Chinese Electronic Periodical Services CEPS, April 2005
8. G. Yang and R. Zhang, "An Emotional Particle Swarm Algorithm", in Advances in Natural Computation: Proceedings of the First International Conference ICNC 2005, part III, page 553, Lecture Notes in Computer Science, Springer Berlin/Heidelberg, 2005
9. H.-P. Schwefel, "Adaptive Mechanismen in der biologischen Evolution und ihr Einfluß auf die Evolutionsgeschwindigkeit", Technical report, Technical University Berlin, 1974
10. D.B. Fogel, "Evolving Artificial Intelligence", Ph.D. Thesis, University of California, San Diego, 1992
11. H.-G. Beyer, "Toward a Theory of Evolution Strategies: Self-Adaptation", in *Evolutionary Computation*, vol. 3, no. 3, pp. 311–347, 1996
12. S. Meyer-Nieberg and H.-G. Beyer, "Self-adaptation in evolutionary algorithms" in F. Lobo, C. Lima, and Z. Michalewicz, editors, *Parameter Setting in Evolutionary Algorithms*, Springer, series: Studies in Computational Intelligence , Vol. 54, 2007
13. J. Kennedy and R.C. Eberhart, "Particle Swarm Optimization", *Proceedings of the 1995 IEEE International Conference on Neural Networks*, pp. 1942–1948, Perth, Australia, 1995
14. M. Clerc, "The Swarm and the Queen: Towards a Deterministic and Adaptive Particle Swarm Optimization", *Proceedings of the 1999 Congress of Evolutionary Computation*, vol. 3, 1951–1957, Washington D.C., USA, IEEE Press, 1999
15. J. Kennedy, "The Particle Swarm: Social Adaptation of Knowledge", *Proceedings of the 1997 International Conference on Evolutionary Computation*, pp. 303–308, IEEE Press, 1997
16. Mendes, R., Kennedy, J., Neves, J., "The fully informed particle swarm: Simple, maybe better". IEEE Transactions on Evolutionary Computation 8, pp. 204–210, 2004
17. S. Mohais, R. Mendes, C. Ward and C. Postho, "Neighborhood re-structuring in particle swarm optimization", AI 2005 – 18th Australian joint conference on artificial intelligence, Sydney, Australia, December 5-9, 2005, in Lecture Notes in Computer Science (ISSN 0302-9743), Springer, 2005

18. R. Storn and K. Price, "Differential Evolution - A simple and efficient adaptive scheme for global optimization over continuous spaces", International Computer Science Institute – *Technical Report* TR-95-012, March 1995
19. M. Clerc et al., Standard PSO 2006, in Particle Swarm Central, http://www.particleswarm.info/, code retrieved Dec. 2006
20. S. Kitagawa and Y. Fukuyama, "Comparison of particle swarm optimizations for optimal operational planning of energy plants", Proceedings of the Swarm Intelligence Symposium 2005, pp. 155–161 June 2005
21. H. Mori and Y. Komatsu, "A Hybrid Method of Optimal Data Mining and Artificial Neural Network for Voltage Stability Assessment", Proceedings of IEEE St. Petersburg PowerTech, Russia, June 2005
22. N. W. Oo and V. Miranda, "Multi-energy Retail Market Simulation with Intelligent Agents", Proceedings of IEEE St. Petersburg Power Tech, St. Petersburg, Russia, June 2005
23. V. Miranda and N. W. Oo, "Evolutionary Algorithms and Evolutionary Particle Swarms (EPSO) in Modeling Evolving Energy Retailers", *Proceedings of PSCC - 15th Power Systems Computation Conference*, Liège, Belgium, Aug 2005
24. N. W. Oo and V. Miranda, "Evolving Agents in a Market Simulation Platform – A Test for Distinct Meta-Heuristics", Proceedings of the 13th International Conference on Intelligent Systems Application to Power Systems ISAP 2005, pp. 482–487, Arlington (VA), USA, Nov. 2005

7

Design of Type-Reduction Strategies for Type-2 Fuzzy Logic Systems using Genetic Algorithms

Woei-Wan Tan and Dongrui Wu

Department of Electrical and Computer Engineering
National University of Singapore
4, Engineering Drive 3, Singapore 117576
Email: wwtan@nus.edu.sg

Summary. Increasingly, research in the field of fuzzy theory is focusing on fuzzy sets (FSs) whose membership functions are themselves fuzzy. The key concept of such type-2 FSs is the footprint of uncertainty. It provides an extra mathematical dimension that equips type-2 fuzzy logic systems (FLSs) with the potential to outperform conventional (type-1) FLSs. While a type-2 FLS has the capability to model more complex relationships, the output of a type-2 fuzzy inference engine is a type-2 FS that needs to be type-reduced before defuzzification can be performed. Unfortunately, type-reduction is usually achieved using the computationally intensive Karnik-Mendel iterative algorithm. In order for type-2 FLSs to be useful for real-time applications, the computational burden of type-reduction needs to be relieved. This work aims at designing computationally efficient type-reducers using a genetic algorithm (GA). The proposed type-reducer is based on the concept known as *equivalent type-1 FSs* (ET1FSs), a collection of type-1 FSs that replicates the input-output relationship of a type-2 FLS. By replacing a type-2 FS with a collection of ET1FSs, the type-reduction process then simplifies to deciding which ET1FS to employ in a particular situation. The strategy for selecting the ET1FS is evolved by a GA. Results are presented to demonstrate that the proposed type-reducing algorithm has lower computational cost and may provide better performance than FLSs that employ existing type-reducers.

7.1 Introduction

Fuzzy logic methodology has been proven to be effective in dealing with complex nonlinear systems containing uncertainties that are otherwise difficult to handle. Technology based on fuzzy theory has been applied to many real-world problems, especially in the area of consumer products. Despite their success, research has shown that fuzzy logic systems (FLSs) employing type-1 fuzzy sets (FSs) whose membership grades are crisp numbers may have limitations in modeling and minimizing the effect of uncertainties [1, 2]. There

W.-W. Tan and D. Wu: *Design of Type-Reduction Strategies for Type-2 Fuzzy Logic Systems using Genetic Algorithms*, Studies in Computational Intelligence (SCI) **66**, 169–187 (2007)
www.springerlink.com

is, therefore, more and more research interest in FLSs constructed using rule bases that utilize at least one type-2 FS. Unlike type-1 FSs, the membership grades of a type-2 FS are themselves FSs in [0, 1]. The union of all the fuzzy membership grades of a type-2 FS models the uncertainties in the shape of a FS and is collectively known as the footprint of uncertainty (FOU). Since the FOU of a type-2 FS provides an extra mathematical dimension, type-2 FLSs have the potential to outperform their type-1 counterparts. To date, type-2 FLSs have been used successfully in many applications, for example, decision making [3, 4], preprocessing of data [5], time-series forecasting [1], survey processing [1, 6], document retrieval [7], speech recognition [8], noise cancelation [9], word modeling [10, 11], control [2, 12–18], wireless communication [19, 20], etc.

Fig. 7.1 shows the structure of a typical type-2 FLS. Unlike type-1 FLSs, the output of the inference engine is a type-2 FS. Type reduction is, therefore, needed to convert the output of the fuzzy inference engine into a type-1 FS before defuzzification can be performed to obtain a crisp output. One of the most popular type-reducer is the Karnik-Mendel (KM) center-of-sets iterative approach [1]. Unfortunately, this type-reduction algorithm is very computationally intensive, especially when there are many membership functions (MFs) and the rule base is large. In order for type-2 FLSs to be useful for real-time applications, the computational burden of type-reduction needs to be relieved. This paper presents a method for using genetic algorithm (GA) to design computationally efficient algorithms for performing type-reduction. The approach utilizes the concept of *equivalent type-1 FSs* (ET1FSs) [21], which enables a type-2 FS to be replaced by a collection of ET1FSs without affecting the input-output relationship. As the role of a type-reducer is to reduce a type-2 FS to a type-1 FS, the type-reduction process reduces to selecting the most appropriate ET1FS to employ in a particular situation if a type-2 FS is viewed as a collection of ET1FSs. Thus, the computational requirement can be reduced and the resulting type-2 FLSs would be more amenable to real-time embedded applications.

The rest of the paper is organized as follows: The theory of type-2 FSs, type-2 FLSs and ET1FSs is first presented in Section 7.2. Section 7.3 then introduces the principle of the proposed type-reducer and how GA may be used to design the type-reducer. Next, a type-reducer is evolved by GA and

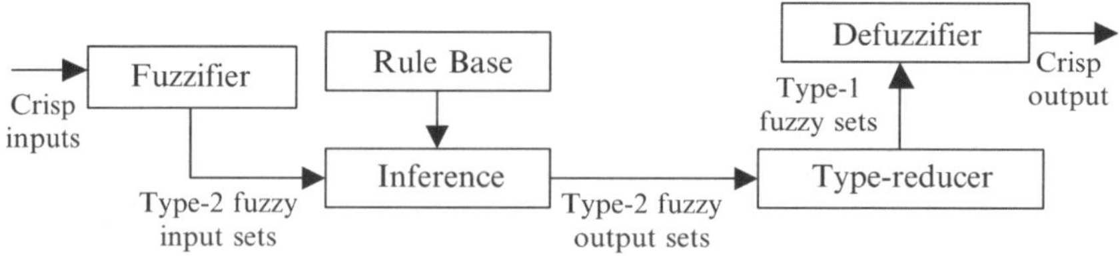

Fig. 7.1. A type-2 fuzzy logic system

its performance is compared with a type-1 FLS and two type-2 FLSs with the KM type-reducer and the uncertainty bound method on a first order time delay system. Finally, conclusions are drawn in Section 7.5.

7.2 Background and Motivations

7.2.1 Type-2 FSs

Type-2 FSs, characterized by membership grades that are themselves fuzzy, were introduced by Zadeh in 1975 [22] to better handle uncertainties. As illustrated in Fig. 7.2, a type-2 FS can be obtained by starting with a type-1 FS and then blurring it. The extra mathematical dimension provided by the blurred area, referred to as the FOU, represents the uncertainties in the shape and position of the type-1 FS. Mathematically, a type-2 FS, $\tilde{A}$, is characterised by the following MF:

$$\tilde{A} = \int_{x \in X} \left[\int_{u \in J_x} f_x(u) \mid x \right] \qquad J_x \subseteq [0,1] \tag{7.1}$$

where $\int \int$ denotes union over all admissible x and u. x is the input (*primary*) variable whose domain is X. u is the *secondary variable* and its domain $J_x \subseteq [0,1]$ is the *primary membership* of x. $f_x(u)$ is a type-1 FS referred to as a *secondary set* that reflects the uncertainties associated with the primary memberships in the FOU. Many types of secondary MF are possible. Since there are no guidelines on how secondary grades should be assigned to the primary memberships, a logical thing to do is to assign the same weight to all primary memberships. Hence, the most frequently used type-2 FSs are *interval type-2 FSs*[1], where each point in the FOU has unity secondary membership grade. For an interval type-2 FS, the FOU is bounded by an upper MF,

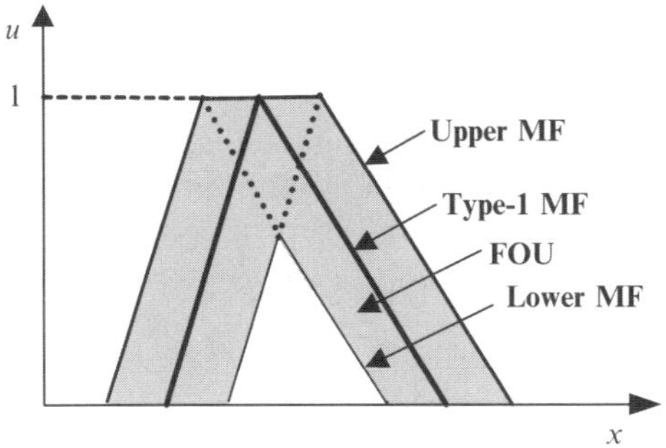

Fig. 7.2. An interval type-2 FS

[1] Some researchers [23–25] call them *interval-valued FSs*.

$\overline{\mu}_{\tilde{A}}(x)$, and a lower MF, $\underline{\mu}_{\tilde{A}}(x)$. Since $\overline{\mu}_{\tilde{A}}(x)$ and $\underline{\mu}_{\tilde{A}}(x)$ are both type-1 FSs, the primary membership, J_x reduces to

$$J_x = \left\{(x,u) : u \in \left[\underline{\mu}_{\tilde{A}}(x), \overline{\mu}_{\tilde{A}}(x)\right]\right\}$$

7.2.2 Interval Singleton Type-2 Fuzzy Logic System (FLS)

In this paper, an interval singleton type-2 FLS [1] is employed. "Interval" means that the input/output domains are characterized by interval type-2 FSs. The term "singleton" denotes that the fuzzifier converts the input signals of the FLS into fuzzy singletons. The inference engine then matches the fuzzy singletons with the fuzzy rules in the rule base. To compute unions and intersections of type-2 FSs, compositions of type-2 relations are needed. Just as the sup-star composition is the backbone computation for a type-1 FLS, the extended sup-star composition is the backbone for a type-2 FLS [1]. To illustrate the extended sup-star operation, consider a rule base that consists of rules with the following structure:

$$R^i: \quad \text{If } x_1 \text{ is } \widetilde{F}_1^i \text{ and } x_2 \text{ is } \widetilde{F}_2^i, \quad \text{then } y \text{ is } \widetilde{G}^i$$

where $i = 1, 2, \ldots, N$ and N is the number of rules. The first step in the extended sup-star operation is to obtain the firing set $\sqcap_{j=1}^{2} \mu_{\widetilde{F}_j^i}(x_j) \equiv F^i(\mathbf{x})$ by performing the input and antecedent operations. As only interval type-2 FSs are used and the meet operation is implemented by the product t-norm, the firing set is the following interval type-1 FS:

$$F^i(\mathbf{x}) = [\underline{f}^i(\mathbf{x}), \overline{f}^i(\mathbf{x})] \equiv [\underline{f}^i, \overline{f}^i] \tag{7.2}$$

where $\underline{f}^i(\mathbf{x}) = \underline{\mu}_{\widetilde{F}_1^i}(x_1) \cdot \underline{\mu}_{\widetilde{F}_2^i}(x_2)$ and $\overline{f}^i(\mathbf{x}) = \overline{\mu}_{\widetilde{F}_1^i}(x_1) \cdot \overline{\mu}_{\widetilde{F}_2^i}(x_2)$. The term $\underline{\mu}_{\widetilde{F}_j^i}(x_j)$ and $\overline{\mu}_{\widetilde{F}_j^i}(x_j)$ are the lower and upper membership grades of $\mu_{\widetilde{F}_j^i}(x_j)$ (See Fig. 7.2). Next, the firing set, $F^i(\mathbf{x})$, is combined with the consequent FS of the ith rule using the product t-norm to derive the fired output consequent sets. The combined output FS may then be obtained using the maximum t-conorm.

Since the output of the inference engine is a type-2 FS, it must be type-reduced before the defuzzifier can be used to generate a crisp output. This is the main structural difference between type-1 and type-2 FLSs. The most commonly used type-reduction method is the center-of-sets type-reducer, which may be expressed as [1]:

$$\begin{aligned} Y_{cos}(\mathbf{x}) &= \int_{y^1 \in Y^i} \cdots \int_{y^N \in Y^i} \int_{f^1 \in F^i(\mathbf{X})} \cdots \int_{f^N \in F^i(\mathbf{X})} 1 \Bigg/ \frac{\sum_{i=1}^{N} f^i y^i}{\sum_{i=1}^{N} f^i} \\ &\equiv 1/[y_l, y_r] \end{aligned} \tag{7.3}$$

where $F^i(\mathbf{X}) = [\underline{f}^i, \overline{f}^i]$ is the interval firing level of the ith rule, $Y^i = [y_l^i, y_r^i]$ is an interval type-1 FS corresponding to the centroid of the interval type-2 consequent set $\widetilde{G}^i$ [1]:

$$C_{\widetilde{G}^i} = \int_{\theta_1 \in J_{x_1}} \cdots \int_{\theta_N \in J_{x_N}} 1 \Bigg/ \frac{\sum_{i=1}^{N} x_i \theta_i}{\sum_{i=1}^{N} \theta_i} \equiv 1/[y_l^i, y_r^i] \tag{7.4}$$

(7.3) may be computed using the KM iterative method [1] as follows:

Set $y^i = y_l^i$ *(or* y_r^i*) for* $i = 1, \ldots, N$;
Arrange y^i *in ascending order;*
Set $f^i = \frac{\underline{f}^i + \overline{f}^i}{2}$ *for* $i = 1, \ldots, N$;
$y' = \frac{\sum_{i=1}^{N} y^i f^i}{\sum_{i=1}^{N} f^i}$;
do
    $y'' = y'$;
    Find $k \in [1, N-1]$ *such that* $y^k \leq y' \leq y^{k+1}$;
    Set $f^i = \overline{f}^i$ *(or* $\underline{f}^i$*) for* $i \leq k$
    Set $f^i = \underline{f}^i$ *(or* $\overline{f}^i$*) for* $i \geq k+1$;
    $y' = \frac{\sum_{i=1}^{N} y^i f^i}{\sum_{i=1}^{N} f^i}$;
while $y' \neq y''$
y_l *(or* y_r*)* $= y'$;

It has been proven that this iterative procedure can converge in at most N iterations [1]. Once y_l and y_r are obtained, they can be used to calculate the crisp output. Since the type-reduced set is an interval type-1 FS, the defuzzified output is:

$$y(\mathbf{x}) = \frac{y_l + y_r}{2} \tag{7.5}$$

7.2.3 Equivalent Type-1 Fuzzy Sets (ET1FSs)

The main difference between type-1 and type-2 FSs is the FOU, and research has shown that the extra mathematical dimension provided by the FOU is mainly responsible for the improved modeling capability of type-2 FLSs [26]. While a type-2 FS is a three dimensional function, the input-output relationship of a type-2 FLS, as with type-1 FLSs, is fixed once its structure and rule base is selected. This characteristic motivates the idea that a type-2 FS can be reduced to a group of type-1 FSs without affecting the output of a type-2 FLS. Formally, the concept of ET1FSs may be defined as:

> *Equivalent type-1 fuzzy sets* is the collection of type-1 FSs that can be used in place of the FOUs in a type-2 FLS.

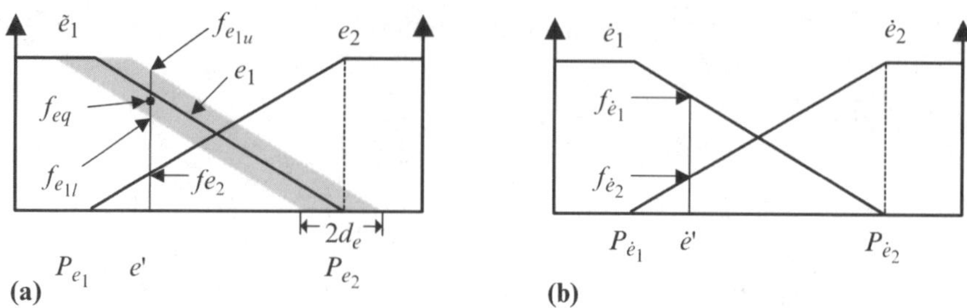

Fig. 7.3. (a) MFs of e domain, (b) MFs of $\dot{e}$ domain

Table 7.1. Rule base of the FLCs

$e\backslash\dot{e}$	$\dot{e}_1$	$\dot{e}_2$
e_1	$\dot{u}_{11}$	$\dot{u}_{12}$
e_2	$\dot{u}_{21}$	$\dot{u}_{22}$

To illustrate the ET1FS concept, consider the example of a two inputs single output type-2 fuzzy logic controller (FLC) and an accompanying base-line type-1 FLC. Both FLCs have two inputs (e and $\dot{e}$) and one output ($\dot{u}$). Each input is characterized by two MFs in its domain. The MFs are shown in Fig. 7.3. The type-2 FS, $\tilde{e}_1$, is obtained by introducing a FOU to the type-1 FLS, shown as the dark thick lines in Fig. 7.3. The type-2 FS used here is an interval one, where each point of the FOU has a unity secondary membership grade. Table 7.1 is the rule base of the FLCs. The entries in Table 7.1 are defined as:

$$\dot{u}_{ij} = K_I \cdot P_{e_i} + K_P \cdot P_{\dot{e}_j} \qquad i, j = 1, 2 \tag{7.6}$$

where P_{e_i} is the apex of the MF for e_i, $P_{\dot{e}_j}$ is the apex of MF for $\dot{e}_j$, as labelled in Fig. 7.3. When the "Product-Sum-Gravity" inference is employed, the resulting type-1 FLC is equivalent to a PI controller with a proportional gain of K_P and an integral gain K_I [27].

Since the baseline type-1 FLC is actually a PI controller, its control surface is linear. The control surface of the type-2 FLC is more complex and nonlinear. For example, the control surface of a type-2 FLC with the following parameters is shown in Fig. 7.4a:

$$P_{e_1} = P_{\dot{e}_1} = -1, \quad P_{e_2} = P_{\dot{e}_2} = 1, \quad d_e = 0.5$$

$$K_I = 1, \quad K_P = 1$$

As it is more complex, the control surface in Fig. 7.4a may not be implemented by a type-1 FLC. However, the control surface can be cutted into numerous slices, each corresponding to a constant $\dot{e}$ value. That is, for a particular input $\dot{e} = \dot{e}'$, a curve representing the relationship between the output $\dot{u}$ and the input e can be obtained. Fig. 7.4b shows the slice corresponding to $\dot{e}' = -1$.

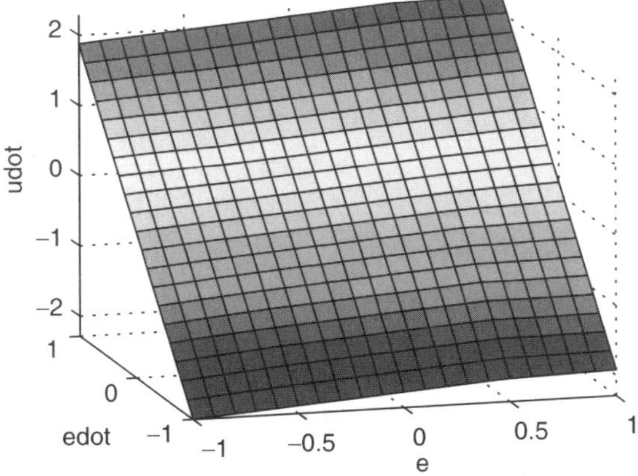

Fig. 7.4a. Control surface of a type-2 FLS

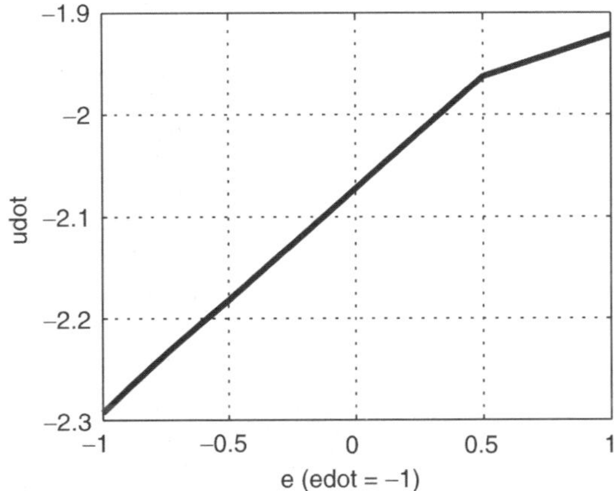

Fig. 7.4b. A slice when $\dot{e} = -1$

As presented in [21], each such slice can be replicated by constructing an ET1FS to replace the type-2 FS $\tilde{e}_1$.

Suppose an input fires the FSs $\tilde{e}_1$, e_2, $\dot{e}_1$ and $\dot{e}_2$ by $f_{\tilde{e}_1} = [f_{e_{1l}}, f_{e_{1u}}]$, f_{e_2}, $f_{\dot{e}_1}$ and $f_{\dot{e}_2}$, respectively, where $f_{e_{1l}}$ and $f_{e_{1u}}$ are the firing strengths on the lower and upper MFs of $\tilde{e}_1$. ET1FSs are identified by first reducing the interval firing set $f_{\tilde{e}_1} = [f_{e_{1l}}, f_{e_{1u}}]$ to a single value, referred as the *equivalent type-1 membership grade* (ET1MG) f_{eq} (Refer to Fig. 7.3), without affecting the crisp output $\dot{u}'$ of the type-2 FLC. Each ET1FS is a set of ET1MGs re-producing one slice in the input-output surface of the type-2 FLS. Hence, the ET1FS of $\tilde{e}_1$ corresponding to $\dot{e} = \dot{e}'$ can be identified by holding $\dot{e}$ constant at $\dot{e}'$ and varying e in discrete steps from $P_{e_1} - d_e$ to $P_{e_2} + d_e$ (the FOU of $\tilde{e}_1$). The remaining slices of the input-output surface can be re-constructed by finding the ET1FSs for other values of $\dot{e}$. The ET1FSs corresponding to the slides in

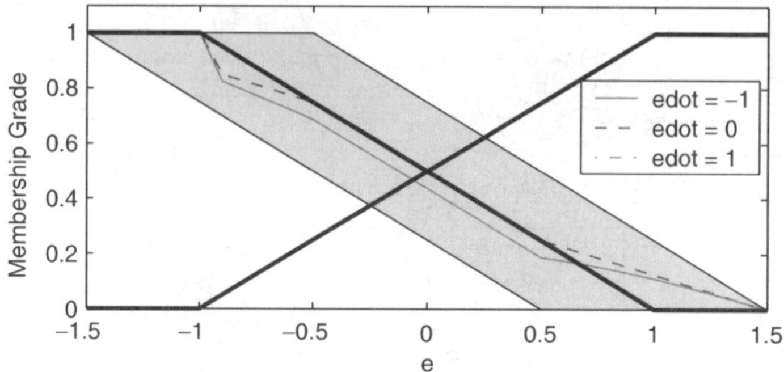

Fig. 7.5. ET1FSs obtained by the KM type-reducer

Fig. 7.4b where $\dot{e}' = \{-1, 0, 1\}$ are shown in Fig. 7.5. Note here the ET1FS corresponding to $\dot{e}' = -1$ coincides with the one to $\dot{e}' = 1$.

7.3 Computationally Efficient Type-Reducer

The schematic structure of a type-2 FLS in Fig. 7.1 shows that type-reduction is performed on the output of the fuzzy inference engine. Consequently, both the inference engine and the type-reducer have to process interval firing strengths. This increases the computational burden and may render type-2 FLSs as unsuitable for certain real-time applications. ET1FSs enable a type-2 FS to be treated as a collection of type-1 FSs. By using this concept, type-reduction may be simplified to finding the ET1FS corresponding to a particular input. More specifically, the type-reducer needs to identify the ET1MG (f_{eq}) for each interval firing strength. Once the ET1MG has been deduced, the firing set of a type-2 FS reduces to a crisp value and a type-1 fuzzy inference engine and defuzzifier can be employed to find the output of the type-2 FLS. In summary, the proposed type-reduction procedure is applied before the inference engine, as illustrated in Fig. 7.6, with the goal of determining the appropriate ET1MGs according to the inputs. Since the inference engine only has to process crisp numbers instead of interval sets in this case, the computational burden is relieved and the calculations may be completed in a shorter time compared to a type-2 FLS based on the KM iterative algorithm.

Although the new approach changes processing order, the proposed type-reducer should not alter the characteristics of a type-2 FLS. From the observations made in [13, 21, 26, 28], a type-reducer in a type-2 FLC should satisfy the following requirements:

1. Since a type-2 FLS reduces to its type-1 counterpart when the FOU disappears, a type-reducer must produce ET1FSs that coincide with the baseline type-1 FSs in this case.

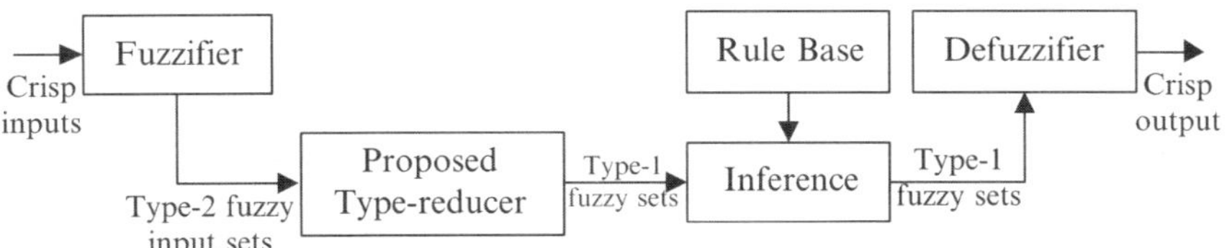

Fig. 7.6. Structure of a type-2 FLS with the proposed type-reducer

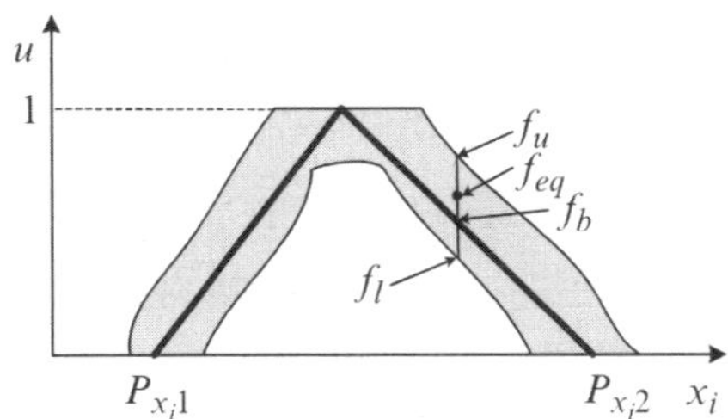

Fig. 7.7. Illustration of the new type-reducer

2. The ET1FSs change with the input. Hence, a type-reducer may be a function of all the input variables.
3. Research on applying type-2 FLS for control [13, 26, 28] demonstrate that the control surface of a type-2 FLC is generally smoother than that of a type-1 FLC, especially around the origin ($e = 0$, $\dot{e} = 0$). The smoother control surface is one factor that makes a type-2 FLC more robust than its type-1 counterpart. The proposed type-reducer should, therefore, also give rise to control surfaces that are smoother.

By taking into account the above requirements, the proposed type-reducer designed using GA for reducing the interval firing set $[f_l, f_u]$ to an ET1MG, f_{eq}, may be defined as:

$$f_{eq} = f_u - \sum_{i=1}^{N} \alpha_i \frac{2|x_i|}{P_{x_i2} - P_{x_i1}} \times (f_u - f_l) \tag{7.7}$$

where: N is the total number of inputs
α_i is a weight evolved using a GA
x_i is the ith input
$P_{x_i2} - P_{x_i1}$ (refer to Fig. 7.7) is the support of the baseline type-1 FS when the FOU disappears.

Analysis verifying that the algorithm defined in (7.7) satisfies the three requirements of a type-reducer will now be presented.

1. Requirement (1) essentially states that f_{eq} must equal f_b in Fig. 7.7 when $f_u = f_l = f_b$. By setting $f_u = f_l = f_b$, it is obvious that (7.7) satisfies this requirement.

2. The term $\sum_{i=1}^{N} \alpha_i \frac{2|x_i|}{P_{x_i2} - P_{x_i1}}$ in (7.7) is motivated by the need for the ET1FSs to vary with the input values. For the FLC described in Fig. 7.3, there are two inputs e and $\dot{e}$. Hence, (7.7) reduces to:

$$f_{eq} = f_u - \left[\alpha_1 \frac{2\,|e|}{P_{e_2} - P_{e_1}} + \alpha_2 \frac{2\,|\dot{e}|}{P_{\dot{e}_2} - P_{\dot{e}_1}}\right](f_u - f_l) \tag{7.8}$$

As (7.8) is a function of both inputs e and $\dot{e}$, the second requirement is fulfilled.

3. To demonstrate that the control surface obtained by using (7.7) is smoother than that of the baseline type-1 FLC, the relationship between the slope of the control surface and the value of f_{eq} is examined. Using the structure in Fig. 7.3 and replacing the interval firing strength $f_{\tilde{e}_1}$ by its ET1MG f_{eq}, the output is:

$$\dot{u} = \frac{f_{eq} f_{\dot{e}_1} \dot{u}_{11} + f_{eq} f_{\dot{e}_2} \dot{u}_{12} + f_{e_2} f_{\dot{e}_1} \dot{u}_{21} + f_{e_2} f_{\dot{e}_2} \dot{u}_{22}}{f_{eq} f_{\dot{e}_1} + f_{eq} f_{\dot{e}_2} + f_{e_2} f_{\dot{e}_1} + f_{e_2} f_{\dot{e}_2}} \tag{7.9}$$

The first derivative of $\dot{u}$ with respect to f_{eq} is:

$$\ddot{u} = \frac{f_{e_2}[f_{\dot{e}_1}(\dot{u}_{11} - \dot{u}_{21}) + f_{\dot{e}_2}(\dot{u}_{12} - \dot{u}_{22})]}{(f_{eq} + f_{e_2})^2 (f_{\dot{e}_1} + f_{\dot{e}_2})} \tag{7.10}$$

Substitute (7.6) into (7.10):

$$\ddot{u} = \frac{K_I f_{e_2} (P_{e_1} - P_{e_2})}{(f_{eq} + f_{e_2})^2} \tag{7.11}$$

(7.11) shows that the slope of $|\dot{u}|$ will decrease as f_{eq} increases. To achieve fast and robust control, the control surface should have a small slope near the origin (steady state) and a large slope far from the origin. Consequently, the ET1MGs (f_{eq}) should be large when e and/or $\dot{e}$ are far from zero, and small when e and/or $\dot{e}$ are around zero. The ET1FSs corresponding to $\dot{e} = \{-1, 0, 1\}$ are plotted in Fig. 7.8. Note here the ET1FS

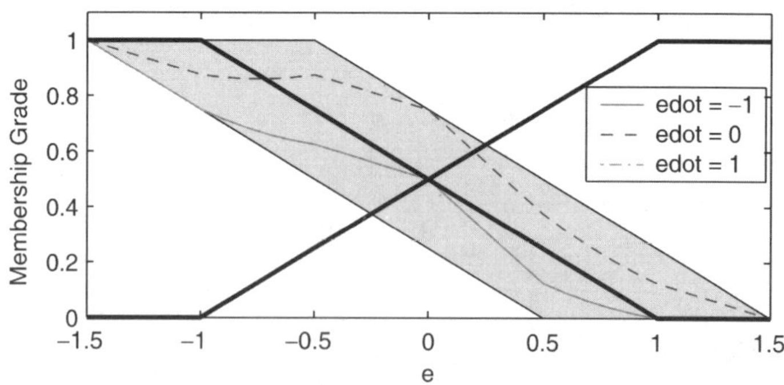

Fig. 7.8. ET1FSs obtained by the proposed type-reducer

corresponding to $\dot{e} = -1$ coincides with the one to $\dot{e} = 1$. Fig. 7.8 shows that the ET1FSs have the desired characteristics as f_{eq} is large when e and $\dot{e}$ are approximately zero and small when the inputs are far away from the origin. Hence, it may be concluded that the control surface has the desired characteristics.

7.4 Simulation Example

In this section, the feasibility of the proposed method for using GA to design a type-reducer is investigated. The performance is analyzed by using type-2 FLC that employs the type-reducer to control the following first order plus dead-time plant:

$$G(s) = \frac{Y(s)}{U(s)} = \frac{K}{\tau s + 1} e^{-Ls} \tag{7.12}$$

where K, τ and L are the static gain, time constant and transportation delay, respectively. It is assumed that the nominal plant is $\frac{1}{10s+1} e^{-2.5s}$. To ensure reasonable control performance is obtained for the nominal plant, the PI parameters used to design the consequent sets of both FLCs are selected by the integral of time-weighted absolute error (ITAE) setpoint tracking tuning rule [29]:

$$K_P = \frac{0.586}{K} \left(\frac{L}{\tau} \right)^{-0.916} = 2.086 \tag{7.13}$$

$$K_I = \frac{1.03 - 0.165 \frac{L}{\tau}}{\tau K_P} = 0.206 \tag{7.14}$$

The MFs of the baseline type-1 FLC used in the study are shown in Fig. 7.9 as the dark thick lines. The FOU of the type-2 MFs used to construct the type-2 FLCs are the shaded regions in Fig. 7.9. Substituting the PI parameters shown above into (7.6), the consequent sets[2] are found and are given in Table 7.2.

From the definition of the proposed type-reducer (Equation 7.7), the weights α_i for reducing the interval firing set to a crisp value are evolved by a GA. Since the type-2 FLC analysed had two inputs and each input domain is characterized by two type-2 MFs, a total of eight weights needed to be tuned by the GA. The GA employed to evolve the weights has a population of 100 chromosomes and is terminated after 50 generations. Chromosomes of the first generation were randomly initialized. To ensure that the fitness increases monotonically, the best chromosome in each generation entered the next generation directly. In addition, a generation gap of 0.8 was used during the reproduction operation so that 80% of the chromosomes in the new generation were determined by the rank reproduction method, while the remaining

[2] Note here crisp values are used for simplicity.

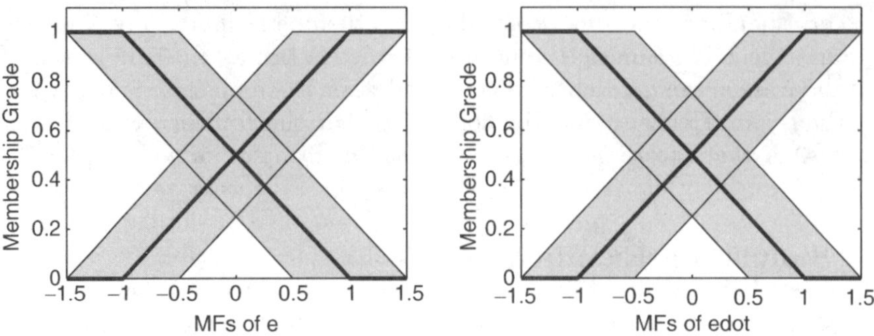

Fig. 7.9. MFs of the FLCs

Table 7.2. Rule base of the FLCs

$e \backslash \dot{e}$	$\dot{e}_1$	$\dot{e}_2$
e_1	−2.2923	1.8797
e_2	−1.8797	2.2923

20% were selected randomly. This strategy helps to prevent premature convergence. The crossover rate was 0.8. Mutation was performed by generating a random number, m, for each gene in the chromosome. If m was smaller than 0.05, non-linear mutation, as defined in (7.15), occurs.

$$x(i+1) = x(i) + \delta(x, i) \tag{7.15}$$

where

$$\delta(x,i) = \begin{cases} (R_x - x) \cdot \left[1 - \lambda^{\left(1 - \frac{i}{1+i_{max}}\right)}\right], \text{ if } rand(1) > 0.5 \\ (L_x - x) \cdot \left[1 - \lambda^{\left(1 - \frac{i}{1+i_{max}}\right)}\right], \text{ otherwise} \end{cases} \tag{7.16}$$

and $x(i)$ was the value of gene x in i^{th} generation, i_{max} was the maximum number of generations, $[L_x, R_x]$ was the interval in which x lied, λ was a random number in [0, 1]. This mutation method enabled finer adjustment to occur as i became larger.

In order to evolve a type-reducer that can cope well with modeling uncertainties, five plants with the parameters shown in Table 7.3 are used to tune the type-reducer. The fitness of a chromosome was evaluated based on the sum of the ITAEs of the five plants. The equations designed by GA for type-reducing the interval firing levels of the type-2 FLC are as follows:

$$f_{e_1} = f_{e_{1u}} - \left[1.4347 \frac{2\,|e|}{P_{e_2} - P_{e_1}} - 3.8964 \frac{2\,|\dot{e}|}{P_{\dot{e}_2} - P_{\dot{e}_1}}\right] (f_{e_{1u}} - f_{e_{1l}}) \tag{7.17a}$$

$$f_{e_2} = f_{e_{2u}} - \left[1.7605 \frac{2\,|e|}{P_{e_2} - P_{e_1}} - 2.6043 \frac{2\,|\dot{e}|}{P_{\dot{e}_2} - P_{\dot{e}_1}}\right] (f_{e_{2u}} - f_{e_{2l}}) \tag{7.17b}$$

Table 7.3. Parameters of the five plants

Parameter\Plant	I	II	III	IV	V
K	1	1	1	0.5	2
τ	10	5	20	10	10
L	2.5	2.5	2.5	2.5	2.5

$$f_{\dot{e}_1} = f_{\dot{e}_{1u}} - \left[0.9601\frac{2\,|e|}{P_{e_2} - P_{e_1}} + 0.2290\frac{2\,|\dot{e}|}{P_{\dot{e}_2} - P_{\dot{e}_1}}\right](f_{\dot{e}_{1u}} - f_{\dot{e}_{1l}}) \quad (7.17c)$$

$$f_{\dot{e}_2} = f_{\dot{e}_{2u}} - \left[0.9041\frac{2\,|e|}{P_{e_2} - P_{e_1}} + 0.1169\frac{2\,|\dot{e}|}{P_{\dot{e}_2} - P_{\dot{e}_1}}\right](f_{\dot{e}_{2u}} - f_{\dot{e}_{2l}}) \quad (7.17d)$$

7.4.1 Comparative Results

To evaluate the feasibility of the type-2 FLC that employs the GA-evolved type-reducer, a comparative study using the following four FLCs were performed:

- Type-1: A type-1 FLC realizing a PI controller with $K_P = 2.086$ and $K_I = 0.206$;
- KM TR: A type-2 FLC using the KM type-reducer;
- UB TR: A type-2 FLC using the uncertainty bound type-reducer [30];
- GA TR: A type-2 FLC using the type-reducer designed by a GA (see Equation 7.17d).

The input MFs are shown in Fig. 7.9 and the rule base in Table 7.2. The ability of the four FLCs to handle modelling uncertainties were compared using the 5 plants given in Table 7.3 as testbeds. Figs. 7.10–7.14 show the respective step responses. The following three performance indices were employed as quantitative measures for comparing the control performances of the four FLCs:

- Integral of the absolute error (IAE): IAE=$\int_0^{100} |e(t)|dt$.
- Integral of the squared error (ISE): ISE=$\int_0^{100} e^2(t)dt$.
- Integral of the time-weighted absolute error (ITAE): ITAE=$\int_0^{100} t|e(t)|dt$.

Table 7.4 shows the performance indices for the four FLCs in the comparative study. When their overall performances are compared, GA TR is the best. This means the type-2 FLC with the proposed type-reducer can handle modelling uncertainties better than the type-1 FLC and the type-2 FLC with KM type-reducer or uncertainty bound method. Another interesting observation is all three type-2 FLCs outperform the type-1 FLC when the response is fast,

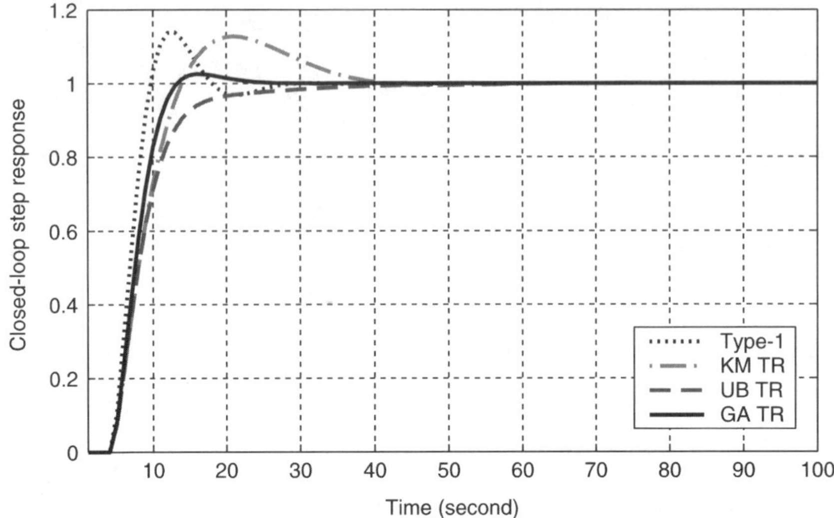

Fig. 7.10. Step response when $K = 1$, $\tau = 10$

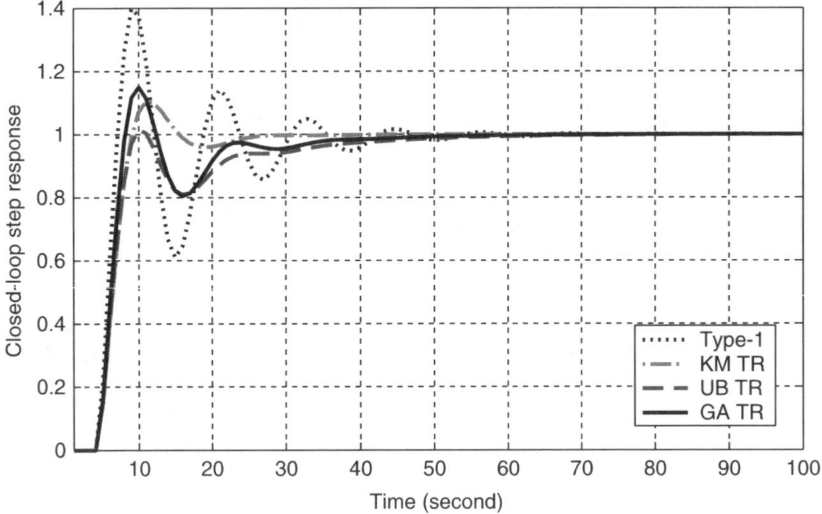

Fig. 7.11. Step response when $K = 1$, $\tau = 5$

i.e. when τ is small (Plant II, Fig. 7.11) or K is large (Plant V, Fig. 7.14). On the other hand, when the response is slow, i.e., τ is large (Plant III, Fig. 7.12) or K is small (Plant IV, Fig. 7.13), the type-1 FLC outperforms all three type-2 FLCs.

An advantage of the proposed type-reducer is its low computational cost. Compared with the KM type-reducer which requires several iterations and

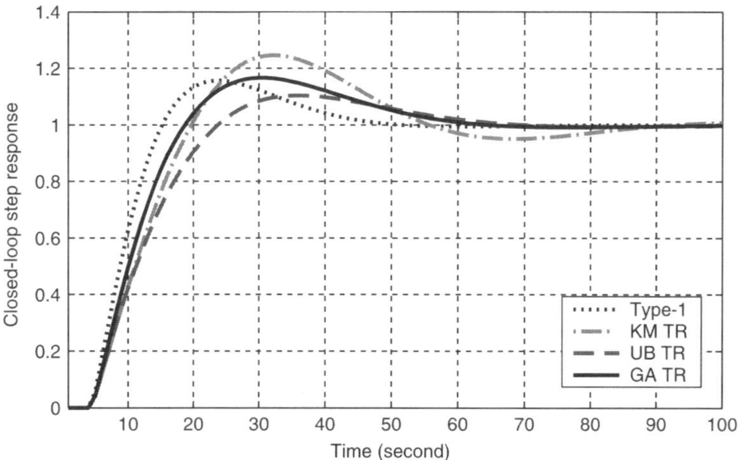

Fig. 7.12. Step response when $K = 1$, $\tau = 20$

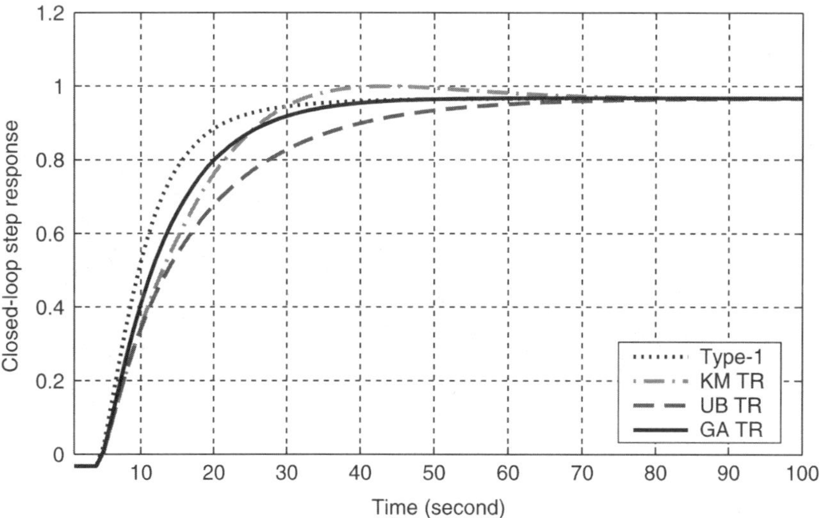

Fig. 7.13. Step response when $K = 0.5$, $\tau = 10$

the number of iterations may vary from run to run, the proposed type-reducer is straight forward. The computational burden is fixed and is much less. Without loss of generality, a comparison of the computational load was performed by assuming that N equally spaced MFs were used to partition each of the two $[-1, 1]$ input domains. The FOU of every type-2 MF was defined as $d_e = d_{\dot{e}} = \frac{1}{N-1}$, i.e. half of the distance between the two adjacent apexes. This study was conducted by first generating 101 points, $e_i = 2(i-1)/100 - 1 (i = 1, \ldots, 101)$, that divided e domain into 100

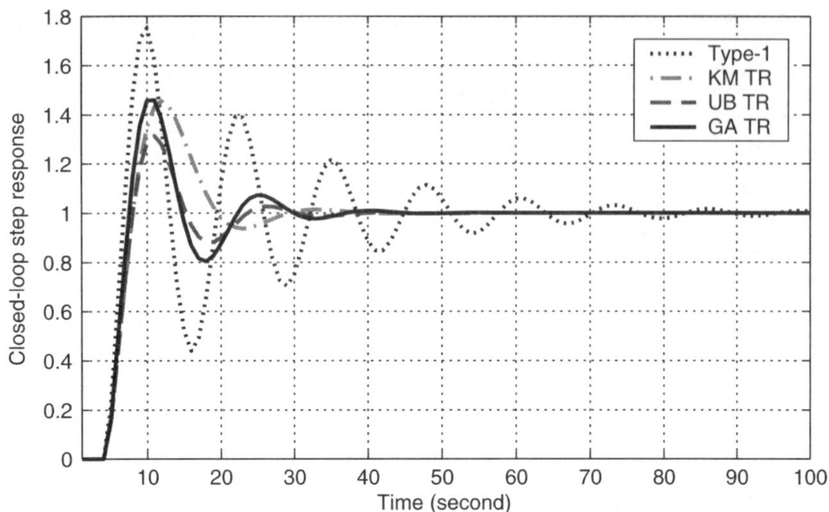

Fig. 7.14. Step response when $K = 2$, $\tau = 10$

Table 7.4. Performances of the four FLCs

(a) IAE of the four FLCs

FLC\Plant	I	II	III	IV	V	Sum
Type-1	6.2690	8.7621	10.5164	9.6941	15.3482	50.5898
KM TR	8.7302	5.6858	16.2223	13.0920	8.2086	51.9389
UB TR	7.7965	7.6447	13.2338	15.7551	6.8963	51.3263
GA TR	6.4481	7.0047	13.1466	11.8570	8.0851	46.5415

(b) ISE of the four FLCs

FLC\Plant	I	II	III	IV	V	Sum
Type-1	4.5960	4.8172	6.2603	6.4649	7.6374	29.7758
KM TR	5.5494	4.2626	8.4429	8.4130	5.1966	31.8645
UB TR	5.4474	4.4632	7.8100	9.1500	4.4992	31.3697
GA TR	5.0406	4.2950	7.3074	7.6748	4.9216	29.2395

(c) ITAE of the four FLCs

FLC\Plant	I	II	III	IV	V	Sum
Type-1	38.0564	104.2077	129.1421	85.3527	307.8952	664.6542
KM TR	84.9387	32.8934	335.7619	157.1687	70.3821	681.1449
UB TR	58.2806	91.3164	200.4673	237.7789	52.7201	640.5633
GA TR	34.8456	70.4150	215.6216	121.7163	72.5361	515.1346

equally-spaced intervals. Another 101 points in $\dot{e}$ domain were generated in a similar manner. By combining these points in all possible ways, 10201 input vectors were generated. Computational cost was evaluated by comparing the time needed to calculate the outputs corresponding to these 10201 input vectors. The platform was an Intel Pentium III 996MHz computer with 256M RAM and Windows XP running MATLAB 6.5. The computation times for

Table 7.5. Comparison of Computational Cost

N\FLC	Type-1	KM TR	UB TR	GA TR
2	1.0 sec	11.9 sec	2.5 sec	1.4 sec
3	1.2 sec	12.8 sec	2.8 sec	1.5 sec
5	1.6 sec	13.3 sec	3.4 sec	2.0 sec
7	2.3 sec	15.8 sec	4.9 sec	2.7 sec
9	3.2 sec	19.6 sec	7.0 sec	3.8 sec

the four FLCs are shown in Table 7.5. Observe that a type-2 FLS with the proposed type-reducer has similar computational cost as a type-1 FLS. Compared to a type-2 FLS with KM type-reducer, the computational burden is greatly reduced. Though the uncertainty bound method is specially designed for reducing computational cost, it is still about two times slower than that of the proposed type-reducer. Thus, the proposed type-reducer may be more suitable for certain types of real-time applications.

In [28] a simplified type-2 FLS structure is proposed to reduce the computational cost. However, that computational cost is also higher than the results here with the same number of input MFs. Besides, the simplified structure is a subset of the type-2 FLS with KM type-reducer. Thus, its best performance is bounded by the type-2 FLS based on the KM type-reducer. On the other hand, the idea in this paper enables one to design different type-reducers, and the performance of the resulting type-2 FLS may be better than a traditional type-2 FLS with KM type-reducer.

There are, however, some limitations to the proposed type-reducer. The proposed type-reducer transforms a type-2 FLC into a type-1 one before the inference engine. This gives rise to minimum computational cost, but it does not allow uncertainties to flow through the inference engine. To overcome this limitation, an alternative strategy may be to evolve a function to replace the KM type-reducer. This function uses the type-2 output FSs from the inference engine as its input and outputs a type-1 FS which will be used by the defuzzifier. This kind of type-reducers may have heavier computational cost than the GA TR proposed herein as the inference engine also has to process type-2 FSs. However, its computational cost is still much less than that of a KM type-reducer, because it calculates the type-1 FS directly without the iterations in the KM type-reducer. The advantage of this approach is the flow of uncertainties is the same as that in a traditional type-2 FLS. The uncertainty bound method [30] can be viewed as an example of this idea.

7.5 Conclusions

In this work, a computationally efficient type-reducer has been proposed. Simulation results demonstrated that it is much faster than the widely used KM iterative method, while at the same time providing better performance. The results are promising and indicate that GAs can be used to evolve faster and better type-reducers according to the specific requirements of a problem.

References

1. Mendel, J.M. (2001): Rule-Based Fuzzy Logic Systems: Introduction and New Directions. Prentice-Hall Upper Saddle River New Jersey
2. Hargras, H.A. Hagras (2004): A hierarchical type-2 fuzzy logic control architecture for autonomous mobile robots. IEEE Trans. on Fuzzy Systems, **12**(4), 524-539.
3. Yager, R.R. (1980): Fuzzy subsets of type-2 in decisions. J. Cybernetics, **10**, 137–159
4. Ozen, T., Garibaldi, J.M. (2004): Effect of type-2 fuzzy membership function shape on modelling variation in human decision making. Proc. FUZZ-IEEE, **2**, 971–976.
5. John, R.I., Innocent, P.R., Barnes, M.R. (2000): Neuro-fuzzy clustering of radiographic tibia images using type 2 fuzzy sets. Information Sciences, **125**, 65–82
6. Auephanwiriyakul, S., Adrian, A., Keller, J.M. (2002): Type-2 fuzzy set analysis in management surveys. Proc. FUZZ-IEEE, 1321–1325
7. Bouchachia, A., Mittermeir, R. (2003): A neural cascade architecture for document retrieval. Proc. Intl Joint Conf. Neural Networks, **3**, 1915–1920
8. Zeng, J., Liu, Z.-Q. (2006): Type-2 fuzzy hidden markov models and their applications to speech recognition. IEEE Trans. on Fuzzy Systems, **14**(3), 454–467
9. Castillo, O., Melin, P. (2004): Adaptive noise cancellation using type-2 fuzzy logic and neural networks. Proc. FUZZ-IEEE, **2**, 1093–1098
10. Mendel, J.M., Wu, H. (2004): Centroid uncertainty bounds for interval type-2 fuzzy sets: Forward and inverse problems. Proc. FUZZ-IEEE, **2**, 947–952
11. Wu, H., Mendel, J.M., (2004): Antecedent connector word models for interval type-2 fuzzy logic systems. Proc. FUZZ-IEEE, **2**, 1099–1104
12. Melin, P., Castillo, O. (2003): A new method for adaptive model-based control of non-linear plants using type-2 fuzzy logic and neural networks. Proc. FUZZ-IEEE, **1**, 420–425
13. Wu, D., Tan, W.W. (2004): A type-2 fuzzy logic controller for the liquid-level process. Proc. FUZZ-IEEE, **2**, 953–958
14. Lee, C.-H., Lin, Y.-C. (2004): Control of nonlinear uncertain systems using type-2 fuzzy neural network and adaptive filter. Proc. 2004 IEEE Intl Conf. Networking, Sensing and Control, **2**, 1177–1182
15. Castillo, O., Melin, P. (2003): Evolutionary computing for optimizing type-2 fuzzy systems in intelligent control of non-linear dynamic plants. Proc. North American Fuzzy Info. Processing Society (NAFIPS), 247–251
16. Lynch, C., Hagras, H.A., Callaghan, V. (2005): Embedded type-2 FLC for real-time speed control of marine & traction diesel engines. Proc. FUZZ-IEEE, 347–352
17. Figueroa, J., Posada, J., Soriano, J., Melgarejo, M., Rojas, S. (2005): A type-2 fuzzy controller for tracking mobile objects in the context of robotic soccer games. Proc. FUZZ-IEEE, 359–364
18. Lin, P.-Z., Hsu, C.-F., Lee, T.-T. (2005): Type-2 fuzzy logic controller desgin for buck DC-DC converters. Proc. FUZZ-IEEE, 365–370

19. Liang, Q., Wang, L. (2005): Sensed signal strength forecasting for wireless sensors using interval type-2 fuzzy logic system. Proc. FUZZ-IEEE, 25–30
20. Shu, H. and Liang, Q. (2005): Wireless sensor network lifetime analysis using interval type-2 fuzzy logic systems. Proc. FUZZ-IEEE, 19–24
21. Wu, D., Tan, W.W. (2005): Type-2 FLS modeling capability analysis. Proc. FUZZ-IEEE, 242–247.
22. Zadeh, L.A. (1975): The concept of a linguistic variable and its application to approximate reasoning-1. Information Sciences, **8**, 199–249
23. Burillo, P., Bustince, H. (1996): Entropy on intuitionistic fuzzy sets and on interval-valued fuzzy sets. Fuzzy sets and systems, **78**, 305–316
24. Gorzalczany, M.B. (1987): A method of inference in approximate reasoning based on intervalvalued fuzzy sets. Fuzzy Sets and Systems, **21**, 1–17
25. Bustince, H. (2000): Indicator of inclusion grade for interval-valued fuzzy sets. Application to approximate reasoning based on interval-valued fuzzy sets, Intl Journal of Approximate Reasoning, **23**(3), 137–209
26. Wu, D., Tan, W.W. (2006): Genetic learning and performance evaluation of type-2 fuzzy logic controllers. Engineering Applications of Artificial Intelligence, **19**(8), 829-841
27. Mizumoto, M. (1995): Realization of PID controls by fuzzy control methods. Fuzzy Sets and Systems, **70**, 171–182
28. Wu, D., Tan, W.W. (2006): A simplified architecture for type-2 FLSs and its application to nonlinear control. ISA Transactions, **45**(4), 503-516
29. Persson, P. (1992): Towards autonomous PID control. Ph.D. dissertation, Lund Institute of Technology
30. Wu, H., Mendel, J.M. (2002): Uncertainty bounds and their use in the design of interval type-2 fuzzy logic systems. IEEE Trans. on Fuzzy Systems, **10**(5), 622–639

8

Designing a Recurrent Neural Network-based Controller for Gyro-Mirror Line-of-Sight Stabilization System using an Artificial Immune Algorithm

Ji Hua Ang, Chi Keong Goh, Eu Jin Teoh, and Kay Chen Tan

Department of Electrical and Computer Engineering National University of Singapore 4 Engineering Drive 3, Singapore 117576.

Summary. The gyro-mirror line-of-sight stabilization platform used to maintain the line-of-sight of electro-optical sensors mounted on moving vehicles is a multivariate and highly nonlinear system. The system is also characterized by a peculiar phenomenon in which a movement about one axis will trigger off a coupled movement in the other axis. Furthermore, uncertainties such as noise, practical imperfections and additional dynamics are often omitted from the mathematical model of the system thus resulting in a non-trivial control problem. In order to handle the complex dynamics of the gyro-mirror as well as to optimize the various conflicting control objectives, a multi-objective artificial immune system framework which combines the global search ability of evolutionary algorithms and immune learning of artificial immune systems is proposed in this chapter for the design of the gyroscope recurrent neural network controller. In addition, a new selection strategy based on the concepts of clonal selection principle is used to maintain the balance between exploration and exploitation of the objective space. Simulation results demonstrate the effectiveness of the proposed approach in handling noise, plant uncertainties and the coupling effects of the cross-axis interactions.

8.1 Introduction

Gyro-mirror line-of-sight (LOS) stabilization systems form the basis of many applications for purposes of navigation, targeting and sighting [23], [26]. In particular, the system is capable of maintaining the LOS of electro-optical sensors mounted on moving vehicles such as helicopters, aircrafts and tanks. However, due to the multivariate, highly nonlinear and tightly coupled nature of the system, the control of gyro-mirror LOS stabilization platforms is a non-trivial problem.

The rapid advancement of control theory has seen the application of various control approaches ranging from the simple proportional-integral-derivative (PID) controller to the more sophisticated optimal and adaptive

J.-H. Ang et al.: *Designing a Recurrent Neural Network-based Controller for Gyro-Mirror Line-of-Sight Stabilization System using an Artificial Immune Algorithm*, Studies in Computational Intelligence (SCI) **66**, 189–209 (2007)
www.springerlink.com

control algorithms being applied to LOS stabilization systems. Lee et al. [10] put forward a composite variable structure augmented adaptive control scheme to provide robust control performance of gyro-mirror LOS system in face of modeling errors and extraneous disturbances. Furthermore, the control scheme guarantees a stable closed-loop response with appropriate selection of control parameters. However, its benefits are limited by virtue of the control chatter that is introduced by the variable structure control. In a similar vein, Tan et al. [26] proposed a composite of PID and adaptive control to handle both parametric and non-parametric uncertainties of the gyro-mirror LOS system. In this arrangement, the PID controller is responsible for handling feedback control while the adaptive controller performs the necessary update to the plant model. Instead of utilizing conventional control techniques, Ge et al. [5] presented a direct adaptive neural network control by means of a radial basis function network (RBF) for effective and robust control over the highly nonlinear gyro-mirror LOS system.

However, these control schemes are often restricted to a particular problem or addresses only a subset of possible performance metrics, for it should be noted that a control design problem involves several design tradeoffs which might not be arbitrated explicitly. Intuitively, it may also be more practical to design a controller that satisfies the different requirements and constraints simultaneously, instead of satisfying a specific control scheme [11], [12], [16], [21]. With this in mind, multi-objective (MO) evolutionary techniques provide a general platform for the simultaneous optimization of competing control requirements. Using an evolutionary approach for MO, a variety of control scheme capable of addressing practical time and frequency domain performance criteria have been proposed in the recent years [4], [23], [27].

The use of artificial neural networks (ANNs) in control has also been receiving increasing attention due to its ability to learn and handle nonlinearities and uncertainties present in the control system [6], [9], [15], [17], [28]. The main ANN control techniques that have been proposed include inverse dynamics, indirect and direct control as well as the more widely used network, Recurrent Neural Networks (RNNs). RNNs, through its unconstrained synaptic connectivity and resulting state-dependent nonlinear dynamics, offer a greater level of computational ability as compared to regular feedforward neural networks (FNNs) architectures. However, a major drawback of the higher complexity is that gradient-based backpropagation learning algorithms and real-time recurrent learning [29] for RNNs are more likely to be trapped in local optima resulting in sub-optimal solutions. Thus the concept of incorporating evolutionary elements into neural networks has lead to the emergence of evolutionary artificial neural networks (EANNs) in recent years. Driven by an evolutionary search process that provides a global perspective to the adaptation process, EANNs have demonstrated to be less sensitive to initial choices of weights and are also capable of dealing with the problem of premature convergence that is usually associated with conventional gradient-based approaches [3], [13], [31].

With this in mind, artificial immune systems (AISs) [2], [8], [14], [32] are a relatively new paradigm in the domain of computational intelligence that have been inspired by the human immune system's ability to remember, recognize and eliminate non-self entities. AISs possess several attractive immune properties such as pattern recognition, uniqueness, self-identity, diversity, immune learning, self-organization and integration with other systems. In particular, features such as predator-prey pattern of response and natural integration with other systems suggest that an AIS might be very fruitful for the development of computational hybrid systems. Intuitively, these features are useful in the design of neural networks.

Presently in the literature, there is a lack of studies in extending AIS for the adaptation of ANNs; not to mention the construction of an AIS-RNN controller to account for the unpredictable parameter variations, disturbances and nonlinear dynamics of the gyro-mirror LOS stabilization system. One of the core objectives of this paper is to extend AIS to the multi-objective design of a RNN controller for the gyro-mirror LOS system. Apart from its capability of optimizing multiple conflicting control objectives simultaneously, the multi-objective artificial immune system (MOAIS) implementation of the RNN controller provides the means of handling high-order system dynamics. Such an automated design methodology allows the MOAIS to identify controllers that are not only capable of maintaining and tracking the reference position but also remain robust in the presence of noise and parameter uncertaintics.

This paper is organized as follows. In section 8.2, an overview of the gyro-mirror stabilization system is given followed by the principles of gyroscopic and formulation of the mathematical model. Description of the RNNs architecture is given in section 8.3. The multiobjective problem as well as the proposed MOAIS for RNN controller design will be discussed in section 8.4. Section 8.5 states the simulation setup while section 8.6 validates and discusses the performance of the MOAIS control design in the presence of plant uncertainties and noise perturbations. Conclusion of the paper is given in section 8.7.

8.2 Gyro-Mirror Line-of-Sight Stabilization System

The following sections have been similarly discussed in [23].

8.2.1 System Overview

The schematic diagram of the gyro-mirror stabilization system is shown in Figure 8.1. It consists of three main components where the first component consists of the flywheel and its motor and controller. The second component consists of two gimbal structure that is responsible for providing two DOF to the flywheel; the inner and outer gambal provides movement along the yaw and the pitch axis, respectively. The third component is a mirror that is

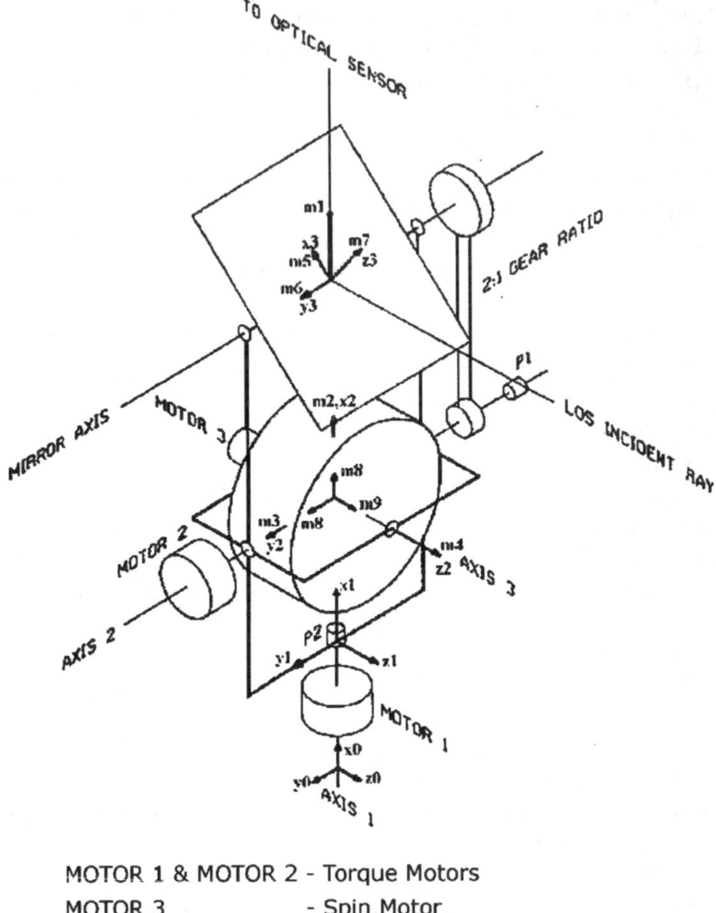

Fig. 8.1. Schematic diagram of gyro-mirror stabilization system

geared to the gimbals through a 2:1 reduction drive mechanism. Specifically, the inner gimbal provides movement along the yaw axis while the outer gimbal provides movement along the pitch axis. As can be seen, the controllers in the gyroscope drive the torque motor in the gimbals for tracking applications. The flywheel is driven by an AC motor and the angular velocity is kept as constant as possible. To provide position/coordinate information, two sensors are used to measure the yaw and pitch angle in a two-degree-of-freedom configuration.

8.2.2 Principles of Gyroscopic

The gyro-mirror LOS stabilization platform is a well-known multivariable and highly coupled nonlinear system that has been extensively used for stabilizing

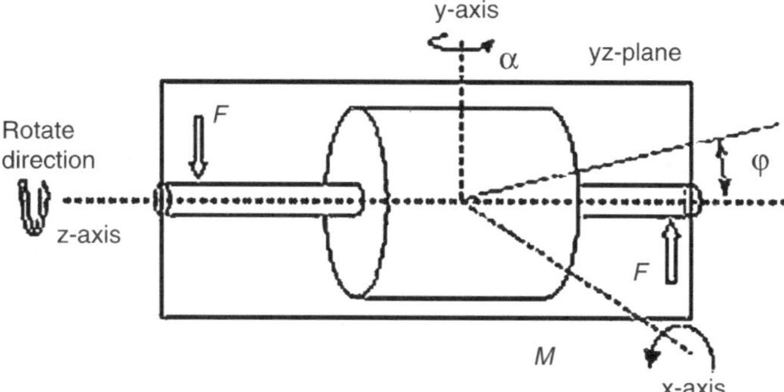

Fig. 8.2. Characteristics of the attribute of precession

and tracking purposes. The term "gyroscope" refers to a variety of mechanisms within a particular system with each having in common a rotating mass that provides an angular momentum. The gyroscope thus serves as a stabilizer for a tracking pointer in both yaw and pitch axes based on the principle of gyroscopic behavior – it has two important attributes namely, precession and directional stability [19], [30]:

(i) Precession: This physical phenomenon happens when a rotating mass is disturbed by a force. The reaction of the system appears quite deceiving since the reaction force's dimension is not confined by the dimension of the action force. Figure 8.2 shows a mass rotating about the z-axis. Supposing two forces F parallel to the y-axis on the yz-plane are applied to the system, a coupling force M about the x-axis would be generated. However, this will only hold true if the mass is not rotating. Instead, the system will react by rotating along the xz-plane about the y-axis through an angle of theta.
(ii) Directional stability: A flywheel spinning at a high-speed possesses an ability to stay in its orientation in inertia space despite the presence of a large external force. This is due to the large momentum generated by spinning the flywheel. Figure 8.3 shows a flywheel mounted in gimbal rings, spinning at high speed. The orientation of the inner gimbals will be fixed in space for any external disturbance along axes 1 and 2 due to the constant momentum incurred. If the magnitude of the momentum is constant, the directional orientation is maintained. This directional stability characteristic is usually employed in numerous inertia and stabilization systems.

The method of passive stabilization [22] is employed in our proposed approach to a stabilized platform for the control system of a nonlinear gyroscope. This method essentially depends on the stability of the axis of a high-speed gyroscope and the extremely slow rate in which it deviates from its original orientation if small torque is applied. It utilizes the large angular momentum of the flywheel to directly stabilize the platform, that is, the gyro

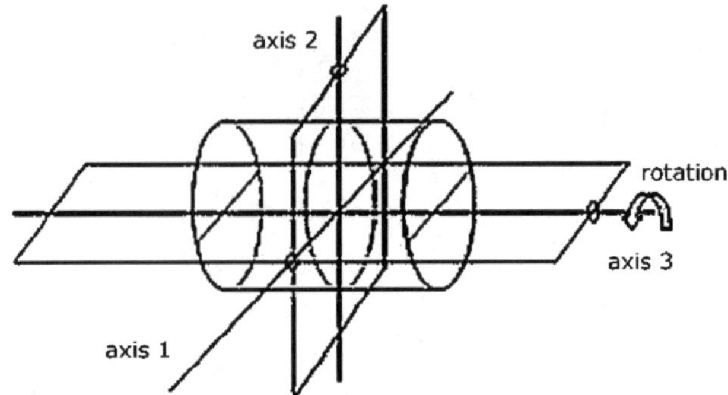

Fig. 8.3. A two-degree-of-freedom gyroscope system

itself provides the muscles for the stabilization. The free gyro stabilized mirror is a system that uses passive stabilization to achieve directional stability, which is subsequently demonstrated in this paper.

8.2.3 Mathematical Model

Based on the coordinate frame defined in Figure 8.1, the dynamic nonlinear model of the gyro-mirror stabilization system can be derived by means of the Lagrangean equations of motion. The system equations can be written as [23], [26],

$$M_1(\theta)\ddot{\theta}_1 + F_1\left(\theta,\dot{\theta}\right) + G_1\left(\theta,\dot{\theta}\right) = \tau_1 \quad (8.1)$$

$$M_2(\theta)\ddot{\theta}_2 + F_2\left(\theta,\dot{\theta}\right) + G_2\left(\theta,\dot{\theta}\right) = \tau_2 \quad (8.2)$$

where $\theta = [\theta_1, \theta_2]^T$, θ_1 and θ_2 are the angles of rotation about the outer and inner gimbal axis respectively. $\dot{\theta} = [\dot{\theta}_1, \dot{\theta}_2]^T$ is the velocity vector corresponding to θ vector while τ_1 and τ_2 are the actuator torques.

Equations (8.1) and (8.2) are defined as,

$$M_1 = (A+D) + (B-D+H)\cos^2\theta_2 + \frac{1}{2}(E+G) + \frac{1}{2}(G-E)\sin\theta_2 \quad (8.3)$$

$$F_1 = -(B-D+H)\dot{\theta}_1\dot{\theta}_2\sin 2\theta_2 + \frac{1}{2}(G-E)\dot{\theta}_1\dot{\theta}_2\cos\theta_2 + J\dot{\theta}_1\dot{\theta}_2\sin\theta_2\cos\theta_2 \quad (8.4)$$

$$G_1 = J\dot{\theta}_2\dot{\theta}_3\cos\theta_2 \quad (8.5)$$

$$M_2 = C + \frac{F}{4} + H \quad (8.6)$$

$$F_2 = \frac{1}{2}(B-D+H)\dot{\theta}_1^2\sin 2\theta_2 - \frac{1}{4}(G-E)\dot{\theta}_1^2\cos\theta_2 + J\dot{\theta}_1^2\sin\theta_2\cos\theta_2 \quad (8.7)$$

$$G_2 = -J\dot{\theta}_1\dot{\theta}_3\cos\theta_2 \quad (8.8)$$

Table 8.1. System Parameters

Parameter	Value	Parameter	Value
A	0.004 kg m^2	F	0.0025 kg m^2
B	0.00128 kg m^2	G	0.00125 kg m^2
C	0.00098 kg m^2	H	0.0032 kg m^2
D	0.02 kg m^2	J	0.0025 kg m^2
E	0.0049 kg m^2	$\dot{\theta}_3$	3000 rpm

Referring to Figure 8.1, A denotes the moment of inertia of the outer gimbal about $g_1 = v_1$, while B, C, and D refer to the moment of inertia about r_1, r_2 and r_3, respectively. E, F, and G denote the moments of inertia of the mirror about m_1, m_2 and m_3 respectively; and H and J denote the moments of inertia of the rotor of flywheel about r_1 and r_3, respectively. Here, v, g, r and m stand for the coordinate axes for the frames of the vehicle/system, outer gimbal, inner gimbal (rotor) and mirror, respectively. $\dot{\theta}_3$ is the velocity of the spin wheel. The method of time domain closed-loop identification has been applied in [23] to identify the various system parameters as listed in Table 8.1. This has been similarly described in [23].

8.3 Recurrent Neural Network Architecture

In control applications, the use of RNNs for dynamical plant identification and controller design has the following advantages:

- An RNN does not require any *a priori* internal understanding of the plant, or any *linearity* assumptions. Necessarily, for successful and more accurate modeling, more data for training the system must be available to offset the lack of internal (first principle) system understanding.
- An RNN approximates plant dynamics through adjustment of the number of hidden nodes and the values of weights. The number of adjustable free parameters of the network, which in this case are the hidden nodes, determines the complexity of the system. Thus, selecting a model structure is equivalent to simply changing the number of hidden nodes. The real states of the plant are not necessary.
- Determining the values of internal weights, referred to as training of the RNN, is a particular form of nonlinear regression, for which effective distributed algorithms are available [1], [18], [29]. It should be stressed that the approximation capabilities of the RNN are limited by the number of training sets available, as well as the complexity of the architecture. However, as previously highlighted, the increase in the complexity of a system typically introduces more local minima, such that during the learning phase, the dynamics of the resultant training algorithm are likelier to converge onto.

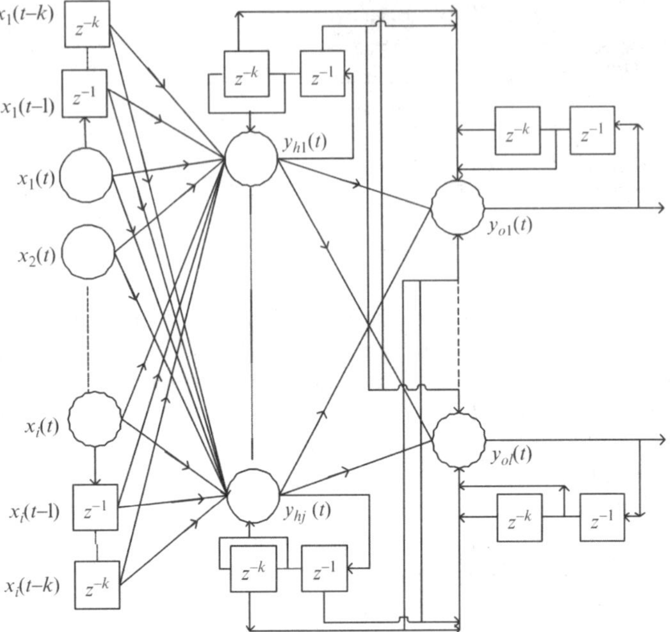

Fig. 8.4. Recurrent neural network structure

The RNN structure adopted in this paper is shown in Figure 8.4. This permits state dependencies, allowing previous states to influence the activation of subsequent activations through this recurrency. The recurrent nature of the signal flow in an RNN however, increases the complexity of the resulting network, and as a result complicating the modeling and subsequent training of the free parameters (synaptic weights) of the network.

For RNNs, we have both *structural* as well as *temporal* state dependencies. Structural dependencies arise from the synaptic weights that are present between neurons (this is common in feedforward architectures), whereas temporal dependencies are a consequent of the fact that the present state of a neuron depends on the state of a neuron a few time-steps (say k time-steps) ago. This being said, for k-step state dependencies, the current state depends on the states of the neurons from 1, 2, ..., k time-steps ago. For each time-step dependency, a matrix W^k defines the set of synaptic weights leading from the states of the neurons k time-steps ago to the present state. Thus for a system with k-step dependency, we can have up to k weight matrices W. Adopting a mathematical expression, the present state at time t is a function of its k previous states weighted by W^k (neglecting the bias terms b), $x(t) = f\left(W^1 x(t-1) + W^2 x(t-2) + \cdots + W^k x(t-k)\right)$. As an illustrative example, assuming w_{ij} denotes the synaptic weight leading from the jth neuron to the ith neuron, w_{ij}^k then represents the synaptic weight leading from the state of the jth neuron k time-steps ago to the ith neuron. In other words, this says that the current activation of the ith neuron depends on the state of the jth neuron k time-steps ago.

Assuming an affine form, the state update of a typical RNN is mathematically described by $x(t) = f(Wx(t-1)+b)$ for a synchronous (parallel) update, where all neurons in the system are updated simultaneously, and by $x_i(t) = f(W_i x(t-1)+b_i)$ for an asynchronous (serial) update where only the ith neuron that is selected, is updated.

8.4 Design of MOAIS-RNN for the Gyroscope Control System

An AIS algorithm attempts to replicate the adaptive nature of the human immune system's ability to recognize and subsequently respond in an appropriate manner to specific threats and in the context of a computational paradigm, these threats are analogous to the problem at hand and the goal of the AIS is to identify solutions to the problem. As was previously mentioned, the actual control system design problem involves the minimization of several conflicting objectives, such as rise-time, overshoot and settling-time. Consequently, MO optimization techniques are more suitable for the design of the resultant controller. This section describes the proposed multi-objective AIS framework for the design of the gyroscope recurrent neural network controller. The multi-objective optimization problem and multi-objective design specifications of the controller will be discussed in section 8.4.1 and 8.4.2, respectively. While the general framework for the MOAIS design of RNN controller is presented in section 8.4.3 and finally the algorithmic flow MOAIS will be described in section 8.4.4.

8.4.1 Multi-Objective Optimization Problem

In a general sense, many real-world applications involve complex optimization problems, most of which are saddled with various competing objectives. Let us consider, without any loss of generality, a minimization problem within decision space, X, of a subset of real numbers. For the minimization problem, we need to find a parameter set P for which

$$\underset{P \in X}{\mathrm{Min}}\, \boldsymbol{F}(\boldsymbol{P}), \mathbf{P} \in \mathbf{R}^{D}, \tag{8.9}$$

Where the parameter set $P = \{p_1, p_2, \ldots, p_D\}$ is a vector with D decision variables and $F = \{f_1, f_2, \ldots, f_M\}$ are M objectives to be minimized.

The solutions to a MO optimization problem exists in the form of an alternate tradeoff set commonly described as a Pareto optimal set where each objective component of any non-dominated solution in the set can only be improved by degrading at least one of its other objective components. Pareto dominance is used to assess the relative strength or fitness between any two

Table 8.2. Design Objectives

Objectives	Design Specifications	Goal
1	Torque saturation (Max(τ_1), Max(τ_2))	<1 Nm
2	Rise-time (T_{rise})	<2s
3	Settling time ($T_{settling}$)	<2s
4	Steady-state error (E_{SS})	<0.001
5	Overshoot (O_{shoot})	<0.1

candidate solutions in MO optimization and this concept has been widely adopted in the research of MO optimization since it was proposed by Pareto. Specifically, a vector $\boldsymbol{F}_a$ is said to dominate another vector $\boldsymbol{F}_b$, denoted as

$$\begin{aligned} &F_a \prec F_b, iff\ f_{a,i} \leq f_{b,i} \forall i = \{1, 2, \ldots, M\} \\ &\text{and } \exists j \in \{1, 2, \ldots, M\}\ where\ f_{a,i} \prec f_{b,i}, \end{aligned} \tag{8.10}$$

8.4.2 Multi-Objective Gyro-Mirror LOS Stabilization System Design Specifications

As mentioned earlier, the gyroscopic principle has two important attributes namely, precession and directional stability. However, the physical phenomenon of precession is such that the application of a torque to one axis will result in the rotation in the direction of another axis. Consequently, the designed controller must be capable of handling such an effect since the movement about one axis will trigger off a coupled movement in the other axis. In addition, the controller must also be able to cope with the multivariate and highly nonlinear nature of the gyro-mirror stabilization system. Although the mathematical model of the system may be available, it should be noted that uncertainties such as noise, practical imperfection and additional dynamics are often omitted. It is for this reason that the design of a controller should be robust in the face of noise and parameter uncertainties, as under many practical conditions, it is almost impossible to obtain a perfectly accurate model. Furthermore, the controller must be capable of maintaining and tracking the desired position despite the hard limited input (of 1 Nm) for the input torque to the plant. The desired design objectives, serving as a guideline for the development of our controller, are listed in Table 8.2. Note that the MO optimization problem considered here is free from constraints and there are five objectives, of which there are several conflicting objectives, e.g., having lower overshoot would usually be at the expense of slower rise time.

8.4.3 Multi-Objective Artificial Immune System Design for Recurrent Neural Network

The proposed immune system controller design methodology for the gyro-mirror stabilization platform composed of four different modules, the RNN

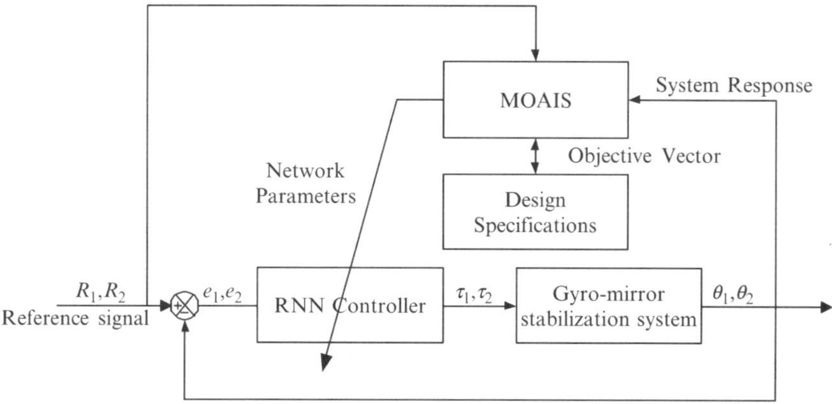

Fig. 8.5. MOAIS automated RNN control design methodology for gyro-mirror stabilization system

controller, the plant model, the MOAIS algorithm and the design specification module as shown in Figure 8.5. Since it is not appropriate to conduct online training of the RNN controller, the response of the control system is simulated based on the plant model. The error feedback of the two channels serves as the input to the RNN controller which generates the appropriate control signals. The candidate controllers identified by the immune system are evaluated based on the desired system response defined in the design specification module. Subsequently, the evaluation results of controllers are used to guide the optimization process of the design procedure. This approach utilizes the given control specifications in addition to the required constraints in an automated manner, intelligently searching for the controller parameters that is best in a Pareto sense.

8.4.4 Algorithmic Flow of MOAIS

The different AIS algorithms that have been proposed thus far in the literature can typically be distinguished by the methods of affinity measure, selection, and the cloning mechanism employed as well as the presence of memory cells. Despite these differences, the different algorithms follow a standard framework which can be extended to handle multi-objectivity as shown in Figure 8.6.

In contrast to single-objective optimization, the solution to an MO optimization problem exists in the form of alternate tradeoffs known as a Pareto optimal set. Each objective component of any non-dominated solution in the Pareto optimal set can only be improved by degrading at least one of its other objective components [20]. Placed in our context, the objective components of the individuals will correspond to the performance requirements specified in Table 8.2. Therefore, the concept of Pareto optimality features prominently in the proposed algorithm. Furthermore, Pareto-based techniques have been

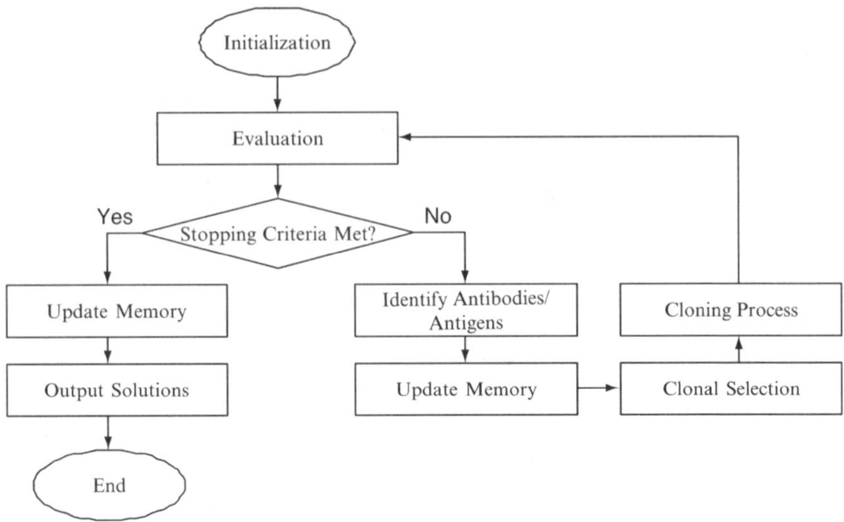

Fig. 8.6. The framework for MO artificial immune systems

employed successfully in MO problems by many researchers and it is also the most appropriate way to generate an entire Pareto front in a single simulation run [24].

The use of MOAIS has the capability of searching for a set of Pareto optimal solutions for MO optimizations. The MOAIS maintains a population of individuals and each of which is assigned a fitness value based on the different objectives to be satisfied. The individuals then go through a simulated evolution process which is described below.

i) Representation: Similar to the evolutionary optimization techniques, the AIS process a set of encoded parameters providing the flexibility to design an appropriate representation of the potential solutions. In order to facilitate the ease of manipulating the embedded information within the individuals, the genotype encodes the weights of the RNN which are represented by real numbers. Chromosome representation could be in any form as long as the architecture of RNN remains after application of variation operators.
ii) Archiving: In our algorithm, elitism is implemented in the form of a fixed-size archive – this is done to prevent the loss of good particles due to the stochastic nature of the optimization process. The size of this archive is usually adjusted based on the desired number of particles distributed along a tradeoff curve (known as a Pareto front) in the objective space. The archive is subsequently updated at each cycle, e.g., if the candidate solution is not dominated by any members in the archive, it will be added to the archive. In a similar manner, any archive members dominated by this solution will hence be removed from the archive. Thus, in order to

maintain a set of uniformly distributed non-dominated particles in the archive, a dynamic niche sharing scheme [25] is employed. When a predetermined archive size is reached, a recurrent truncation process [7] based on a niche count is used to eliminate the most crowded archive member.

iii) Clonal Selection: Similar to the work presented in [2], the principle of clonal selection forms the core of the MOAIS. In this paper, we define memory cells as an archive of non-dominated solutions to preserve the best antigens found by the immune system and the antibodies as the candidate solutions in the AIS population. At every generation, the population of antibodies is evaluated. Fitness evaluation is based on affinity measurement. The affinity between the memory cells and an antigen is calculated based on their euclidean distance in the objective space. The smaller the euclidean distance between the memory and the antibody, the higher the affinity, i.e., higher fitness value. The affinity of an antibody is given in Equation 8.11.

$$Affinity(k) = 1/\sum_{j=1}^{N}\sum_{i=1}^{M} ||memory_i(j) - antibody_i(k)|| \tag{8.11}$$

where M = archive size, N = number of objectives and k = index of individuals.

In contrast to existing techniques, the clonal selection of memory cells is based on the degree of its representation in the evolving population.

iv) Antibody cloning and variation: In this paper, binary tournament selection is implemented to select the appropriate memory cells for the cloning and hyper-mutation process. The number of individuals selected is based on *clone_rate*. After the selection of the memory cells, these non-dominated antibodies will be cloned such that the number of individuals is equal to the population size. In cloning, exact replication of the cell is done. Subsequently, uniform mutation will be applied to the clones. The mutation strength is adapted as,

$$\sigma_w = 0.1 \cdot (uppbd_w - lowbd_w) \tag{8.12}$$

where $lowbd_w$, $uppbd_w$ corresponds to the minimum, maximum of the associated weights in the population. $uppbd_w$ and $lowbd_w$ might differ in every generation however the lower and upper limits of the RNN weights are -1 and 1 respectively.

8.5 Simulation Setup

The proposed MOAIS RNN design methodology is implemented using the MATLAB technical software and the simulations are performed on an Intel Pentium 4 2.8 GHz computer. Thirty independent runs are performed for each of the case studies in Section 8.6 so as to obtain statistical information such as consistency and robustness of the algorithms.

8.5.1 RNN Setup

The recurrent neural network architecture that is used in this article is restricted in its synaptic connectivity. Figure 8.4 gives an illustrative example of the proposed architecture. In the proposed architecture, internal recurrency is limited to local signal flows – in other words, only *self-coupling/feedback* is used for both the hidden and output layer of neurons. Recurrency does not occur between nodes of the output and hidden layer, nor does it occur between nodes of the same layer. External recurrency on the other hand, refers to the feedback that occurs between the output and input layer neurons.

The number of hidden layers and hidden neurons used in the simulation is restricted to one and three, respectively. For both internal and external recurrency, the time delay that is afforded is between 1 to 3 time steps. There are two output neurons, each corresponding to *Torque* 1 and *Torque* 2 mentioned in section 8.2. The initial weights of the RNN are randomly assigned in the range $[-1, \ldots, 1]$.

8.5.2 MOAIS Parameters

The stopping criteria used in this paper is when either all the objectives have been met or the maximum number of generations has been reached. The values of the various parameter settings in the algorithm are tabulated in Table 8.3. During the adaptation of the RNN controller, the reference signals comprise 200 samples as given by equation 8.13 and 8.14.

$$R_1 = 0.2 \quad 0 < t < 20 \tag{8.13}$$

$$R_2 = \begin{cases} 0.0 & 0 < t < 10 \\ 0.2 & 10 \leq t < 20 \end{cases}$$

For the simulation studies, the test signals are given as,

$$R_1 = \begin{cases} 0.2 & 0 < t < 10 \\ 0.1 & 10 \leq t < 20 \end{cases} \tag{8.14}$$

Table 8.3. MOAIS Parameter Setting

Parameter	Values
Population, memory size	20
Mutation method	Normal
Mutation strength	Adaptive
Mutation rate	0.1
Clone rate	0.2
Niche radius	Dynamic
Evaluations	6000

$$R_2 = \begin{cases} 0.0 & 0 < t < 5 \\ 0.2 & 5 \leq t < 15 \\ 0.1 & 15 \leq t < 20 \end{cases} \tag{8.15}$$

8.6 Results and Discussions

In order to validate the proposed approach, experimental studies are carried out to investigate the controller robustness in the presence of plant uncertainties and noise perturbations in section 8.6.1 and 8.6.2 respectively.

8.6.1 Robustness Validation of Plant Uncertainties

As mentioned, the actual model of the gyro-mirror stabilization system may be different from the mathematical model due to practical imperfection and parameter drifts. Therefore, the system controller design based on the mathematical model described in section 8.2 must be capable of maintaining control under these uncertainties. Experiments are conducted at normally distributed noise variation of $\sigma^2 = 30\%$ with zero mean on plant transfer function parameters to simulate the conditions of plant uncertainties. The system response and control signal based on the original plant model is shown in Figure 8.7 and

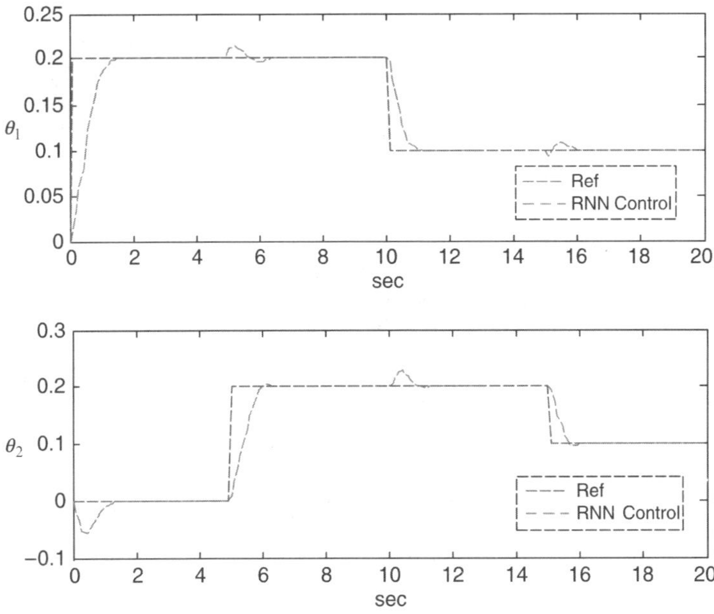

Fig. 8.7. Gyro-mirror stabilization system response based on original plant model

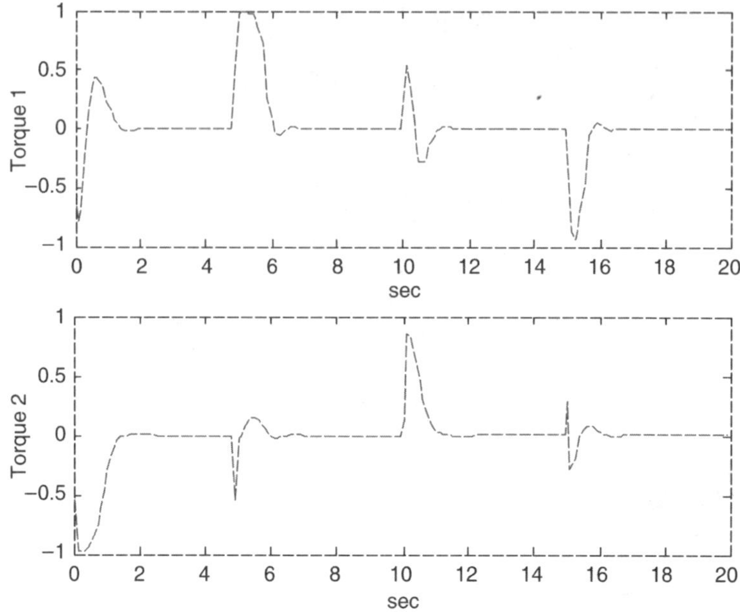

Fig. 8.8. RNN controller output signal based on original plant model

Figure 8.8, respectively. The system responses for 30% parameter variation are plotted in Figure 8.9 while the control signals are plotted in Figure 8.10.

From Figure 8.7, it is clear that the design RNN controller is capable of satisfying the nominal system specifications defined previously. Furthermore, it can be observe from Figure 8.9 that the RNN controller remains robust despite the variations made to the plant model. By comparing Figure 8.8 and Figure 8.10, the adaptation of the RNN controller to uncertain plant parameters is evident by the control signals produced by the controller in response to the plant parameter variation. In general, it can be seen that the main effects of model variation is the longer time required to seek the new reference signals as well as being more susceptible to the coupling effects of the axis. However, it should be noted that the controller is still capable of sustaining good performance in terms of steady state error.

8.6.2 Robustness Validation of Noise Perturbation

Many real-world applications are also characterized by a certain degree of noise, manifesting itself in the form of signal distortion or information uncertainty. Noise in the sensor readout of the feedback loop has a disruptive presence resulting in the failure of the gyro-mirror stabilization system to maintain LOS. In this study, noise is implemented as an additive normally distributed perturbation of $\sigma^2 = 30\%$ with zero mean on the error feedback signal. The

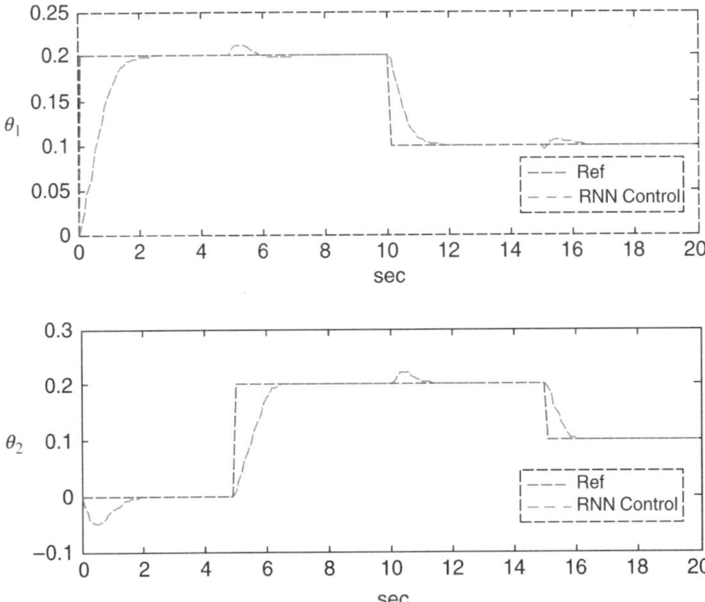

Fig. 8.9. Gyro-mirror stabilization system response under 30.0% parameter variation

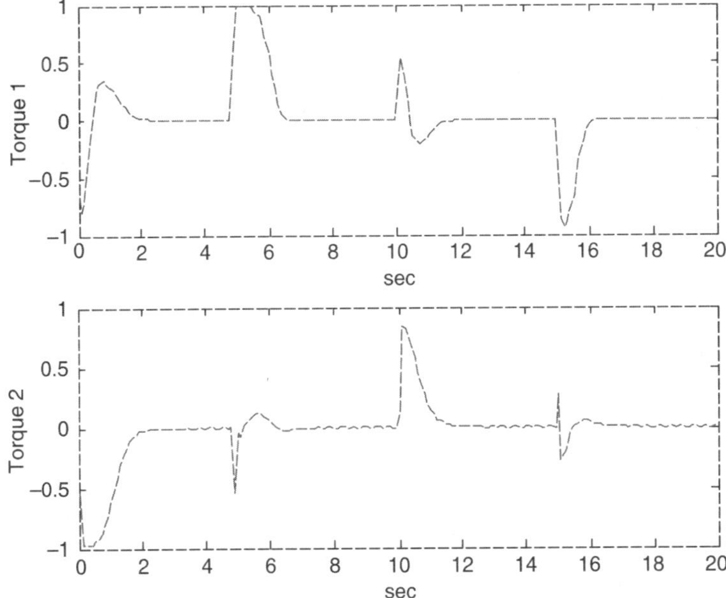

Fig. 8.10. RNN controller output signal under 30.0% parameter variation

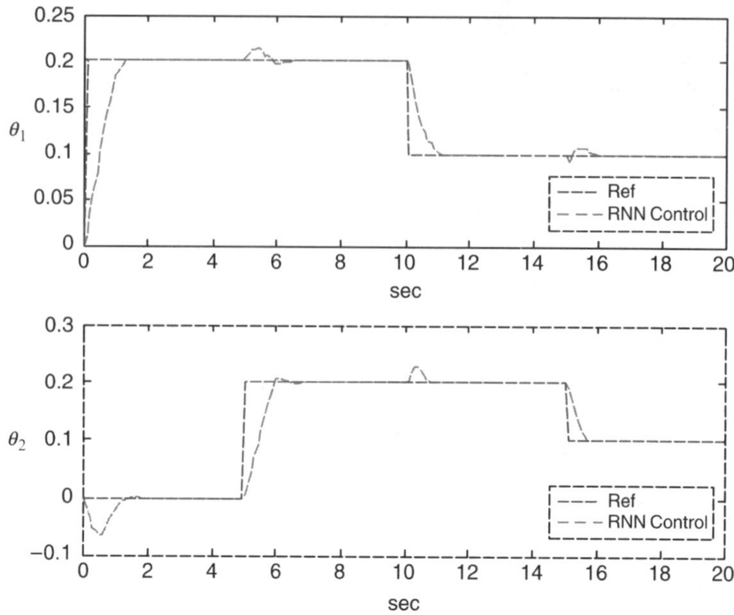

Fig. 8.11. Gyro-mirror stabilization system response under 30.0% noise

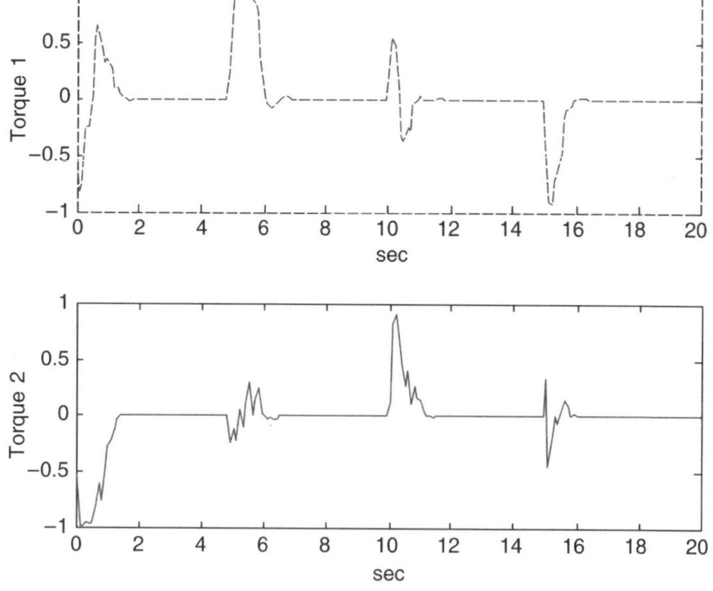

Fig. 8.12. RNN controller output signal under 30.0% noise

system responses with noise implemented are plotted in Figure 8.11 while the control signals are plotted in Figure 8.12.

By comparing the system performance in Figure 8.7 and Figure 8.11, it is apparent that the tracking ability of the controller is still good despite the introduction of 30% noise. In fact, the presence of noise did not prevent the controller from maintaining the desired reference position as evident from the good steady-state error in Figure 8.11. This is probably due to the recurrent nature of the RNN controller which creates a low pass filtering effect. The main difference observed in system performance is the transient response of the system – which in this case exhibits slower rise times and higher overshoots, as well as a more pronounced coupling effect. The resulting control signal produced by the controller, as shown in Figure 8.12, shows evidence of higher levels of fluctuations, which we believe is to correct the sudden onset of unknown dynamics that have been introduced into the system.

8.7 Conclusions

Recurrent neural networks (RNNs), through their unconstrained synaptic connectivity and resulting state-dependent nonlinear dynamics, offer a greater level of computational ability compared against regular feedforward neural network architectures. A corollary of this is a higher degree of complexity, which in turn leads to gradient-based learning algorithms for RNNs being more likely to be trapped in local optima, thus resulting in sub-optimal solutions. This motivates the use of evolutionary computation methods which center about the use of population-based global-search techniques as an optimization scheme. In this article, a new algorithm based on an Artificial Immune System (AIS) paradigm to train a recurrent neural network controller applied to a gyro-mirror line-of-sight stabilization system has been presented. We have demonstrated the viability of AIS in a dynamic setting, as epitomized by the gyro-mirror line-of-sight control problem, and supported by a number of standard simulation tests, highlighting its performance and robustness. Experimental results obtained from simulations of the evolved controller demonstrate good performance in terms of satisfying the specified control specifications as well as robust behavior in the presence of noise.

References

1. Almeida, L.B.: Backpropagation in nonfeedforward networks, *Neural Computing Architectures*, I. Aleksander, Ed. Cambridge, MA: M.I.T. Press, 1989.
2. Coello Coello, C.A., Cortes, N.C. "Solving multiobjective optimization problems using an artificial immune system", *Genetic Programming and Evolvable Machines*, vol. 6, no. 2, pp. 163–190, 2005.
3. Fogel, D.B., Wasson, E.C., Boughton, E.M. "Evolving neural networks for detecting breast cancer", *Cancer letters*, vol. 96, no. 1, pp. 49–53, 1995.

4. Fonseca, C.M., Fleming, P.J. “Multiobjective optimal controller design with genetic algorithms”, in *Proceedings on IEE Control*, pp. 745–749, 1994.
5. Ge, S.S., Lee, T.H., Zhao, Q. “Real-time neural network control of a free gyro stabilized mirror system”, in *Proceedings of the American Control Conference*, pp. 1076–1080, 1997.
6. Herrmann, G., Ge, S.S., Guo, G. “Practical implementation of a neural network controller in a hard disk drive”, *IEEE Transactions on Systems Technology*, vol. 13, no. 1, pp. 146–154, 2005.
7. Khor, E.F., Tan, K.C., Lee, T.H., Goh, C.K. “A study on distribution preservation mechanism in evolutionary multi-objective optimization”, *Artificial Intelligence Review*, vol. 23, no. 1, pp. 31–56, 2005.
8. Kurpati, A., Azarm, S. “Immune Network Simulation with Multiobjective Genetic Algorithms for Multidisciplinary Design Optimization”, *Engineering Optimization*, vol. 33, pp. 245–260, 2000.
9. Kuljaca, O., Swamy, N., Lewis, F.L., Kwan, C.M. “Design and Implementation of Industrial Neural Network Controller Using Backstepping”, *IEEE Transactions on Industrial Electronics*, vol. 50, no. 1, pp. 193–201, 2003.
10. Lee, T.H., Tan, K.K., Mamun, A., Lee, M.W., Khoh, C.J. “Composite Control of a gyro Mirror Line-of-Sight Stabilization Platform-Design and Auto-tuning”, in *Proceedings of the 3rd World Congress on Intelligence Control and Automation*, pp. 3150–3155, 2000.
11. Li, Y., Tan, K.C., Ng, K.C., Murray-Smith, D.J. “Performance based linear control system design by genetic evolution with simulated annealing”, in *Proceedings of the 34th IEEE Conference on Decision and Control,* New Orleans, LA, pp. 731–736, 1995.
12. Li, Y., Tan, K.C., Marionneau, C. “Direct design of linear control systems from plant I/O data using parallel evolutionary algorithms”, in *Proceedings of the International Conference on Control’96,* UK, pp. 680–686, 1996.
13. Liu, Y., Yao, X., Higuchi, T. “Evolutionary Ensembles with Negative Correlation Learning”, *IEEE Transactions on Evolutionary Computation*, vol. 4, no. 4, pp. 380–387, 2000.
14. Luh, G.C., Chueh, C.H., Liu, W.W. “MOIA: Multi-Objective Immune Algorithm”, *Engineering Optimization*, vol. 35, no. 2, pp. 143–164, 2003.
15. Narendra, K.S., Parthasarathy, K. “Identification and control of dynamical systems using neural networks”, *IEEE Transactions on Neural Networks*, vol. 1, pp. 4–27, 1990.
16. Nye, W.T., Tits, A.L. “An application-oriented, optimization-based methodology for interactive design of engineering systems”, *International Journal on Control*, vol. 43, pp. 1693–1721, 1986.
17. Oh, W.S., Bose, B.K., Cho, K.M., Kim, H.J. “Self Tuning Neural Network Controller for Induction Motor Drives”, in Proceedings of the IEEE Annual Conference of the Industrial Electronics Society, vol. 1, pp. 152–156, 2002.
18. Pearlmutter, B.A. “Learning state space trajectories in recurrent neural networks”, *Neural Computation*, vol. 1, no. 2, pp. 263–269, 1989.
19. Perry, J.: Spinning Tops and Gyroscopic Motion, *Dover Publications*, New York, 1957.
20. Srinivas, N., Deb, K. “Multiobjective optimization using non-dominated sorting in genetic algorithms,” *Evolutionary Computation*, vol. 2, no. 3, pp. 221–248, 1994.

21. Tan, K.C., Li, Y. "Performance-based control system design automation via evolutionary optimization", *Engineering Applications of Artificial Intelligence*, vol. 14, pp. 473–486, 2001.
22. Tan, K.C., Ang, D.C., Tan, K.K. "Genetic algorithm based fuzzy control of gyro-mirror line-of-sight stabilization platform", *Fifth International Conference on Control, Automation, Robotics and Vision*, pp. 1543–1547, 1998.
23. Tan, K.C., Lee, T.H., Khor, E.F., Ang, D.C. "Design and real-time implementation of a multivariable gyro-mirror line-of-sight stabilization platform", *Fuzzy Sets and Systems*, vol. 128, issue 1, pp. 81–93. 2002.
24. Tan, K. C., Lee, T.H., Khor, E.F. "Evolutionary algorithms for multi-objective optimization: performance assessments and comparisons", *Artificial Intelligence Review*, vol. 17, no. 4, pp. 251–290, 2002.
25. Tan, K.C., Khor, E.F., Lee, T.H., Sathikannan, R. "An evolutionary algorithm with advanced goal and priority specification for multi-objective optimization", *Journal of Artificial Intelligence Research*, vol. 18, pp. 183–215, 2003.
26. Tan, K.K., Mamum, A., Lee, T.H., Khoh, C.J. "PID-Augmented Adaptive Control of a Gyro Mirror LOS System", in *Proceedings of the 3rd World Congress on Intelligent Control and Automation*, pp. 3144–3149, 2000.
27. Thompson, H.A., Fleming, P.J. "An integrated multidisciplinary optimisation environment for distributed aero-engine control system architectures", in *Proceedings of the 14th World Congress of International Federation of Automatic Control*, pp. 407–412, 1999.
28. Tian, L., Collins, C. "A dynamic recurrent neural network-based controller for a rigid-flexible manipulator system", *Mechatronics*, vol. 14, pp. 471–490, 2004.
29. Williams, R.J., Zipser, D. "Experimental analysis of the real-time recurrent learning algorithm", *Connection Science*, vol.1, no. 1, pp. 87–111, 1989.
30. Willems, P.Y.: Gyrodynamics, *Springer*, Berlin, 1974.
31. Yao, X., Liu, Y. "A new evolutionary system for evolving artificial neural networks", *IEEE Transactions on Neural Networks*, vol. 8, no. 3, pp. 694–713, 1997.
32. Yoo, J., Hajela, P. "Immune network simulations in multicriterion design", *Structural Optimization*, vol. 18, pp. 85–94, 1999.

9

Distributed Problem Solving using Evolutionary Learning in Multi-Agent Systems

Dipti Srinivasan and Min Chee Choy

Department of Electrical and Computer Engineering
National University of Singapore
4 Engineering Drive 3, Singapore 117576

Summary. This chapter presents a new framework for solving distributed control problems in a cooperative manner via the concept of dynamic team building. The distributed control problem is modeled as a set of sub-problems using a directed graph. Each node represents a sub-problem and each link represents the relationship between two nodes. A cooperative ensemble (CE) of agents is used to solve this problem. Agents are assigned to the nodes in the graph and each agent maintains a table of link relationship with all the other nodes of the problem. In the cooperative ensemble, each agent generates three sets of outputs iteratively based on the input variables it receives. They are, namely, the need for cooperation, the level of cooperation and the control directives. These outputs are used for dynamic team building within the cooperative ensemble. Agents within each team can issue a collaborative control directive and they take into account the mistakes of all the members in the team. In addition, each agent has a neuro-biologically inspired memory structure containing the addictive decaying value of all its previous errors and it is used to facilitate the dynamic update of the agent's control parameters. The cooperative ensemble has been implemented in the form of distributed neural traffic signal controllers for the distributed real-time traffic signal control. It is evaluated in a large simulated traffic network together with several existing algorithms. Promising results have been obtained from the experiments. The cooperative ensemble is seen as a potential framework for similar distributed control problems.

9.1 Introduction

A variety of definitions have been offered for cooperation in previous published works and from [1], the definition is as follows: "Cooperation occurs when two or more agents work together in a common environment to more effectively reach the maximal union of their goals". The neurobiological basis of cooperative behaviours has been shown in some recent research works [2]. This further reinforces the possibility of implementing cooperative behaviours

D. Srinivasan and M.C. Choy: *Distributed Problem Solving using Evolutionary Learning in Multi-Agent Systems*, Studies in Computational Intelligence (SCI) **66**, 211–227 (2007)
www.springerlink.com

when designing intelligent entities which are derived from neurobiological concepts (e.g. artificial neural networks, multi-agent systems). In fact, in recent years, the concept of cooperation has been adopted by many research works to build hybrid neural networks [3, 4], collaborative agents [5] and systems with cooperation mechanisms. With the onset of distributed problem-solving paradigm, many researches are looking into the feasibility of developing a distributed platform for cooperative problem solving.

Cooperative distributed problem solving (CDPS) research investigates the areas in which the fields of artificial intelligence and distributed processing overlap. It concerns the development of knowledge and reasoning techniques that a network of problem-solving nodes or agents uses to cooperate effectively in solving a distributed, complex problem. In many cases, each node in the network does not have sufficient expertise, resources and information to solve a problem. On the other hand, different nodes might have the expertise to solve different parts of the problem. In addition to this, different nodes might have different information or viewpoints on a problem due to the nature of the network architecture.

In this paper, we present a new framework called the cooperative ensemble (CE). The cooperative ensemble comprises of distributed cooperative agents which possess the ability to form teams dynamically. Agents in each team can work together to generate a collaborative control directive. In addition, a new cooperative parameter update algorithm has been developed to complement the dynamic team building feature of the cooperative ensemble.

The following section will describe the modeling of the distributed control problem for CDPS. Section 9.3 covers the overall architecture of the CE while Section 9.4 describes the actions of each agent in the CE. Section 9.5 presents the dynamic team building feature of the CE and Section 9.6 touches on the cooperative parameter update algorithm in each agent. Section 9.7 presents an actual application of CE to solve a real-world distributed control problem, namely, the problem of real-time traffic signal control. Section 9.8 gives a brief description of the simulation results. Section 9.9 discusses other possible applications of the CE for distributed control and concludes the research work presented in this paper.

9.2 The Distributed Control Problem

The problem of distributed control can be modeled as a set of sub-problems using the concept of a directed graph. Define the following:

$G = (N, L)$	directed graph with a set N of n nodes and a set L of l links describing the traffic network
A_i	the CE agent acting at node $i \in N$
T	denotes the T iteration where T is a positive integer.

$I_i(T)$	local input information vector of A_i at iteration T
$c_i(T)$	vector of control parameters of A_i at iteration T
$c_i(T)^o$	optimal vector of control parameters of A_i at iteration T
$l_{ij}(T)$	the link relationship between node i and node j as seen from node i at iteration T(note that $l_{ij} \neq l_{ji}$)
$l_{ij}(T)^o$	the optimal link relationship between node i and node j as seen from node i at iteration T
$\{L_i\}$	the set $\{L_i\}$ contains all l_{ij}, for $j = 1, 2, \ldots .n$, and $j \neq i$, where n is the total number of nodes in G
$u_i(T) = F_{iT}[I_i(T), c_i(T)]$	control function of A_i at iteration T
$u_i(T)^o = F^o_{iT}[I_i(T), c_i(T)^o]$	optimal control function of A_i at iteration T
$s(T)$	overall state vector of the distributed control problem at iteration T
C	cost function for the distributed control problem

The distributed control problem can thus be stated as:
Distributed Control Problem for the CE: At each iteration T, find the set of control function $u_i(T)$ and $\{L_i\}$ for each A_i, $i \in N$ that minimizes the cost function C.

The choice of control function for each A_i is entirely up to the individuals, depending on the type of distributed control problem and the availability of well known control functions. The CE essentially presents a platform for dynamic team building as well as parameter updating for the agents once the choice of the control function is known.

9.3 The Cooperative Ensemble

The CE is a set of cooperative intelligent agents which are used to solve the problem as described in Section II. A general schematic of an agent in the CE is as shown in Fig. 9.1. The agent essentially generates three outputs based on the inputs it received. The first output will be the problem-specific control directive/s. The second output is that of the need for cooperation and the third output is the level of cooperation. For some problems where the choice of the control function is known, the agent will have to include an additional module known as the cooperative inference engine so as to generate the two cooperation-related outputs (as shown in Fig. 9.1). Otherwise, the three outputs can also be generated by a single, customized control function (e.g. a knowledge base system such as a fuzzy neural control function using fuzzy rule base).

The values of the need for cooperation as well as the level of cooperation can take the form of crisp outputs or fuzzified outputs (which are then further mapped to an additional set of fuzzy rules). In any case, the need for cooperation represents the need for team building of an agent while the level

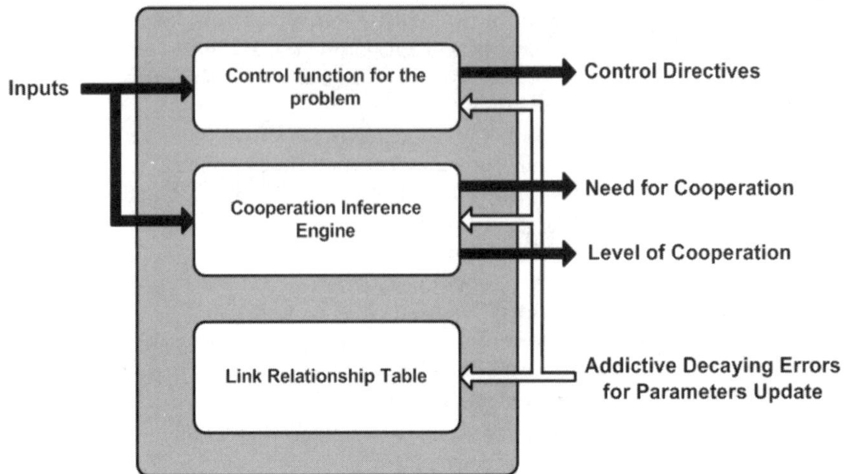

Fig. 9.1. General Schematic of an agent in the CE

of cooperation represents the degree of collaborative action as desired by the agent (i.e. if an agent desires a high degree of collaborative action, it will seek to form a bigger team). Besides the control function and the cooperation inference engine, each CE agent also maintains a table of link relationship (as shown in Fig. 9.1). The link relationship table describes the relationship between the agent and all the other agents in the CE. This table is used together with the two cooperation-related outputs for dynamic team building.

Finally, as shown in Fig. 9.1, addictive decaying errors are used to update the parameters in the three modules of the agent. More details of this process will be given in Section VI. The following section describes the actions of each agent in the CE.

9.4 Actions of Each Agent

The actions of each agent in the CE are shown in Fig. 9.2. The initialization process takes place first. After receiving the inputs for a particular iteration T, each agent will proceed to generate the three outputs shown in Fig. 9.1. After all the agents have generated their outputs, an agent will be selected randomly for dynamic team building (details of which will be given in the following section). The random selection process will repeat itself until all the agents have been considered. This means that all the agents have either:

1) Been selected to be part of others' teams.
2) Formed their own teams.
3) Chosen not to be part of any team.

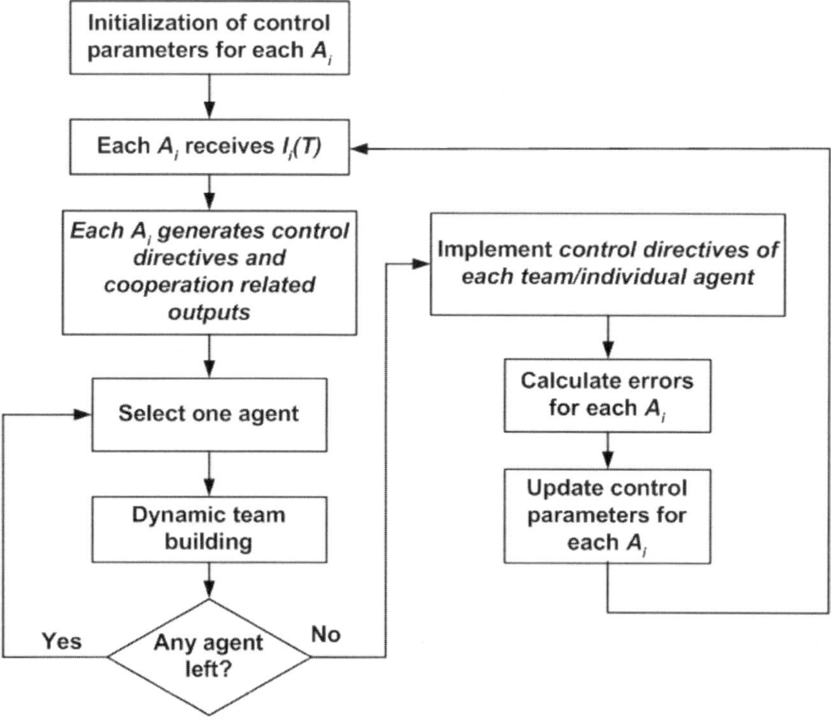

Fig. 9.2. Actions of each agent in the CE

Agents in a team can choose to make a group decision pertaining to their respective outputs. This means that they can choose to adopt a common control directive using various means (such as the aggregate of all control directives, or taking the control directive of the most important agent in the team, etc.). The agents who have chosen not to be part of any team will continue to stick to their initial set of control directives. After the control directives have been implemented into the problem, the respective local errors of each agent are computed and the control parameters of each A_i will be updated using the cooperative parameters update algorithm (details of which will be given in Section VI). For, the next iteration $T+1$, a new set of inputs $I_i(T+1)$ will be read in for each A_i and the process will repeat itself again. This process continues until a stopping criterion is met or when the agents are deactivated. The following section will cover the dynamic team building feature of the CE.

9.5 Dynamic Team Building

Fig. 9.3 shows the process of dynamic team building. In this process, an agent A_i is first randomly selected (via a discrete uniform distribution) to form its own team. A_i will process the various values of l_{ij} in its table of link relationships or the set $\{L_i\}$. For a chosen target node j, A_i will determine if A_j and itself desires cooperation or not. If either of them do not desire to cooperate, A_i will have to change its target node and select another l_{ij} from $\{L_i\}$. If cooperation is desired by both agents and l_{ij} is greater than an empirical correlation threshold value, the respective levels of cooperation desired by the two agents will then be compared. If A_i desires a higher level of cooperation compared to A_j, A_j will be added as part of the team of A_i.

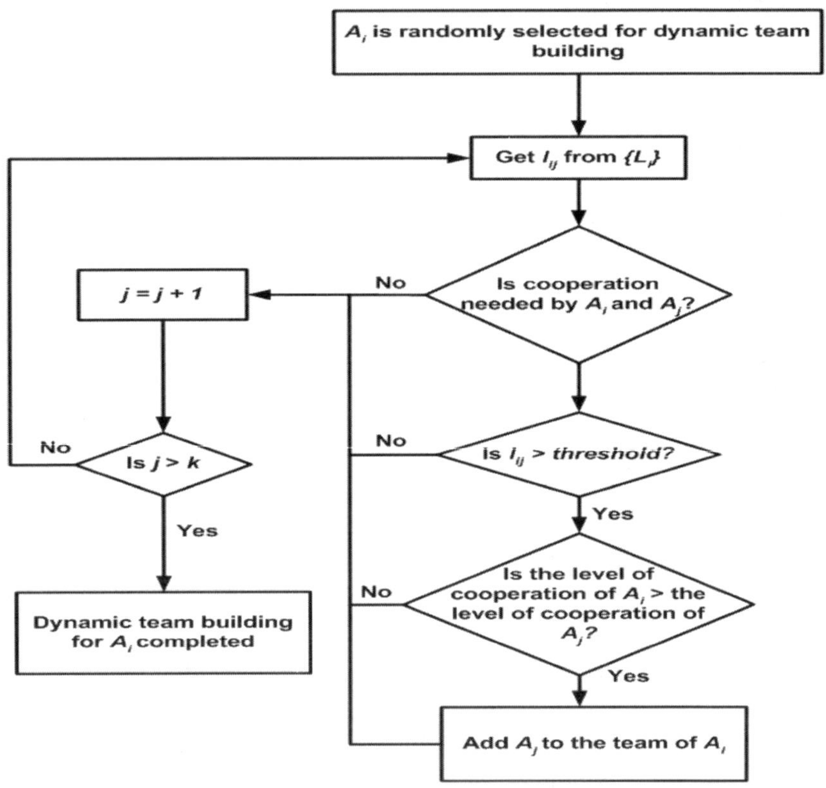

Fig. 9.3. Dynamic Team Building

Also, A_j will not be further considered in the next round of random selection (which means that each agent can only be a member of not more than one team). The process of dynamic team building for A_i is complete once every member in the set $\{L_i\}$ has been processed. Hence, in the situation where either one of the following is true for each l_{ij} in the set $\{L_i\}$,

1) $l_{ij} < threshold$
2) The level of cooperation desired by A_i is less than all the other agents'.

The agent A_i will not be able to form its own team. However, it still can be included in another agent's team as long as it wants to cooperate in the first place. If A_i does not want to cooperate with others during iteration T, it will not be able to form any team and it will have to implement its own control directives for that iteration.

Given that the link relationship table, the cooperation inference engine and the control function are being updated iteratively, it can be seen that the process of team building is a dynamic one where new teams can be formed, old teams can be dissolved, and the size of existing teams can be changed. The next section presents a cooperative parameters update algorithm that leverages on the feature of dynamic team building.

9.6 Cooperative Parameters Update Algorithm

In this section, we describe how the parameters of the various modules in the agent (Fig. 9.1) can be updated in a cooperative manner. The main concept that is being introduced in this algorithm is an addictive decay method to take into account all the previous errors of an agent, which are in turn decaying iteratively. This concept is neuro-biologically inspired and it is largely similar to related design issues of Organizational Memory Systems (OMS) [6, 7] in which an OMS has to store all its previous experiences in memory [8] while being appropriately forgetful [9] at the same time. In the context of the CE (analogous to an organization), its previous experiences are stored in each agent in the form of the addictive errors. At the same time, an appropriate forgetful mechanism is implemented in the form of decay functions.

The number of errors that an agent receives during each iteration is dependent on the size of its team. Fig. 9.4 and 9.5 illustrate the concept of the cooperative parameters update algorithm. As shown in the figures, the number of local errors which Agent A considers changes as it adds an additional member to its team. Given this, it can be seen that this cooperative parameters update algorithm and the dynamic team building process are complementary in nature.

Assume the CE is activated at iteration 1. Let $e_{li}(T)$ be the local error of the node n_i at iteration T. It is assumed that the local error $e_{li}(T)$ can be reasonably approximated/calculated after the main distributed control problem has been subdivided into several sub problems (each represented by a

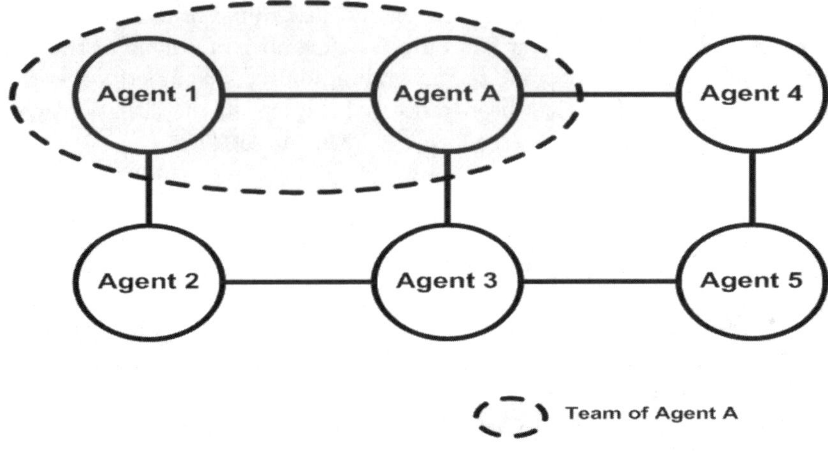

Fig. 9.4. Cooperative Parameters Update (part 1)

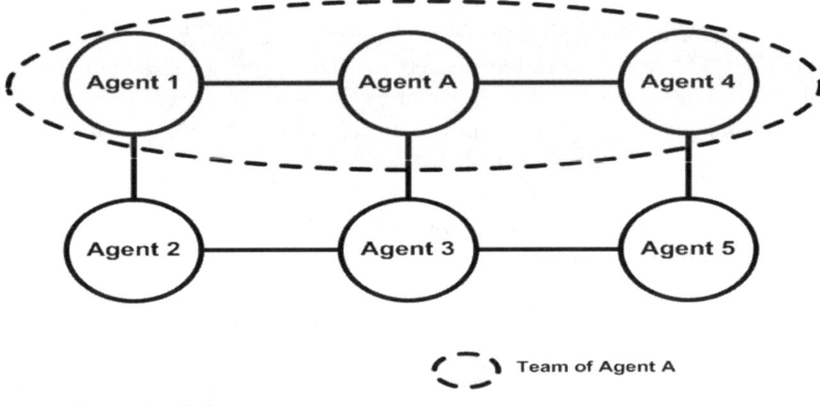

Fig. 9.5. Cooperative Parameters Update (part 2)

node). The final error of the node n_i, $E_{fi}(T)$ that will be considered by A_i at iteration T will be:

$$E_{fi}(T) = (1 - \gamma) \sum_{x=1}^{M} e_{lx}(T) + \gamma Ci(T) \tag{9.1}$$

where $0 < \gamma < 1$ and it represents the relative importance of historical errors and current errors, M is the total number of members in the team of A_i at iteration T, $C_i(T)$ is the addictive decaying errors term of A_i at iteration T and it is defined as follows:

$$C_i(T) = \sum_{t=1}^{T-1} \sum_{j=1}^{N(t)} D\left(e_{lj}(t), T, t\right) \tag{9.2}$$

where $N(t)$ is the total number of team members which A_i has at iteration t, and $D()$ is the decay function (a non-increasing $D(x) >= 0$ for $x >= 1$), which takes in $e_{lj}(t)$, T and t as the inputs. $e_{lj}(t)$ is the local error which A_i considered during iteration t. During iteration T, the decay function D computes the decayed version of $e_{lj}(t)$ by considering the difference between iteration T and t (i.e. the longer the error has been around, the more it would have decayed). The choice of the decay function can vary depending on applications. Some possible choices are exponential decay functions, sliding windows decay functions, polynomial decay functions, etc. Due to practical reasons, the decayed version of $e_{lj}(t)$ will be stored in the agent's memory as long as $|e_{lj}(t)| > T_D$, where T_D is the decay threshold. If the decayed version of $e_{lj}(t)$ is smaller than T_D, it will be removed from the agent's memory (or "forgotten" by the agent).

The final error obtained in (1) can be used to update the control parameters of the agent using a variety of standard parameter update algorithm. The distinct advantages of the addictive decay method for computing the error for each agent are:

1) For some parameter update algorithms (e.g. the gradient descent algorithm), the tendency for the algorithms to converge to a local minima is high. This will be an extremely critical issue if such algorithms are used to solve dynamic problems. However, in this case, the error of each agent is computed as shown in (1) and its value is dependent on the number of team members which the agent has at each iteration as well as the addictive decayed versions of all its previous errors. Hence, even when a local minima is reached in the previous iteration, the agent may still be able to get out of it in the next iteration due to changes in its team structure as well as the inclusion of the addictive decayed errors.
2) Given the addictive nature of the error computation method, recurring errors can be reinforced iteratively even though its previous versions have decayed. Non-recurring errors will vanish once its value is reduced beyond T_D. An appropriate choice of decay functions can thus ensure that the

historical errors remain in memory for an appropriate length of time, depending on the dynamics of the problem.

9.7 Cooperative Ensemble for Real-Time Distributed Traffic Signal Control

The CE is designed to solve complex distributed control problem as defined in Section II. The control of traffic signal at multiple signalized intersections in a traffic network can be modeled as one such problem. In this case, each agent in the CE is a local traffic signal controller. Given the unavailability of good control functions for real-time traffic signal control, the concepts of fuzzy neural networks are used to design the control modules (including the cooperation inference engine) of each agent for this problem (refer to Fig. 9.6).

The control directives generated by each agent are essentially traffic signal control policies which are used to adjust the length of the traffic signals. At the end of each iteration, the local error for each agent is obtained by computing the average delay of the traffic intersection which the agent is controlling. The cooperative parameter update algorithm is then used to update the neural weights as well as the link table of each agent. In each agent, the decay function used is that of an exponential decay function. Also, from equation (9.1), γ is set to 0.5 to signify the equal importance of current errors and historical decayed errors.

The agents in this cooperative ensemble are implemented to control the traffic signal of 25 signalized intersections in a simulated traffic network of

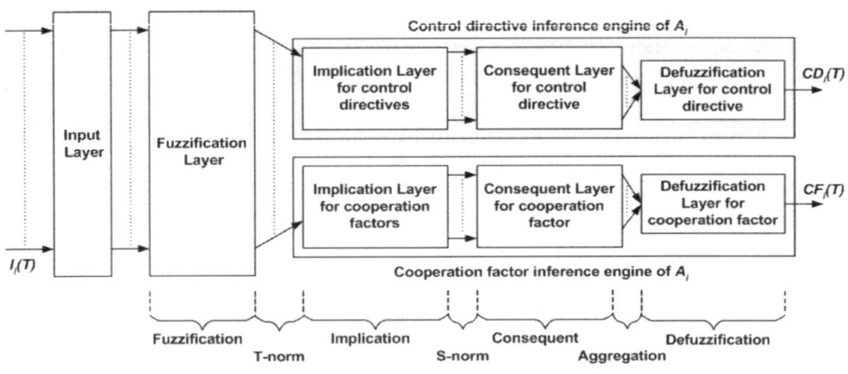

Fig. 9.6. Fuzzy Neural Networks Implementation of CE Agent for Traffic Signal Control

a section of the Central Business District (CBD) region of Singapore. The traffic network is simulated using a microscopic traffic simulator known as PARAMICS [10]. The performance of the CE is compared with the performances of two other hybrid multi-agent systems, namely the hierarchical multi-agent system (HMS) [11] and the stochastic approximation based (SPSA-NN) multi-agent system [12, 13]. In addition, a simulated version of a well known proprietary traffic signal control program known as GLIDE is also used for benchmarking.

9.8 Simulation Results

Two simulation scenarios have been designed to evaluate the performances of the CE and to compare it with the performances of the other techniques. They are namely the typical scenario with morning and evening peak periods (24 hours) and the extreme scenario with multiple peak periods (24 hours). The former depicts a typical day-to-day scenario for that section of the CBD while the latter is a fictitious scenario which is designed to test the performance of these techniques under extreme traffic conditions (with high, fluctuating traffic volumes). The performance measure used is the current mean speed of the vehicles. Given that the simulation platform is microscopic in nature, we are able to obtain the mean speed of each vehicle when the simulation is running. For each method, six separate simulation runs are carried out and the corresponding aggregated results are plotted in the graphs shown in Fig. 9.7 and 9.8.

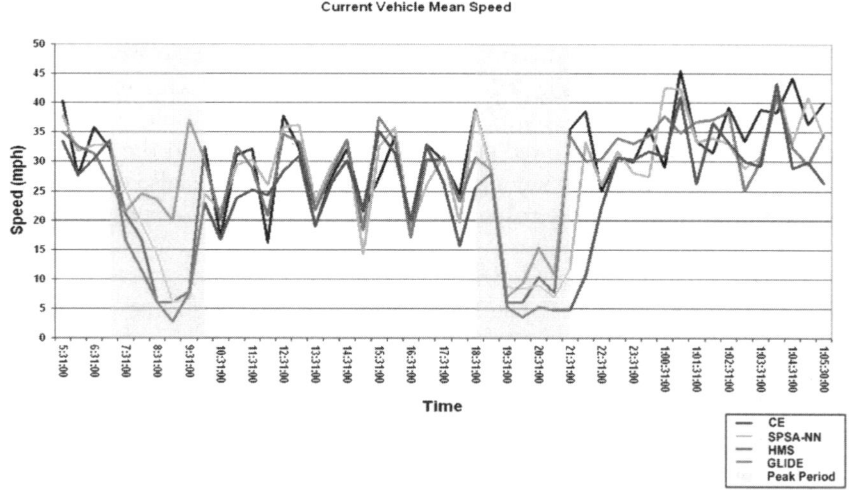

Fig. 9.7. Current Vehicle Mean Speed for the typical scenario with morning and evening peaks (24 hours)

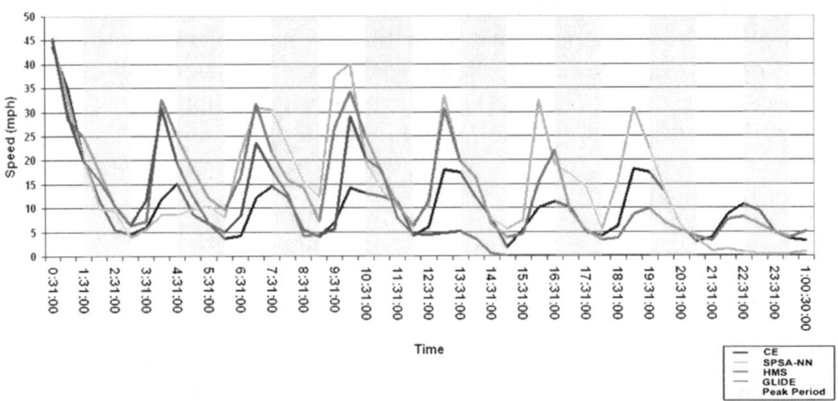

Fig. 9.8. Current Vehicle Mean Speed for Long Extreme Scenario with Multiple Peaks (24 hrs)

Table 9.1. Current Vehicle Mean Speed for the typical scenario with morning and evening peaks (24 hours)

Control technique	Current Vehicle Mean Speed (mph)		
	1st Peak Period	2nd Peak Period	End of Simulation
CE	**32.5**	**35**	**40**
SPSA-NN	**25**	**12.5**	**35**
HMS	**31**	**35**	**35**
GLIDE	**23**	**5**	**26.5**

As can be seen from Fig. 9.7, for the typical scenario with morning and evening peak periods (24 hours), the CE yields the best result at the end of the simulation. This is followed closely by the SPSA-NN and the HMS while GLIDE performs the worst among the four methods. This is also evident from the tabulated results of the respective average current mean speed shown in Table 9.1.

For the extreme scenarios with multiple peak periods (24 hours), the graph in Fig. 9.8 shows that the HMS yields the best performance, followed very closely by the CE. The SPSA-NN as well as GLIDE degenerate and on the average, the current vehicle mean speed drops to zero before the end of the simulation runs. For each method, the average values of the current mean speed at the end of each peak period is given in Table 9.2. It can be seen that the current mean speed for GLIDE drops to zero typically at the 5th peak period while the current mean speed for SPSA-NN typically drops to zero at the 7th peak period.

Table 9.2. Current Vehicle Mean Speed for the long extreme scenario with multiple peaks (24 hours)

Control technique	Current Vehicle Mean Speed (mph)							
	1st Peak Period	2nd Peak Period	3rd Peak Period	4th Peak Period	5th Peak Period	6th Peak Period	7th Peak Period	8th Peak Period
CE	5	5	5	5	2.5	5	4	4
SPSA-NN	5	10	15	7.5	4	5	4	0
HMS	7	10	14	7.5	6	4	4	4
GLIDE	7	5	5	5	0	0	0	0

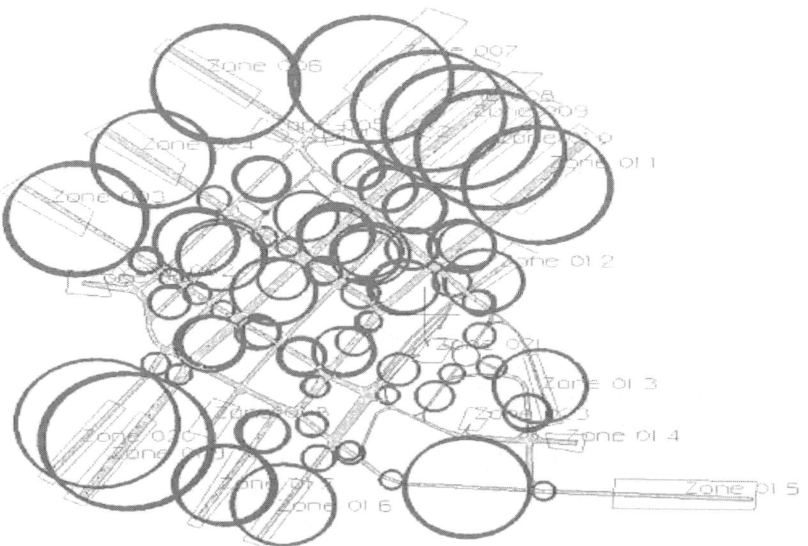

Fig. 9.9. Traffic Network controlled by the GLIDE benchmark after 24 hours (long extreme scenario with multiple peaks)

In order to better illustrate the conditions of the traffic network at the end of the long extreme scenario, two-dimensional top down screenshots of the traffic network are taken from PARAMICS Modeler to show the state of the traffic network when the traffic signals are controlled by GLIDE (Fig. 9.9) and Cooperative Ensemble-based multi-agent system (Fig. 9.10). The screenshots are captured at 0030 hrs, at the end of the 24 hours for the extreme scenario. The PARAMICS modeling environment was preset to denote 13 stopped or queued vehicles with a hotspot or red circle. As can be seen from the Fig. 9.9, the traffic network evolves into a pathological state with over saturation at 0030 hrs when GLIDE was implemented. This is likely to be the result of the steady increase of total mean delay and the drop in the current vehicle mean speed to zero as early as after the 5th peak period when the GLIDE

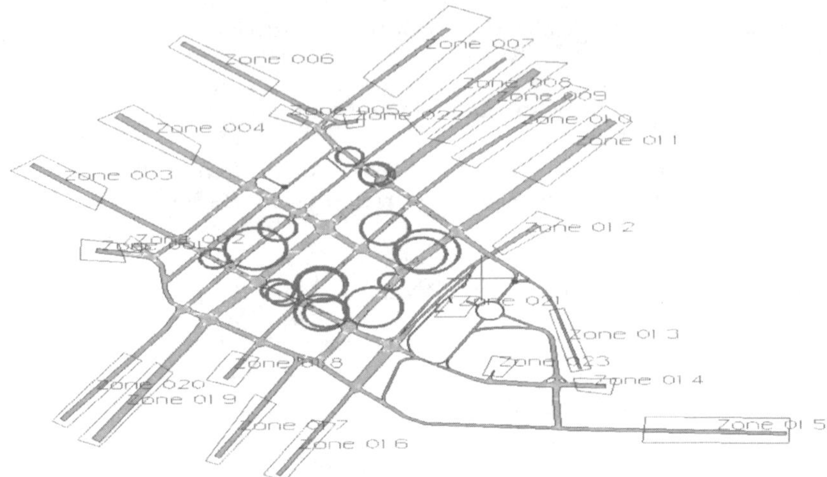

Fig. 9.10. Traffic Network controlled by the CE after 24 hours (long extreme scenario with multiple peaks)

benchmark was implemented. The number of congested road sections is well over thirty in number.

On the other hand, using the CE, congestions are confined to the central region of the traffic network and the number of congested links is reduced to less than ten (as shown in Fig. 9.10). Despite the presence of these hotspots, the traffic continues to flow at a slow speed even at the end of the extreme scenario simulation.

These experiments establish the superior performance of the proposed cooperative ensemble multi-agent system in providing effective real-time distributed traffic signal control for a large complex traffic network. Also, it can be seen that the CE manages to maintain a robust performance even under extreme conditions.

Ref [14] proposed a multi-agent system based on game theory for control of traffic signals. The cooperative ensemble approach proposed in our paper, however, aims to leverage on the collaborative actions of the agents to attain a smooth traffic flow in this complex network. The application of game theory [14] does not yield complete cooperation since individual agents have to compete to maximize their own goals, rather than working towards a common goal. Additionally, our ensemble approach presents a dynamically changing cooperative neighborhood and consequently a dynamically updating algorithm, which has been successfully tested on a large traffic network using real-time information from an actual traffic network.

9.9 Discussion and Conclusions

A new framework for cooperative distributed problem solving, known as the cooperative ensemble (CE), has been presented in this chapter. This distributed, cooperative problem solving approach has been implemented in the form of multiple interacting, autonomous hybrid agents to provide effective signal optimization strategies for real-time traffic management of an arterial network. At the same time, new hybrid multi-agent system based methods for effective cooperation, online learning, policy formulation as well as knowledge acquisition have been conceptualized and demonstrated in this study. Although the approach presented in this paper has been applied to a traffic control problem, this hybrid multi-agent approach can be used for solving dynamic distributed control problems in general. The approach incorporates several advanced, humanized features such as cooperation and online learning in the large-scale hybrid multi-agent systems. Extensive simulations and testing has been done to validate this approach using the complex problem of providing real-time traffic signal control in a large, real world traffic network using state-of-the-art traffic network simulator.

From the positive results, it can be concluded that although cooperative multi-agent systems are attractive solutions to the Distributed Control Problem, various computational intelligence techniques such as neural networks, fuzzy logic and evolutionary computation can play an integral and significant role in intelligent, autonomous multi-agent systems. These roles include the facilitation of more effective online learning, knowledge representation, acquisition, and decision making, etc.

Comparison of results with various intelligent multi-agent approaches shows that systems with enhanced cooperative behaviours yield better performances when they are being implemented in extremely difficult and complex scenarios. This is shown by the better performances of the cooperative ensemble approach in the extreme simulation scenarios. The enhanced cooperative behaviours via properly implemented cooperation mechanisms can therefore be used to allow agents to gain a better perspective (which can be a higher level perspective, a more diverse perspective, etc.) of the overall problem and not be constrained to its local perspective when solving dynamic problems.

The hybrid multi-agent approach has been applied in a wide spectrum of application domains concerning commerce, manufacturing, defense, etc. by many researchers in recent years [5]. This trend, coupled with existing successful applications, indicates the great potential of adopting the distributed problem solving approach (often in the form of multi-agent systems) for solving complex problems, which are often ill-defined and dynamic in nature. The promising results obtained in this study reinforce this notion. In addition, the results show how complex cooperation mechanisms can be used in a meaningful manner in multi-agent systems to solve real world, complex problems.

The proposed cooperative ensemble approach can be potentially applied to solve other distributed control problems that can be modeled in a similar manner as presented in Section II. Examples include intelligent agents for traffic monitoring and intrusion detection in computer networks, intelligent agents for distributed sensor networks as well as distributed decision support systems.

Future work will involve fine-tuning certain aspects of the algorithm such that it can be used as a form of meta-heuristic for solving distributed control problems that are dynamic in nature. Other forms of cooperation mechanisms (e.g. cooperation by imitation) can also be conceptualized and used to improve the performance of multi-agent systems as well as other control systems.

Acknowledgement

The authors would like to thank the Land Transportation Authority of Singapore (LTA) for providing data necessary for the simulation modeling.

References

1. W. Zachary, S. Robertson, "Introduction," *In Cognition, Computing, and Cooperation*, S. Robertson, W. Zachary, and J. Black, Eds. Norwood, NJ: Ablex, 1990.
2. J.K. Rilling, D.A. Gutman, T.R. Zeh, G. Pagnoni, G.S. Berns, and C.D. Kilts, "A Neural Basis for Social Cooperation", *Neuron*, Vol. 35, Page(3) 395–405, July 2002.
3. G. Auda, M. Kamel, and H. Raafat, "A New Neural Network Structure with Cooperative Modules", In Proceedings of IEEE World Congress on Computational Intelligence (ICNN), Vol. III, Page(s) 1301–1306, 1994.
4. G. Auda, and M. Kamel, "CMNN: Cooperative modular neural network", In Proceedings of IEEE Conference on Neural Networks, vol. 1, Page(s) 226–331, 1997.
5. L. Panait and S. Luke, "Cooperative Multi-Agent Learning: The State of the Art", Autonomous Agents and Multi-Agent Systems, Volume 11, Number 3, Page(s) 387–434, November 2005.
6. Argyris, C. & Schön, D.A. Organizational learning: A theory of action perspective, Addison-Wesley, Reading, MA, 1978.
7. Churchland, P.S. & Sejnowski, T.J. The computational brain, The MIT Press, Cambridge, MA, 1993.
8. Stein, E.W. & Zwass, V. "Actualizing organizational memory with information systems," Information systems research, 6(2), 85–117, 1995.
9. Huber, G.P. "Organizational learning: The contributing processes and the literatures," Organizational Science, 2(1), 88–115, 1991.
10. Quadstone. *PARAMICS Modeller v4.0 User Guide and Reference Manual.* Quadstone Ltd., Edinburgh, U.K., 2002.
11. M.C. Choy, D. Srinivasan, R.L. Cheu, "Cooperative, Hybrid Agent Architecture for Real-time traffic control", IEEE Transactions on Systems, Man and Cybernetics, Part A, Vol. 33, Issue 5, Page(s) 597–607, Sept. 2003.

12. M.C. Choy, D. Srinivasan, R.L. Cheu, "Simultaneous Perturbation Stochastic Approximation based Neural Networks for Online Learning", In the Proceedings of the 7th IEEE International Conference on Intelligent Transportation Systems, Page(s)1038–1044, Oct. 3–6, 2004.
13. M.C. Choy, D. Srinivasan, R.L. Cheu, "Neural Networks for Continuous Online Learning and Control", to appear in the IEEE Transactions on Neural Networks, 2006.
14. Ana L. C. Bazzan, "A Distributed Approach for Coordination of Traffic Signal Agents", Autonomous Agents and Multi-Agent Systems, Volume 10, Number 2, Page(s) 131–164, March 2005.

10

Evolutionary Computing within Grid Environment

Ashutosh Tiwari, Gokop Goteng, and Rajkumar Roy

Decision Engineering Centre, Manufacturing Department, School of Applied Sciences, Cranfield University, Cranfield, Bedfordshire, MK43 0AL, UK
{*a.tiwari, g.l.goteng, r.roy*}*@cranfield.ac.uk*
http://www.cranfield.ac.uk/sas/mem/decisionengineering/

Summary. Evolutionary computing (EC) techniques such as genetic algorithms (GA), genetic programming (GP), evolutionary programming (EP) and evolution strategies (ES) mimic nature through natural selection to perform complex optimisation processes. Grid-enabled environment provides suitable framework for EC techniques due to its computational and data capabilities. In addition, the semantic and knowledge Grids aid in the design search and exploration for multi-objective optimisation tasks. This chapter explores some problem solving environments such as Geodise (Grid-Enabled Optimisation Design Search for Engineering), FIPER (Federated Intelligent Product Environment), SOCER (Service-Oriented Concurrent Environment), DAME (Distributed Aircraft Maintenance Environment) and Globus toolkit to demonstrate how EC techniques can be performed more efficiently within a Grid environment. Service-oriented and autonomic computing features of Grid are discussed to highlight how EC algorithms can be published as services by service providers and used by service requestors dynamically. Grid computational steering and visualisation are features that can be used for real-time tuning of parameters and visual display of optimal solutions. This chapter demonstrates that grid-enabled evolutionary computing marks the future of optimisation techniques.

10.1 Introduction

The use of Evolutionary computing (EC) to solve many optimisation problems is still hindered by the inadequate computational power and data manipulation platform for computational and data intensive problems. Examples of such computational and data intensive applications are engineering design optimisation, bioinformatics, pharmaceutical and particle physics simulations and many others. Additionally, since EC techniques mimic nature through natural selection, knowledge-driven problem solving environments (PSEs) are ideal frameworks for optimisation problems using EC. Grid computing (GC) promises to provide computational resources and services for optimisation

A. Tiwari et al.: *Evolutionary Computing within Grid Environment*, Studies in Computational Intelligence (SCI) **66**, 229–248 (2007)
www.springerlink.com

processes using EC techniques. Besides, the service-oriented architecture (SOA) in Grid-based deployments provide easy and seamless access to heterogeneous distributed optimisation codes to users in a dynamic and coordinated manner. Service-oriented problem solving environments (SO-PSEs) such as Geodise (Grid-Enabled Optimisation Design Search for Engineering), FIPER (Federated Intelligent Product Environment) and SORCER (Service-Oriented Concurrent Environment) are good examples of Grid-based PSEs that use semantic Grid and ontology to intelligently drive optimisation processes. Globus toolkit which serves both as a middleware as well as a PSE has been the major innovative contribution to the second generation Grid by the Global Grid Forum (GGF) which has now been merged with the Enterprise Grid Alliance (AGA) to become the Open Grid Forum (OGF).

This chapter will focus on how evolutionary computing techniques can exploit Grid-based framework to provide seamless access to optimisation resources and services for users involved in multidisciplinary optimisation applications. The first part of this chapter will discuss evolutionary computing and related application areas. The second part will look at intelligent Grid environments such as semantic Grid, knowledge Grid and autonomic computing. The third part will concentrate on service-oriented and problem solving environments for Grid-based problems. The fourth part will delve on problem solving environment case studies. The last part of this chapter will talk about collaborative evolutionary computing.

10.2 Evolutionary Computing

Evolutionary computing uses techniques such as genetic algorithms (GAs), genetic programming (GP), evolutionary programming (EP) and many others. These techniques work on the principles of natural selection based on Darwin's theory of evolution. This principle works on the composition of genetic traits called chromosomes, in which successive operations through crossover or mutation give rise to better performing off-springs (population) due to successive refinement of these hereditary traits. In the same way, evolutionary computing techniques mimic this phenomenon of natural selection to improve upon classical methods of optimisation such as preference-based and generating methods that work well only on single-objective optimisation problems [10]. With evolutionary computing techniques, multi-objective optimisation problems can be solved, producing multiple optimal solutions in a single simulation run. This is not possible with classical optimisation methods. However, the evolutionary computation can be computationally expensive for complex optimisation problems. Imagine an optimisation problem of an underground system with hundred of thousands of optimal solutions. This is where Grid-based framework comes in. Grid computing is a distributed computing infrastructure which aims to overcome large-scale computation within the fields of science, engineering and business with its secured, pervasive, persistent, dynamic and coordinated computational resources and services [13].

Evolutionary programming which was introduced by Fogel is an attempt to create artificial intelligence (AI) using finite state machines (FSM) to predict future events on the basis of previous behaviours of elements during successive transitions [3]. FSM is an abstract machine which converts input symbols into sequence of output results depending on some finite set of states and finite state of transition rules. With this in mind, the Semantic Grid, which is an extension of the Grid that gives well-defined meaning to processes to ensure reuse in subsequent similar applications, is a good tool to capture optimisation processes for future use within evolutionary computing domains. Other evolutionary algorithms such as GA and evolutionary strategies work in similar ways that may need semantic Grid as knowledge based framework for intelligent application of evolutionary computing used in various real world problems and experimental simulations of random and optimised stimulus sequences over a range of design parameters [38]. This chapter will take a look at intelligent Grid environments and later discuses service-oriented and problem solving environments as they relate to evolutionary computing applications. Figure 10.1 is a simple diagram of GA process.

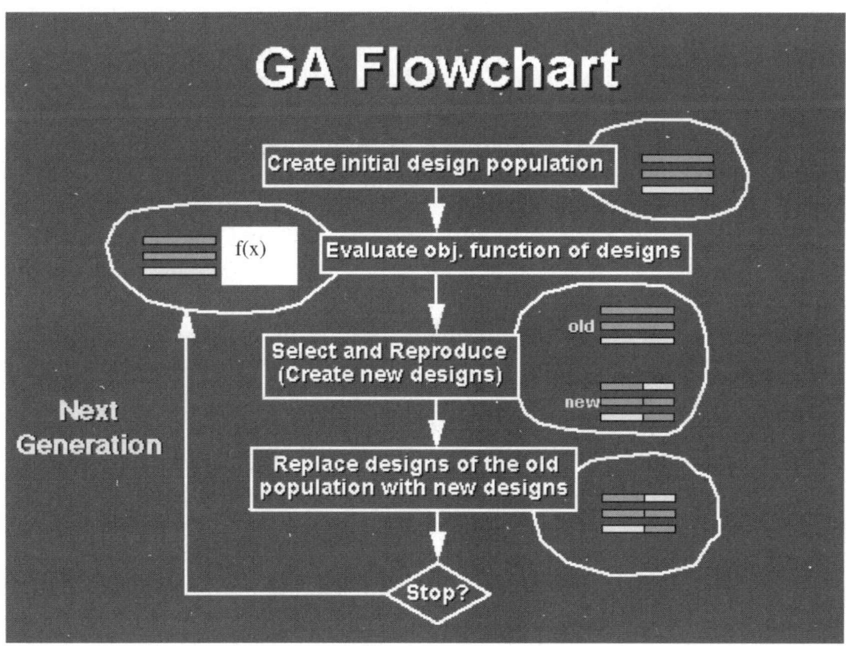

Fig. 10.1. Genetic Algorithm (GA) Flowchart

10.3 Intelligent Grid Environment

Evolutionary computing algorithms produce explosion of solutions in each iteration which require intelligent analysis to select the optimum solution. Again, it is important to capture optimisation processes and reuse them for new optimisation tasks or to serve as a guide to new optimisation engineers. The use of ontology in semantic and knowledge Grids is important in creating an intelligent problem solving environment for evolutionary computing applications. Ontology is the structured representation of the concepts, functions and relationships of the building blocks of a special field of knowledge using high-level syntax common to the users of that field of knowledge.

10.3.1 Semantic and Knowledge Grids

Research activities usually involve multidisciplinary experts working together for collaboration. Dynamic coordination and resource sharing is very important in this arrangement. Also of great importance is how each expert will understand what others are doing and how this affects the overall aim of the group. However, dynamic coordination and resource scheduling is a big challenge coupled with knowledge representation issues that can be understood by all cooperating experts. Semantic Grid aims at capturing the processes of activities of domain experts and publishing it for reuse by others in an intuitive manner. Knowledge Grid on provides management services for processes produced from semantic Grid using Ontologies through rich information service protocols [32]. Knowledge Grid enables knowledge reuse through description, discovery and access to resources using protocols such as Grid Services Description Language (GSDL) and Universal Description, Discovery and Integration (UDDI) services.

Evolutionary computing creates population of solutions with repeated iterations. Thus, more data is produced which requires intelligent method of analysis to select most suitable optimum solution using multiple trade-offs. This process of obtaining the optimum solution in the presence of multiple trade-offs is better achieved through semantic Grid which uses ontology to explicitly specify conceptualisation where definitions associate concepts, functions, context, taxonomies and relationships with high-level (human-readable) text as well as low-level (machine-readable) axioms [20]. The high-level text serves as an intuitive inference point and knowledge reuse for collaborators can be achieved while the low-level is important for communication among computational resources as well as among the collaborators. The layered architecture of the Grid provides a convenient way of representing human-readable optimisation rules and processes and machine-readable axioms on different layers. For example, to present a transparent view to users, semantic and knowledge-based rules should fall within the application level [33]. [33] demonstrated this way of using intelligent systems in their N2Grid project which uses neural networks to exchange information and exploit available computational resources using a 3-layered architecture. The layers are knowledge layer which uses 2-dimensional information for analysis and mining of data to provide problem solving mechanisms, information layer which uses 1-dimensional information for semantic interpretation and data layer which provides data administration.

Intelligent knowledge-based layered architecture for the Grid is the next-generation of future Grid implementation using service-oriented models [6]. Service-oriented architectures will be discussed later. Additionally, the semantic Grid seeks to incorporate its predecessor, the semantic Web, which is an extension of the Web in which information is given well-defined meaning [5]. Just like the semantic Web, the semantic Grid needs an expressive and extensible way of describing Grid functionalities at all levels using flexible mechanisms to explore trade-offs in Grid's complex decision space that incorporates heuristics and constraints into optimisation and computational processes of evolutionary computing techniques [15].

10.3.2 Autonomic Computing

Grid computing is a global infrastructure for coordinated and dynamic platform for large-scale science and business activities [13]. This global infrastructure will grow at a rate not imaginable with many heterogeneous interacting hardware, software, services and people. In this case, managing the infrastructure may become almost impossible with conventional system management and scheduling algorithms. Autonomic computing is a concept to address this complexity in managing hundreds of thousands of heterogeneous Grid components with each accessing and using other services and resources as well as others using it as their resource or service in a symbiotic set up [27]. The ultimate goal is to build a Grid resource management system within every Grid service that can automatically manage itself and adapt to its changing environment even in the presence of errors or other harsh conditions through autonomic configurations and reconfigurations. An agent-based or master-worker paradigm is an innovative way to view the autonomic framework [16]. Arguably, security issues concerning the protection of each service provider in the Grid may hinder this autonomic goal of resource management from a global point of view. The most innovative way to overcome this is through the single-sign on proxy protocol found in middleware service such as Globus. This will allow resources for evolutionary computation to be used from multiple administrative domains among multiple workloads as the middleware is called upon to authenticate and authorise the integration of services through some adaptive autonomic concepts [22]. [36] demonstrated that autonomic computing can allow resource providers to collaborate while maintaining autonomy over their individual codes. [36] used genetic algorithm optimisation logic as Grid service at Southampton University in UK to drive the design search process while the analysis code was located in Singapore to evaluate the objective functions. The synergistic effect of computational powers at both locations also helped in speeding up the identification of an optimum minima using evolutionary algorithm such as the GA. Dynamic composition of these resources is also very important to synthesise available heterogeneous resources and services based on conflicting objectives and constraints during optimisation processes [2].

10.4 Grid-Based Architectures

The layered architecture of the Grid provides a convenient way for service providers to publish services based on high-level (human-dependent) or low-level (machine dependent) requirements regardless of the middleware used. An extension to the conventional architecture is the service-oriented architecture (SOA). SOA together with Grid-enabled problem solving environments (PSEs) provide a collaborative framework for virtual organisations to work together and have seamless access to resources and services dynamically in a coordinated manner. SOA will allow programmers to provide different EC codes as services that can be used by design engineers for optimisation processes regardless of location. An evolutionary algorithm may work well for a particular optimisation problem and performs poorly for another problem [10]. In this case, SOA provides a rich pool of different evolutionary algorithms as services. Autonomic computing helps to automate the selection process of which algorithm is best suited for a particular optimisation problem based on the objective functions and constraints. PSEs offer platform for easy access to EC tools and services for computation, data, collaboration and resource management. This is important because EC techniques are meant to solve a

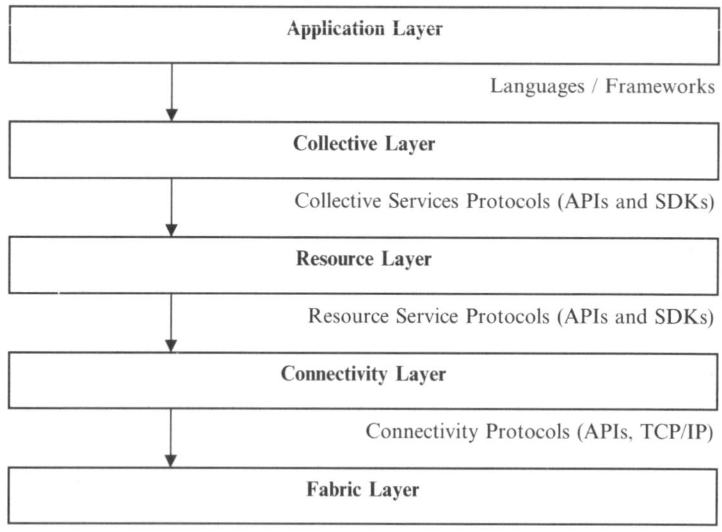

Fig. 10.2. Grid Computing Architecture [1]

variety of optimisation problems and this gives many users who need different optimisation algorithms can seamlessly have access to the services.

10.4.1 Service-Oriented Architecture

To understand service-oriented architectures (SOAs), it is important to trace the history of the electricity power Grid from which the computational Grid derived its name. Before 1910, there was electricity, but the supply and usage was in crude forms. To use electricity power whether for domestic or industrial purpose, one needed to construct, own and maintain a power generating plant. With this form of electricity generation, there was not enough power to drive heavy machineries and this hampered the invention and production of many devices that needed electricity power to operate and function. Intermittent power failures and the cost of constructing and maintaining the generating plant added to this problem. This trend changed after 1910 when the electricity power Grid was invented. With the electricity power Grid, users do not need to construct, own and maintain a generating plant. All users need to do is to subscribe for the utility (electricity power service) and 'plug and use' the power transmitted from any source (hydro, solar, etc). In this way, electricity became cheaper and more reliable. This gave incentive to scientists and technologists to invent and fabricate many devices that can use electricity. The electricity power sector became a big multi-billion dollar industry and contributed greatly to the growth of the economy. This is because it is now a service-oriented electricity power where generation, maintenance, transmission and distribution are essential components of the utility. This means that electricity power was itself not a problem, but the problem was lack of 'Grid' with its associated distribution and transmission features [13].

In the same way, computational scientists having experienced similar bottlenecks with computational power in applications such as CFD (computational fluid dynamics) and engineering design optimisation, are beginning to look towards the 'Grid' to do the same for them just as it did for the electrical power engineers and scientists in 1910. By construing the computational power Grid, computational resources can be published by service providers and consumed by service requestors (users) who are then charged per usage. This means that with Grid, application users may not need to own software systems in the future but subscribe to them and 'plug and use'. Figure 10.3 shows a simple arrangement of how dynamic Grid coordination takes place between service providers and service requestors.

The above analogy between electricity utility and computational utility may sound good. However, service-oriented computational Grid cannot be as easily achievable as service-oriented electricity Grid. This is because the vision of the computational Grid is ambitious due to heterogeneous and dynamic coordination of hardware, software, sensors, expensive instruments, visualisation systems and people. The charge rate per service may not be as static as in electricity; it will include properties such as quality of service (QoS), time

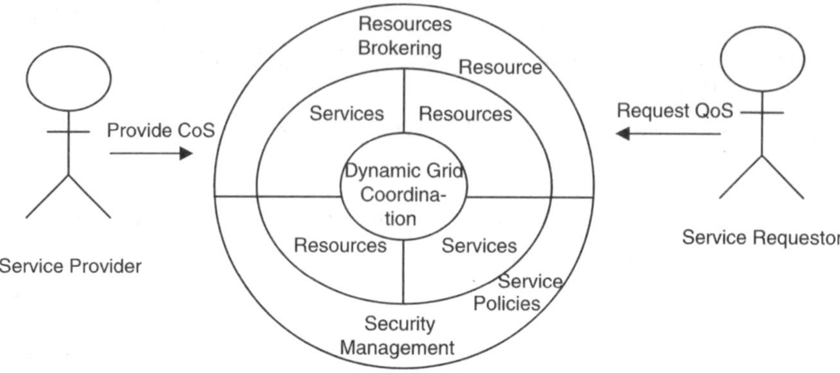

Fig. 10.3. Resource Brokering between Service Providers and Service Requestors. Note: Cost of Service = CoS, Quality of Service = QoS

of services (ToS) and cost of service (CoS) may be dynamic in nature. The middleware issues and security issues involved in sharing knowledge related resources and information are complicated. These issues were not part of electricity power Grid. This means that computational scientists need to do much more than what was done to realise the electricity power Grid. Ownership and deployment of legacy codes such as GAs within SOA involve third generation Grids using OGSA (Open Grid Services Architecture) and WSRF (Web Services Resource Framework) platforms [18]. Grid computing aims to satisfy not only static physically distributed users but even mobile users can access resources dynamically. The DAME (Distributed Aircraft Maintenance Environment) project uses this concept to troubleshoot and maintain aircrafts having engine problems while flying through knowledge-based Grid services [23]. This makes SOA a dynamic interacting for reorganising software applications to adapt to static and mobile application coordination [24]. This location neutrality and technology neutrality between tightly coupled and loosely coupled systems are presenting a big challenge to SOA implementation. Evolutionary computing stands to benefit from this architecture because many different algorithms can be published as services and users can subscribe for the optimisation algorithm that best fits their needs. In addition, one does not need to own the optimisation codes, just subscribes for it and pay per usage. Total cost of ownership (TCO) of evolutionary computing tools which involves development, maintenance and upgrades is drastically reduced. There will be services for evolutionary computations, data management of evolutionary results, middleware for handling interoperability among different objective functions for multi-objective optimisation processes, visualisation services for output, collaboration services where optimisation expertise are distributed across geographical locations and modeling and search services. [26] developed a service-oriented system for design optimisation using genetic algorithm (GA), simulated annealing (SA) algorithm and Tabu Search

(TS) algorithm. The system has data and parametric model located in different sites of the Grid connected by a shared knowledge repository which allows a particular algorithm to be selected at a time.

10.4.2 Problem Solving Environments

To better understand and appreciate the role of Grid-enabled problem solving environments (PSEs) within multidisciplinary context, two scenarios will be described here. The first scenario is in the automobile industry. Consider that an automobile company is to launch a new car into the market soon and has concluded the design phase and is in the process of contacting its suppliers to bid for the components. It is important to note here that no single car manufacturer produces all the components required to manufacture a complete car from engine, wheels and light-fittings to windscreen. Most components are sub-contracted to suppliers. So, each competing supplier is likely to make changes or alterations to the original design sent to it because is not feasible or too costly or offer a better design at a cheaper rate or give other reasons to the original equipment manufacturer (OEM). The OEM will now look at the observations made by the suppliers and take a decision. This process of refining the design will be repeated until both the OEM and its suppliers finally agree. This process is not as straight-forward as described. It takes a lot of documentation, meetings, approvals, decision-making and many others. It is time consuming. Remember the earlier a new product gets to the market the better for the manufacturer because there other competitors who are equally working hard to get a better share of the same product in the market. Imagine that the OEM has 100 suppliers and these suppliers are scattered in different geographical locations around the world. This means spending a lot of time attending to suppliers. It is in this light that PSEs can help bring together both the OEM and its suppliers together within a Grid-enabled virtual organisation (VO) to work together in cooperation. This type of Grid platform is called Extraprise Grid [1].

One may ask the question why Grid since there are PDM (product data management) systems and collaboration packages around. The need for Grid-enabled PSE in this case is because it requires computational steering (making changes on one part of the design dynamically) and observing the effects on the whole designs using Grid visualisation systems. Every supplier can see the effect of changing a parameter or requirement and each supplier will better appreciate the goal of the manufacturer. When all stakeholders in the business have a common goal, the chance of rolling-out a good product is enhanced. Besides, the PSE provides easy access to both the OEM and suppliers with design optimisation codes and other computational services through Grid portals to allow seamless access regardless of geographical location. The computation of the different objective functions are carried out at different supplier's sites, making it faster. Different evolutionary algorithms are also used for the different components and the results shared. Figure 10.4 shows a hypothetical

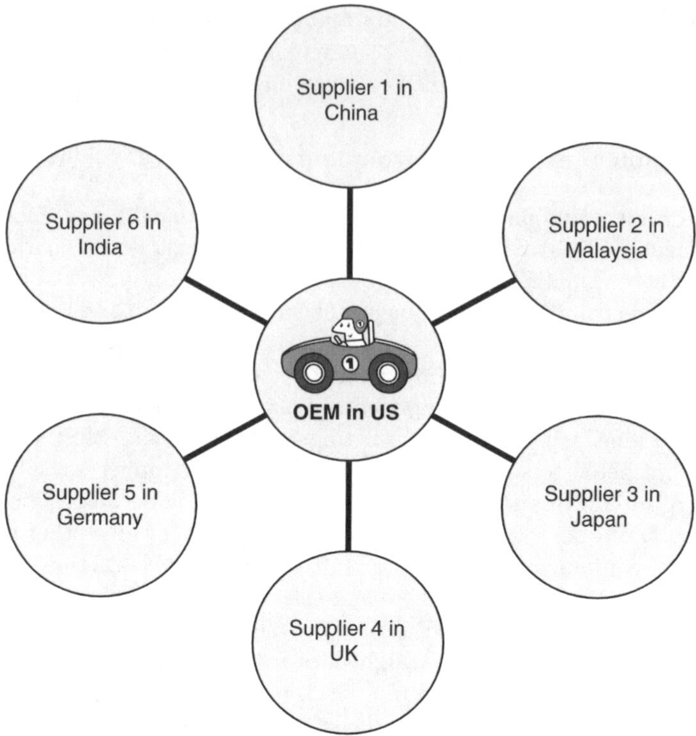

Fig. 10.4. A Hypothetical Car Manufacturer (OEM) with many Suppliers of its Car Parts Located around the World

car manufacturer located in the U.S. with several of its car parts supplied by its partners distributed around the world. In this arrangement, Grid platform is a good environment for coordinated collaboration.

Using the OEM/supplier scenario, evolutionary computation for different objective functions can be carried out at different geographical locations (suppliers) and the results converged at the centre of decision making (OEM). In this way, the computation will be faster than using conventional methods. The Grid middleware provides dynamic coordination and interoperability for the heterogeneous objective functions in case of multi-objective optimisation. In addition, design experts have seamless access to evolutionary algorithms through the Grid portal which enhances their efficiency. There is also reliable security feature which allows legacy codes to be shared, yet ownership and intellectual property are preserved. Decision making for optimal solution is enhanced as different optimisation codes can be used for a particular problem without much delay and results compared. An added advantage in using PSE for EC is the use of Grid visualisation instruments to display solutions in a graphical way. This aid understanding of the fitness functions.

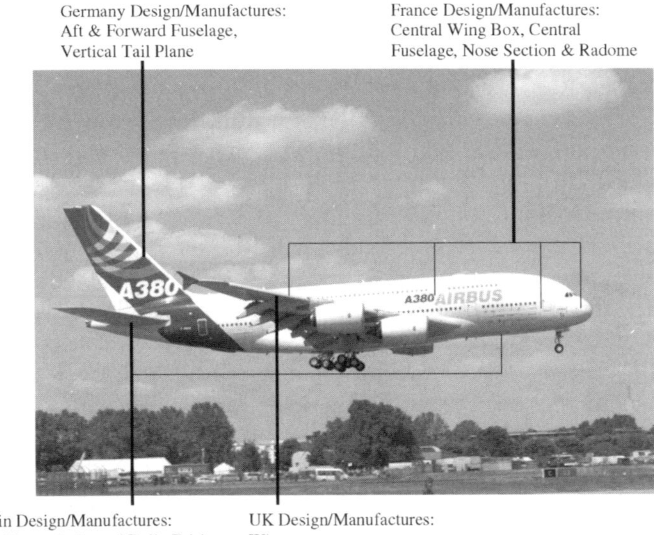

Fig. 10.5. A380 with Different Parts Designed and Manufactured in Different Countries (http://www.bbc.co.uk/southerncounties/community/airbus/parts_330.gif, 10/2/2007)

The second scenario is in aerospace industry. Introducing a new aircraft into the market is a complex task that involves design experts who are hard to gather in one location. For example, Airbus manufactures different parts of its aircrafts in different European countries (Figure 10.5). The wings are manufactured in UK, the forward fuselage and vertical tail in Germany, the centre wing box and centre fuselage in France and the tail planes and central belly in Spain. To integrate these different designs of the different parts of the aircraft requires constant collaboration. Additionally, it requires huge computational power and knowledge sharing. The Grid offers a suitable environment for these design experts to share computational and optimisation algorithms to come up with a good product within the shortest possible time. Semantic and knowledge Grids can play a great role in capturing and reusing the design processes. Visualisation Grid which provides details for all experts involved in the project can aid in decision making and similar to the case of the automobile scenario, make the experts focus on achieving the same goal. The Grid middleware can overcome the heterogeneity of the different hardware and software used by the different experts. This Grid framework which allows experts to collaborate and share resources and services within the same company with different branches located in different countries is called Enterprise Grid [1].

This aerospace scenario shows that the computations involved in the design of the different parts are distributed globally. Grid computing provides a platform upon which evolutionary computations can be subdivided and performed at the different nodes. Different algorithms can be used for the different aircraft parts for optimisation processes and be combined to form a hybrid evolutionary algorithm [14]. [14] used Hooke & Jeeves algorithm for local search and GA for evolutionary optimisation of heat processes and by combining the two algorithms (Hooke & Jeeves and GA), a more robust and efficient solution was obtained.

Computer Aided Design (CAD) systems are used for design of engineering parts within Grid-enabled environment. Evolutionary techniques are used as reverse-engineering mechanisms to produce CAD models so that the interactions of the interfaces of the various parts located at different Grid nodes and boundary conditions can be investigated under different constraints [30]. The Grid middleware can accommodate the heterogeneity in the CAD systems as well as the different objective functions used. Genetic algorithms usually guide the search for solutions using fixed fitness function but revisions to the criteria for the optimal solution involve manually modifying the fitness function [21]. Autonomic Grid computing can provide an agent-based evolutionary techniques that automatically changes the fitness function as the problem space and solution co-evolve in a dynamic manner [21].

Problem solving environments can be applied to all engineering applications because most other engineering disciplines require a range of experts. Therefore, the two scenarios are also applicable to many engineering products. It can be seen that PSEs integrate services, computational resources and people both within and outside organisations through Enterprise and Extraprise Grids. PSEs support remote problem definition that leads to the selection and application of appropriate engineering design search, exploration and optimisation techniques using evolutionary algorithms such as GA and genetic programming (GP) for the optimisation of multidisciplinary engineering processes [26]. The need for workflow management within this cross-domain and dynamic service demand and consumption calls for fuzzy timing techniques as part of the scheduling algorithms in PSEs [7]. Such workflow management strategies become more important when dealing with application areas that require high levels of precision in which case semantic modelling using Ontologies within the PSE is a good idea [4]. A combination of soft computing techniques such as neural networks and fuzzy logic termed as Neuro-Fuzzy gives an intelligent computational and optimisation feature to problem solving environments for complex design optimisation tasks [17]. It is important to note that different PSEs are designed for different optimisation applications because designers generally use approximation techniques to solve problems in order to reduce the otherwise enormous computational effort involved in each particular optimisation strategy [24]. In this regard specialised PSEs could solve either single-objective or multi-objective problems using different options of EC in each case.

Structural design based on parallel finite element analysis in engineering designs is usually parallelised for easy computation. Grid-enabled asynchronous genetic algorithm is parallelised to perform optimisation tasks at different nodes [25]. [25] used Grid problem solving environment to parallelise evolutionary computing techniques parallel algorithms that serve as master-slave for the optimisation of simulation based designs to get better performance than conventional computing facilities offer. Grid computing focuses on large-scale resource allocation though its open service system. [9] used Grid-enabled platform to demonstrate how evolutionary computing resources can be allocated for global optimisation and search tasks. This system uses GA resources to maximise Grid resource allocation and reliability through dynamic security facility. Example of application areas that use Grid-enabled evolutionary computing techniques are engineering design optimisation, particle physics, bioinformatics, pharmaceutical and oil and gas.

10.4.3 Computational Steering and Visualisation Grid

Evolutionary computing techniques for multidisciplinary design optimisation which involves multi-objective functions may require the collaboration of distributed experts located in different geographical locations around the world. Computational steering allows each expert to make changes to particular parameters in the course of the optimisation to observe interesting patterns in the optimal solutions. Visualisation Grid located at the distributed nodes enhances the understanding and decision making during this process. Visualisation Grid can display the graphical representation of the fitness functions and convergence points during optimisation. Interrupting optimisation process for GA to observe patterns in the solutions using EC algorithms can help experts to pinpoint which parameter or parameters have great impact on the process. Advanced visualisation capabilities integrated with simulation based EC algorithms and design processes provide not only valuable design insight and design directions but also facilitate collaboration in design [19]. Close designer interaction with the design process supports exploration involving off-line processing and visualisation of initial results which leads to a redefinition of the solution space by dynamic evolutionary computing techniques [8]. [12] developed an online Grid visualisation system for data analysis. This system gives more meaning to the huge data produced by scientific instruments and evolutionary computing techniques. [11] described a multi-objective visualisation tool using Grid environment to represent engineering design information in multi-dimensional views. Commercial companies such as Boeing and Raytheon are using Grid-enabled Virtual Reality Network (VRN) for computational steering and visualisation on a real-time basis to improve their design processes [11]. Example of application areas that use computational steering are engineering design, physical sciences and life sciences.

10.5 Examples of Intelligent Grid-Enabled Problem Solving Environments

There are Grid projects that have developed prototypes to demonstrate the use of optimisation techniques using EC algorithms and ontological intelligent systems. This section will describe briefly the working features of five of these projects. However, there are many more of such projects that have been implemented within the Grid community. The projects described here are Geodise (Grid-Enabled Optimisation Design for Engineering), FIPER (Federated Intelligent Product Environment), SORCER (Service-Oriented Concurrent Environment), DAME (Distributed Aircraft Maintenance Environment) and Globus Toolkit (GT).

10.5.1 Geodise Project

Geodise is one of the numerous UK e-Science projects which aim to provide Grid-based seamless access to intelligent knowledge repositories, state-of-the-art collection of EC optimisation and search tools, analysis codes, distributed computing services and data resources (http://www.geodise.org/) for multi-disciplinary engineering design optimisation. The project was funded by the UK EPSRC (Engineering and Physical Sciences Research Council) for an initial period of 3 years from 2001 to 2004 with a grant value of £2,872, 450 ($5,084,237). The project is based at the University of Southampton in UK with collaborating research partners at Universities of Oxford and Manchester also in UK. Collaborating companies are BAE Systems, Rolls Royce and Microsoft UK Ltd. The basic components of the Geodise Toolkit are application services, collaboration toolkits, data mining and analysis services, data management services, domain ontology and metadata, workflow services and problem solving environments. Geodise is usually used in conjunction with Matlab for its scripting and visualisation features. Matlab also have EC algorithms such as GA for optimisation and search applications. Geodise has proved to be suitable for engineering processes involving computational fluid dynamics (CFD). The use of Geodise tools by the engineer is facilitated by intelligent design advisors targeted initially at CFD [29]. Geodise is an open source software and is freely downloadable at the Geodise website. Figure 10.6 presents a simple Geodise architecture.

10.5.2 FIPER Project

FIPER is a 4-year project sponsored by the National Institute for Standards and Technology-Advanced Technology Program (NIST-ATP) in the U.S. Its aim is to produce an intelligent system that leverages the emerging web technologies in which engineering tools such as CAD (Computer Aided Design), CAE (Computer Aided Engineering), PDM (Product Data Management), EC

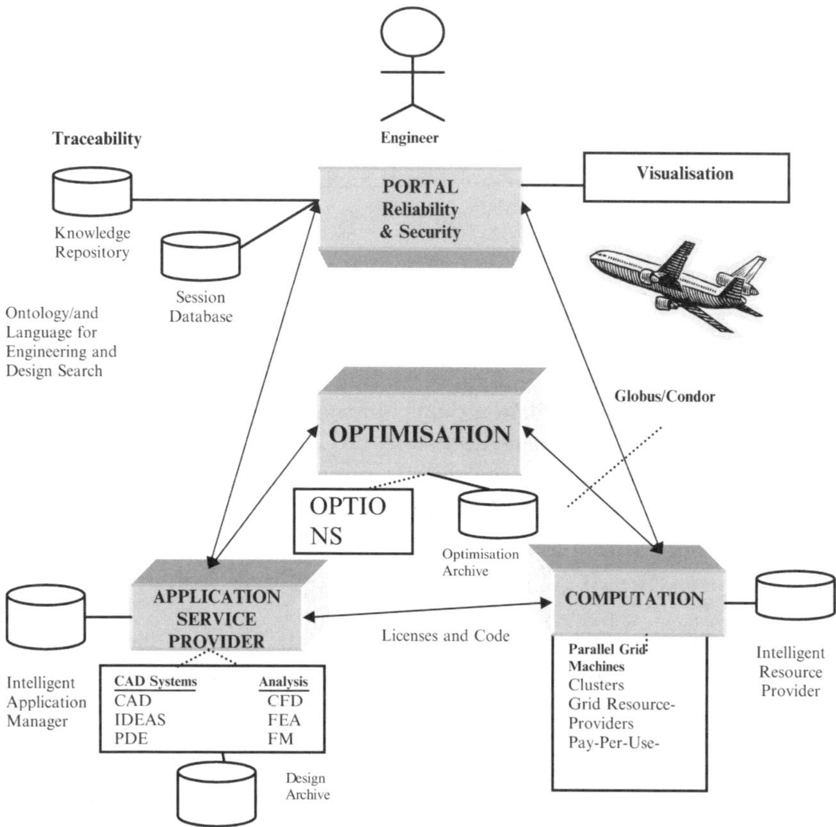

Fig. 10.6. Simple Geodise Architecture

optimisation techniques such as GA, GP and many others act as distributed service providers as well as service requestors communicating through intelligent context models for concurrent engineering design optimisation of FIPER has a grant worth of $21.5 million with Ohio and Stanford Universities as academic partners and GE (General Electric) teaming with Engineous Software, BFGoodrich, Parker Hannifin and Ohio Aerospace Institute as company partners [31]. FIPER uses GE Aircraft Engine functionalities to capture the designer's knowledge in Knowledge Based Engineering (KBE) systems to create Intelligent Master Model (IMM). This IMM contains the 'what', the 'why' and the 'how' of a design using EC algorithms to produce a range of global optimal solutions [31]. FIPER provides a graphical environment that permits interactive click-and-drag Grid programming interface which users can reuse and execute concurrently [35].

10.5.3 SORCER Project

SORCER is an extension of the work of FIPER project. SOCER lab is based at the Texas Tech University; U.S.A. The goal of SORCER is to form Grids of distributed services that provide engineering data, applications, tools and EC optimisation algorithms for concurrent engineering design disciplines [37].

10.5.4 DAME Project

DAME is also a UK project which provides an intelligent Grid demonstrator system for health monitoring and fault trouble-shooting in mobile aircrafts. The need for mobile devices to have secured and seamless access to Grid services while actively working on the field is important in both scientific and business disciplines [23]. In this framework, EC algorithms are used to select optimal diagnostic solutions based on certain constraints and parameters using multiple objective functions for the optimisation.

10.5.5 Globus Tookit

Globus is an open source toolkit considered as the de facto middleware for large-scale computing services. The Globus toolkit is funded by various organisations such as the Defense Advanced Research Projects Agency (DARPA) USA, National Science Foundation (NSF) USA, Globus Alliance and the UK e-Science program. The I-WAY (Information Wide Area Year) projects which succeeded in linking many supercomputing centres was the first initiative that led to the development of the Globus toolkit in 1995. Globus toolkit consists of many services and protocols such as Globus Security Infrastructure (GSI), Globus Information Services (GIS) and Globus Resource Allocation and Management (GRAM). Globus has evolved to the current version 4.0 with many additions to the protocols DAI (Data Access and Integration) developed by the UK e-Science Centre and the standard 'plumbing' features of the OGSA (Open Grid Services Architecture) which provides standard implementation interface for participants (service providers and service requestors) in the Grid so as to ease and facilitate interoperability among heterogeneous resources and services. To join in the Grid, local users must follow the standard protocols to reduce customisation efforts. This philosophy encourages autonomic Grid resource management. This autonomic characteristic of the Globus toolkit is where the use of EC algorithms is useful for dynamic adjustment and adaptation strategies for resource management. Adaptation is the key feature that will make the complex Grid composition a success. This will help software agents to dynamically change states to suit their environments. Multi-objective optimisation will play a great role in scheduling resources and services for requestors through efficient assignment of the resource to the right users based on inference rules within evolutionary context.

10.6 Collaborative Evolutionary Grid Computing

The original intention of the Grid is to link supercomputing stations for collaboration. The sharing and dynamic coordination of computational resources involves policies that govern the activities of service providers and services requestors. A set of individuals and institutions that define rules about what should be accessed and who should access them is called virtual organisation. These types of Grids which share resources and services for common scientific or business goals are called collaboratories [28]. The field of evolutionary computing needs collaboratories to explore more evolutionary algorithms that can tackle more complex optimisation problems. Grid-enabled collaboratories will enhance distributed and secured access to computational power, data management resources and EC algorithms. Real-time solution for evolutionary computations is guaranteed due to the synergy of distributed computational resources within the virtual organizations that made up a collaboratory.

10.7 Summary and Conclusions

This chapter presented how Grid Computing can enhance the functionalities provided by EC techniques through its distributed and dynamically coordinated computational resources and services. Grid resources include hardware, software, instruments, visualisation systems and evolutionary algorithms. Grid-based architectures such as service-oriented architecture (SOA) and problem solving environments (PSEs) were discussed. Examples of intelligent PSEs such as Geodise, FIPER, SORCE, DAME and Globus were briefly explained within the context of evolutionary optimisation applications.

Grid has become a key platform upon which large-scale computations can be performed in a more coordinated way. Evolutionary computing (EC) is suited for Grid application due to its computational and data intensive nature. EC algorithms produce enormous solutions for complex multi-objective optimisation problems and the Grid, using its distributed nodes, can handle the selection of an optimal solution from the feasible set of solutions. Evolutionary computations of multiple objective functions can be physically carried out in different locations around the world and the results converged at the point of need. Data Grid has capabilities for data access and integration of multiple data produced by multiple objective functions during computations. The heterogeneity in different objective functions and constraints can be overcome using Grid middleware during computations done at different locations. In addition, the middleware allows different software and hardware to interoperate which enhances the use of different software for different objective functions. In addition, collaborative evolutionary computing framework within Grid environment can enhance new discoveries in evolutionary computation. Grid collaboratories can allow different experts in optimization processes located at different places to work together and share their expertise. Optimisation processes using evolutionary algorithms are done under confidential environments by companies due to the competitive nature of the economy and the need to safeguard the intellectual property of the company. Grid security feature which has single sign-on proxy encryption protocols provide secured platform for this multidisciplinary collaboration. Semantic and Knowledge Grids provide intelligent problem solving environments for design engineers who can have easy access to computational codes such as GAs. It is important to note that the use of EC techniques within Grid environment is still at its infancy and much work needs to be done in this area. However, Grid promises to bring revolutionary efficiency in the way EC algorithms are used.

References

1. Abbas, Ahmar. Grid Computing: A Practical Guide to Technology and Applications. USA: Charles River Media, Inc. (2004).
2. Agarwal, Manish and Parashar, Manish. Enabling Autonomic Compositions in Grid Environments. Proceedings of the Fourth International Workshop on Grid Computing (GRID'03). USA: IEEE Computer Society. (2003).
3. Back, Thomas; Hammel, Ulrich, and Schwefel, Hans-Paul. Evolutionary Computation: Comments on the History and Current State. IEEE Transactions on Evolutionary Computation. (1997) 13–17.
4. Cannataro, Mario; Comito, Carmela, and Guzzo, Antonella Veltri Pierangelo. Integrating Ontology and Workflow in PROTEUS, a Grid-Based Problem Solving Environment for Bioinformatics. Proceedings of the International Conference on Information Technology: Coding and Computing (ITCC'04). USA: IEEE Computer Society. (2004).
5. Cannataro, Mario and Talia, Domenico. Semantics and Knowledge Grids: Building the Next-Generation Grid. IEEE Intelligent Systems. (2004) 56–63.
6. Cannataro, Mario and Talia, Domenico. Towards the Next-Generation Grid: A Pervasive Environment for Knowledge-Based Computing. Proceedings of the International Conference on Information Technology: Computers and Communications (ITCC'03). USA: IEEE Computer Society. (2003).
7. Cao, Junwei; Jarvis, Stephen A., and Saini, Subhash Nudd Graham R. GridFlow: Workflow Management for Grid Computing. Proceedings of the 3rd IEEE/ACM International Symposium on Cluster Computing and the Grid (CCGRID'03). USA: IEEE Computer Society. (2003).
8. Cvetkovic, Dragan and Parmee, Ian. AgentBased Support within an Interactive Evolutionary Design System. Parmee, I. C. Adaptive Computing in Design and Manufacture. 5th ed. London, UK: Springer-Verlag London Limited, (2002): 355–367.
9. Dai, Yuan-Shun and Wang, Xiao-Shun. Optimal Resource Allocation on Grid Systems for Maximizing Service Relaibility using a Genetic Algorithm. Elsevier: Reliability Engineering and System Safety. (2005).
10. Deb, Kalyanmoy. Multi-Objective Optimisation using Evolutionary Algorithms. England, UK: John Wiley & Sons Ltd. (2001).

11. Eddy, John and Lewis, Kemper. Multidimensional Design Visualization in Multiobjective Optimization. 9th AIAA/ISSMO Symposium on Multidisciplinary Analysis and Optimization. USA: AIAA Inc. (2002).
12. Foster, Ian; Insley, Joseph; Laszewski, Gregor von; Kesselman, Carl, and Thiebaux, Marcus. Distance Visualization: Data Exploration on the Grid. IEEE Computer Society. (1999) 26–43. CODEN: 0018-9162/99.
13. Foster, Ian and Kesselman, Carl. The Grid: Blueprint for a New Computer Infrastructure. San Francisco, USA: Morgan Kaufman Publishers Inc. (1999).
14. Fraga, E. S. and Zilinskas, A. Experience with Hybrid Evolutionary/Local Optimization for Process Design. Parmee, I. C. Adaptive Computing in Design and Manufacture. 5th ed. London, UK: Springer-Verlag London Limited. (2002) 53–64.
15. Gil, Yolanda, Deelman, Ewa, Blythe, Jim, Kesselman, Carl and Tangmunarunkit, Hongsuda. Artificial Intelligence and Grids: Workflow Planning and Beyond. IEEE Intelligent Systems. (2004) 26–33.
16. Goux, Jean-Pierre, Kulkarni, Sanjeev, Yoder, Michael and Linderoth, Jeff. Master-Worker: An Enabling Framework for Applications on the Computational Grid. Kluwer Academic Publishers: Cluster Computing. (2001) 463–70.
17. Joshi, Anupam, Weerawarana, Sanjiva, Ramakrishnan, Narendran, Houstis, Elias N. and Rice, John R. Neuro-Fuzzy Support for Problem-Solving Environments: A Step Toward Automated Solution of PDEs. IEEE Computational Science & Engineering: Neural Networks Problem Solving Environments. (1996) 44–56.
18. Kacsuk, P., Goyeneche, A., Delaitre, T.; Kiss, T., Farkas, Z. and Bocko, T. High-Level Grid Application Environment to Use Legacy Codes as OGSA Grid Services. Proceedings of the Fifth IEEE/ACM International Workshop on Grid Computing (GRID'04). USA: IEEE Computer Society. (2004).
19. Kodiyalam, Srinivas; Yang, R.J., and Gu, Lei. High Performance Computing & Rapid Visualization for Design Steering in MDO. 44th AIAA/ASME/ASCE/AHS Structures, Structural Dynamics, and Materials Conference. USA : AIAA Inc. (2003).
20. Li, Yuhua and Lu, Zhengding. Ontology-Based Universal Knowledge Grid: Enabling Knowledge Discovery and Integration on the Grid. Proceedings of the 2004 IEEE International Conference on Services Computing (SCC′04). USA: IEEE Computer Society. (2004).
21. Masher, M. L. and Garza, A. Gomez de Silva. Adapting Problem Specifications and Design Solutions Using Co-evolution. Parmee, I. C. Adaptive Computing in Design and Manufacture. 5th ed. London, UK: Springer-Verlag London Limited. (2002) 257–271.
22. Naik, Vijay K.; Sivasubramanian, Swaminathan, and Krishnan, Sriram. Adaptive Resource Sharing in a Web Services Environment. IFIP International Federation for Information Processing. (2004) 311–330.
23. Ong, M.; Alkarouri, M.; Ren, X.; Allan, G.; Kadirkamanathan, V.; Thompson, H. A., and Fleming, P. J. Grid-Based Decision Support with Pro-Active Mobile Computing. Proceedings of the 2005 IEEE International Conference on Services Computing (SCC'05). USA: IEEE Computer Society. (2005).
24. Papazoglou, Mike P. Service-Oriented Computing: Concepts, Characteristics and Directions. Proceedings of the Fourth International Conference on Web Information Systems Engineering (WISE'03). USA: IEEE Computer Society. (2003).

25. Park, Jin Woo, Park, Si Hyoung, Moon, Ji Joong, Yoon, Youngha and Kim, Seung Jo. High-Fidelity Simulation Based Optimum Design Utilizing Computing Grid Technology. 46th AIAA/ASME/ASCE//AHS/ASC Structures, Structural Dynamics & Materials Conference. USA: AIAA Inc. (2005).
26. Parmee, I. C., Abraham, J., Shackelford, M., Rana, O. F. and Shaikhali, A. Towards Autonomous Evolutionary Design Systems via Grid-Based Technologies. Proceedings of ASCE 2005 International Conference on Computing in Civil Engineering (IPDPS'03). ASCE. (2005).
27. Pattnaik, Pratap, Ekanadham, Kattamuri and Jann, Joefon. Autonomic Computing and Grid. Berman, Fran, Fox, Geofrey C. and Hey, Anthony J. G. Grid Computing: Making the Global Infrastructure a Reality. England, UK: Jonh Wiley & Sons Ltd. (2003) 351–384.
28. Pearl, Laura, Welch, Von, Foster, Ian, Carl, Kesselman and Tuecke, Steven. A Community Authorization Service for Group Collaboration. Proceedings of the Third International Workshop on Policies for Distributed Systems and Networks (POLICY'02). USA: IEEE Computer Society. (2002).
29. Pound, G.E., Eres, M.H., Wason, J.L., Jiao, Z., Keane, A.J. and Cox, S.J. A Grid-Enabled Problem Solving Environment (PSE) for Design Optimisation within Matlab. Proceedings of the International Parallel and Distributed Processing Symposium (IPDPS'03). USA: IEEE Computer Society. (2003).
30. Robertson, C. and Fisher, R. B. Better Surface Interactions by Constrained Evolution. Parmee, I. C. Adaptive Computing in Design and Manufacture. 5th ed. London, UK: Springer-Verlag London Limited. (2002) 133–142.
31. Rohl, Peter J.; Kolonay, Raymond M.; Irani, Rohinton K.; Sobolewski, Michael; Kao, Kevin, and Bailey, Michael W. A Federated Intelligent Product Environment. AIAA. (2000).
32. Roure, David De; Jennings, Nicholas R., and Shadbolt, Nigel R. The Semantic Grid: A Future e-Science Infrastructure. Berman, Fran//Fox, Geofrey C.//Hey, Anthony J. G. Grid Computing: Making the Global Infrastructure a Reality. England, UK: John Wiley & Sons Ltd. (2003) 437–470.
33. Schikuta, Erich and Weishaupl, Thomas. N2Grid: Neural Networks in the Grid. IEEE. (2004) 1409–1414.
34. Scurr, A. D. and Keane, A. J. The Development of a Grid-Based Engineering Design Problem Solving Environment. Parmee, I. C. Adaptive Computing in Design and Manufacture. 5th ed. England, UK: Springer-Verlag London Limited. (2002) 65–73.
35. Sobolewski, Michael and Kolonay, Raymond M. Federated Grid Computing with Interactive Service-Oriented Programming. Concuurent Engineering: Research and Applications. (2006) 1455–66.
36. Song, Wenbin; Ong, Yew Soon; Ng, Hee Khiang; Keane, Andy; Cox, Simon, and Lee, Bu Sung. A Service-Oriented Approach for Aerodynamic Shape Optimisation Across Institutional Boudaries. 2004 8th International Conference on Control, Automation, Robotics and Vision. USA: IEEE. (2004).
37. Soorianayanan, Sekhar and Sobolewski, Michael. Monitoring Federated Services in CE Grids. Sobolewski, Michael//Cha, Jianzhong. Concurrent Engineering: The Worldwide Engineering Grid. China: Tsinghua University Press and Springer-Verlag. (2004) 89–95.
38. Wager, Tor D. and Nichols, Thomas E. Optimisation of Experimental Design in fMRI: A General Framework using a Genetic Algorithm. IEEE NeuroImage Academic Press. (2003) 18293–309.

11

Application of Evolutionary Game Theory to Wireless Mesh Networks

Athanasios Vasilakos[1] and Markos Anastasopoulos[2]

[1] National Technical University of Athens, GRC 50100, Greece
University of Western Macedonia, Kozani
email: vasilako@ath.forthnet.gr

[2] National Technical University of Athens, GR 15780, Greece
email: mark@telecom.ntua.gr

Summary. Wireless Mesh Networks (WMN) with static nodes, consisting of point-to-point links operating at frequencies above 10GHz are studied in this chapter. The dominant fading mechanism deteriorating the performance of these wireless networks is rain attenuation. Due to spatial-temporal properties of rain fading, routing protocols that usually operate well under clear sky conditions seem to be inefficient. In this chapter, a new routing protocol applying evolutionary game theory is presented. Traffic is being controlled if one considers an infinite number of agents that each is responsible for an infinitesimal load. Their aim is to maximize their individual throughput selfishly without considering the performance of the whole network. The routing strategy of every agent is being revised continuously by sampling another path using input from the physical layer. Finally, the performance of the proposed routing scheme is evaluated through extended numerical simulations for its stability and scalability.

11.1 Introduction

Wireless mesh networking is an emerging technology that uses wireless multi-hop links to provide a cost-efficient way for community or enterprise users to have broadband Internet access and share network resources. Due to its distributed, dynamic, self-configuring, self-optimizing and self-healing architecture, each node in the network is capable of independently adapting its operation based on the current environment according to predetermined algorithms and protocols. In this context, theoretical models for performance evaluation for those kinds of networks are scarce. *Evolutionary Game Theory* (EGT) offers a suite of tools that may be useful in modeling the interaction among independent nodes in a WMN.

One of the most challenging issues in WMNs is the design of robust, scalable and efficient routing protocols [1]. A number of studies related to routing in WMNs, have been recently reported in the literature (see e.g [2–4]), each

A. Vasilakos and M. Anastasopoulos: *Application of Evolutionary Game Theory to Wireless Mesh Networks*, Studies in Computational Intelligence (SCI) **66**, 249–267 (2007)
www.springerlink.com

dealing with routing performance from a different point of view. Particularly, the impact of performance metrics on a routing protocol is studied in [3], where the link quality source routing (LQSR) is proposed on the basis of DSR. LQSR aims at selecting a routing path according to link quality metrics. Three performance metrics, i.e., the expected transmission count (ETX) [2], per-hop RTT, and perhop packet pair are implemented separately in LQSR, while in [4], a new performance metric, called the weighted cumulative expected transmission time (WCETT) is proposed for the routing protocol. WCETT takes into account both link quality metric and the minimum hop-count. Finally, the authors, in [5–7] exploit multi-path routing to perform better load balancing and to provide high fault tolerance.

In this article, emphasis is given in the design of a routing protocol for WMNs operating in frequencies above 10 GHz. The motivation of this study is based on the recently published IEEE 802.16 WirelessMAN [8] and the ETSI BRAN HIPERACCESS [9] air interface standards, where the operation frequency-range 10–66 GHz is suggested. Following this assumption, rain is the dominant fading mechanism and the only atmospheric phenomenon exhibiting significant spatial inhomogeneity within the network area. Due to the fact, that, the aforementioned approaches of multipath routing do not take into account the special characteristics of the access physical propagation medium (spatial-temporal characteristics of rain structure) their performance seems to be in inefficient. In particular, the routing decision is usually taken by a centralized authority based on stale information. Many times that information (sometimes is inaccurate due to time varying channels e.g. rain fading) can lead to undesirable oscillations resulting in poor performance. In this study, traffic is managed by a large number of agents, each responsible for a small amount of traffic (see e.g. [10, 11] for similar approaches). Their aim is to maximize their individual throughput selfishly without considering the performance of the whole network. The routing strategy of every agent is being revised continuously by sampling another path using input from the physical layer. The time intervals at which agents reconsider their strategy are Pareto distributed. The population of agents that uses each strategy is modeled using a certain population dynamics in continuous time- the replicator dynamics- [27] known from Evolutionary Game Theory (EGT).

The remainder of this paper is organized as follows. Section 11.2 presents a brief overview of the preliminaries of Evolutionary Game Theory. An application of EGT in WMNs is discussed in Section 11.3. Then the maximum throughput routing problem and an algorithm that improves convergence is formulated in Section 11.4. Section 11.5 provides an analytical description of the simulation environment along with the numerical results. Finally, several important conclusions are drawn in Section 11.6.

11.2 Background of Evolutionary Game Theory

In a conventional game, the objective of a rational player is to choose a strategy maximizing its payoff. Instead, in the frame of EGT, the game is played continuously by socially conditioned players who are randomly drawn from a large population. In EGT, the players are individual organisms and strategies are heritable phenotypes. Then, EGT deals with the survival of individuals using a variety of alternative strategies.

In general, an evolutionary process combines two significant mechanisms: a *mutation mechanism* that provides variety and a *selection mechanism* that favours some varieties over others. The role of mutation is highlighted by the idea of an Evolutionary Stable Strategy (ESS), while the role of selection by the concept of replicator dynamics. The replicator dynamic assumes that a subpopulation grows when it plays strategies that are better than the average one, whereas a subpopulation declines whenever it employs strategies worse than average. The idea of ESS is to let the equilibrium to intercept the invaders. Note that ESS is actually a refinement of the NE.

11.2.1 Evolutionary Stable Strategies

ESS is a key concept in EGT. A population following such a play strategy is invincible. Specifically, suppose that the initial population is programmed to play a certain pure or mixed strategy profile x. Now let in a small population share ϵ of individuals, where $\epsilon \in (0,1)$, to play some other pure or mixed strategy y. Hence, if an individual is drawn to play the game, then the probability that the opponent plays the mutant strategy y is ϵ, and the probability that the opponent plays the incumbent strategy x is $1-\epsilon$. Under these assumptions, the payoff of such a match is the same as that in the match where an individual plays the mixed strategy $w = \epsilon y + (1-\epsilon)x$. The corresponding payoff to the incumbent and mixed strategy is $u(x,w)$ and $u(y,w)$, respectively. A strategy x is said to be evolutionary stable if and only if

$$u[x, \epsilon y + (1-\epsilon)x] > u[y, \epsilon y + (1-\epsilon)x] \tag{11.1}$$

for sufficiently small positive ϵ.

Definition 1. *A strategy x is said to be evolutionary stable if for every strategy $y \neq x$ there exists some $\overline{\epsilon}_j \in (0,1)$ such that inequality (11.1) holds for all $\epsilon \in (0, \overline{\epsilon}_j)$*

There is also a weaker notion of evolutionary stability, called neutral stability. Instead of requiring that all mutants earn less than the incumbent strategy, neutral stability requires that no mutant prospers in the sense of earning a higher payoff than the incumbent strategy.

Definition 2. *A strategy x is said to be neutral stable if for every strategy $y \neq x$ there exists some $\bar{\epsilon}_j \in (0,1)$ such that inequality*

$$u[x, \epsilon y + (1-\epsilon)x] \geq u[y, \epsilon y + (1-\epsilon)x] \tag{11.2}$$

holds for all $\epsilon \in (0, \bar{\epsilon}_j)$

Using linearity of expected utility in probabilities, equation (11.1) can be written as follows

$$(1-\epsilon)u(x,x) + \epsilon u(x,y) > (1-\epsilon)u(y,x) + \epsilon u(y,y) \tag{11.3}$$

For values of ϵ close to zero the above yields either

$$u(x,x) > u(y,x) \tag{11.4}$$

or

$$u(x,x) = u(y,x) \quad \text{and} \quad u(x,y) > u(y,y) \tag{11.5}$$

Hence, it is obvious that an ESS must be a NE, otherwise inequality (11.4) does not hold. Furthermore, any strict NE is an ESS. The latter holds since strict NE satisfy (11.4) for all other strategies.

11.2.2 The Replicator Dynamics

The replicator dynamics specifies how populations shares, associated with different pure strategies, evolve over time. This model was first proposed by Taylor and Jonker [27] and in contrast to evolutionary stability; every individual is programmed to play a pure strategy. In this section this dynamic will be discussed and it will be investigated how its states are related to the set of Nash equilibria.

To define the replicator dynamics we take into consideration a large but finite population of agents who are all programmed to play pure strategies $i \in K$, where K is the set of strategies set. At any time instant t, let $\lambda_i(t) \geq 0$ be the number of individual who are programmed to pure strategy i. The total population is given by $\lambda(t) = \sum_{i \in K} \lambda_i(t)$. Let $x_i(t) = \lambda_i(t)/\lambda(t)$ be the fraction of players using pure strategy i at time t. The associated *population state* is defined by the vector $x(t) = (x_1(t), x_2(t), ..., x_k(t))$. Then the expected payoff to using pure strategy i, given the population is in state x is $u(e^i, x)$ and the *population average payoff*, that is the payoff to an individual which is drawn randomly from the population, is $u(x,x) = \sum_{i=1}^{k} x_i u(e^i, x)$. Suppose that payoffs are proportional to the reproduction rate of each individual and, furthermore, that a strategy profile is inherited. This leads to the following dynamics for the population shares x_i.

$$\dot{x_i} = x_i(u(e^i, x) - u(x,x)) \tag{11.6}$$

Equation (11.6) denotes the fact that populations associated with worse than average strategies shrink, whereas those with strategies that are better than average ones grow. However, sometimes even a strictly dominated strategy may earn more than average, hence, it is not clear a priori whether such strategies get wiped out in the replicator dynamics. The following theorem gives the answer to that question [26].

Preposition 1 *If a pure strategy i is strictly dominated then $\xi_i(t, x^0)_{t \to \infty} \to 0$, where $\xi_i(t, x^0)$ is the population at time t and x^0 is the initial state.*

Now, in case a pure strategy i is weakly dominated by some strategy y, and the subpopulation programmed to pure strategy i does not vanish over time, then all those pure strategies j in which y is better than i, vanish from the population.

Preposition 2 *Suppose that a pure strategy i is weakly dominated by some strategy y. if $u(y - e^i, e^j) > 0$ then $\xi_i(t, x^0)_{t \to \infty} \to 0$ or $\xi_j(t, x^0)_{t \to \infty} \to 0$(or both) for any x^0*

Finally, it should be noted that the population shares $x_i > 0$ and $x_j > 0$ increases over time if strategy i earns higher payoff than strategy j. The latter results from the following equation

$$\frac{d}{dt}\left[\frac{x_i}{x_j}\right] = [u(e^i, x) - u(e^j, x)]\frac{x_i}{x_j} \tag{11.7}$$

From the above, it becomes obvious that suboptimal strategies have the trend of increasing their share. However, there is a close connection between steady states of replicator dynamics and NE. Thus, due to the fact that in NE, all strategies being played have the same average payoff, it is deduced that, every NE is a steady state. Unfortunately, steady states need not to be a NE, e.g., any state where all agents use the same pure strategy is a steady state, but, it in not stable

11.2.3 Other Dynamics

In this section a brief overview of few continuous-time pure-strategy selection dynamics arising from adaptation by myopic imitation, particularly Imitation Driven by Dissatisfaction and Imitation of Successful Agents, will be discussed [29].

Imitation Driven by Dissatisfaction

In this model each individual selects at random an agent from the population, according to a uniform probability, and adopts his/her pure strategy. The agents with less successful strategies on average review their strategy at a higher rate than agents with more successful strategies. Then a payoff-monotonic selection dynamic arises. More specifically,

$$r_i(x) = \rho\left[u(e^i, x), x\right]$$

where $r_i(x)$ is the average review rate for an agent attached to pure strategy i and ρ is a strictly decreasing function. Under these assumptions, the population dynamics are given by

$$\dot{x}_i = x_i \left(\sum_{j \in K} x_j \left(\rho[u(e^j, x), x] - \rho[u(e^i, x), x]\right) \right) \tag{11.8}$$

where K is the strategy set.

Imitation of Successful Agents: Model 1

Suppose that an i-strategist agent samples a j-strategist and the observed payoff is $u(e^i, x) + \epsilon$ and $u(e^j, x) + \epsilon'$ respectively, where, ϵ, ϵ' are random variables, denoting the observation error, such that their difference has a continuously differentiable probability distribution function $\phi : R \to [0, 1]$. The reviewing agent switches to the sampled agent's strategy if and only if the following inequality holds
$u(e^i, x) + \epsilon > u(e^j, x) + \epsilon'$. The conditional probability that the agent will switch from the strategy i to j is thus $\phi\left[u(e^j, x) - u(e^i, x)\right]$. Since the probability that the agent will sample strategy j is x_j, the resulting probability distribution over the set of pure strategies is

$$p_i^j(x) = \begin{cases} x_j \phi\left[u(e^j - e^i, x)\right] & if i \neq j \\ 1 - \sum_{j \neq i} x_j \phi\left[u(e^j - e^i, x)\right] & otherwise \end{cases} \tag{11.9}$$

Assuming that all review rates are constantly equal to one (in order to isolate the effect of choice probabilities), the following dynamics is derived

$$\dot{x}_i = x_i \left(\sum_{j \in K} x_j \left(\phi\left[u(u^j, x), x\right] - \phi\left[u(u^i, x), x\right]\right) \right) \tag{11.10}$$

Imitation of Successful Agents: Model 2

Let the weight factor that an agent switches from strategy i to strategy j be $\omega_i[u(e^j, x), x] > 0$, where ω_i is a non decreasing function. The probability that an i-strategist agent will adopt strategy j is given by

$$p_i^j(x) = \frac{\omega_i[u(e^j, x), x] x_j}{\sum_{h \in K} \omega_i[u(e^h, x), x] x_h} \tag{11.11}$$

Combining equation (11.11) with the assumption of unit review rates, results in the following selection dynamics

$$\dot{x}_i = x_i \left[\sum_{j \in K} \frac{\omega_j[u(e^j, x), x]x_j}{\sum_{h \in K} \omega_j[u(e^h, x), x]x_h} - 1 \right] \tag{11.12}$$

11.3 Application to Wireless Mesh Networks

Evolutionary game theory is a mathematical tool that has been applied in many different research fields including biology, sociology, finance and philosophy. In this section, an application of EGT in the design of a routing protocol for WMNs will be presented.

11.3.1 Network Topology

A Wireless Mesh Network (WMN) is modeled as a directed graph *G(N,A)* where N is the set of all nodes and A is the set of all directed links $(i, j), i, j \in N$. In this model, a set of nodes, labeled $n = 1, ..., N$, can send, receive, and relay data across communication links. The topology is supposed to be static and each link (i, j) has a fixed finite capacity X_{ij} bits per sec. It is further assumed that, two nodes i, j are connected if they can transmit packets to each other with a packet loss probability less that a specific threshold. The set of links connected to node i by a link are defined as V_i. Finally the bit error ratio in each link is denoted by BER_{ij}. A typical example of an IEEE 802.16 WMN topology is depicted in Fig. 11.1.

11.3.2 Problem Definition

For the routing of data packets across the network the multicommodity flow model is adopted. Such models are widely used in the literature of network routing and optimization (see, e.g., [15, 16]). A commodity $c \in C$ is defined by a set of source nodes (origin nodes labeled O) and a set of destination nodes (destination nodes labeled D). For each origin-destination pair, there is a traffic demand $x^{(c)}$ to be routed from node $o^{(c)} \in O$ to $d^{(c)} \in D$. In the context of routing of data in a communication network, $x^{(c)}$ is the arrival rate of traffic injected into network at node $o^{(c)}$ and ejected at node $d^{(c)}$. Let $P^{(c)}$ be the set of all loop-free paths between source and destination.

The goal, is to find a positive flow vector $\mathbf{x} = \{x^{(c)}\}$ that routes information via a path, $p_i^{(c)} \in P^{(c)}$, with the maximum throughput, under the flow conservation condition

$$\sum_{j \in V_i} \left(x_{ji}^{(c)} - x_{ij}^{(c)} \right) = \begin{cases} x^{(c)} & if\ i = o^{(c)} \\ -x^{(c)} & if\ i = d^{(c)}, \forall i \in N \\ 0, & otherwise \end{cases} \tag{11.13}$$

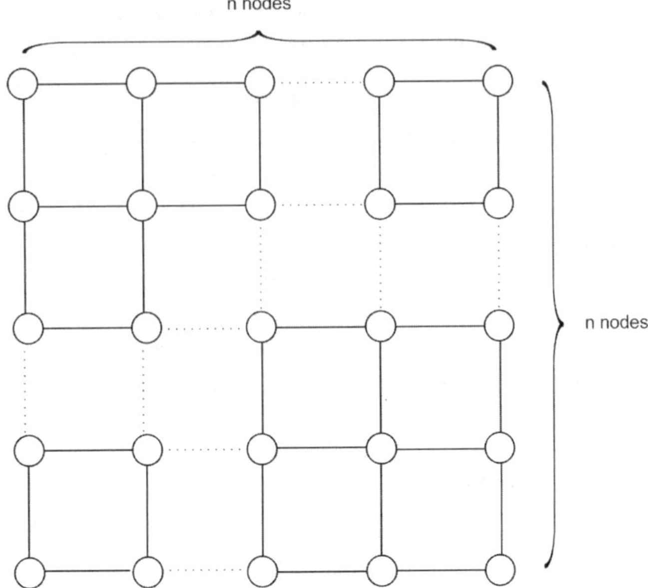

Fig. 11.1. A typical IEEE 802.16 WMN toplology with nxn nodes

and capacity constraints

$$\sum_{c} x_{ij}^{(c)} \leq X_{ij} \tag{11.14}$$

where $x_{ij}^{(c)}$, is the transmission rate of commodity c in link $(i, j) \in A$.

11.3.3 Impact of Physical Layer in Throughput

The WMN considered, is supposed to operate in frequencies above 10 GHz, specifically in Ku, Ka and V bands. However in these frequency bands the main physical layer impairment that degrades the performance of a wireless link is rain attenuation. We notate the rain attenuation induced on microwave path as $A_R(dB)$. For a fixed modulation scheme, coding rate and power level the packet error ratio, PER is an increase function of that could be expressed as

$$PER = f(A_R), f : R^+ \rightarrow [0, 1] \tag{11.15}$$

To ensure reliability in data transfer, TCP protocol is adopted. A measure that indicates how well TCP behaves, over the wireless link, is throughput. Several models exist in the literature for TCP throughput [18–20]. Let RTT denote the average round trip time seen by the connection. From [18], in terms of packets/sec, TCP throughput in link (i, j) can be written as

$$Thrp = min\left(\frac{1}{RTT}\sqrt{\frac{3}{2PER}}, X_{ij}\right) \tag{11.16}$$

From the aforementioned, assuming high values for X_{ij}, maximum throughput is gained if the transmitted information is routed via the most reliable path. That procedure is quoted in what follows.

11.4 Problem Formulation using Evolutionary Game Theory

11.4.1 Description of the Routing Problem

In the routing problem under consideration, the traffic is controlled by a large population of agents, each responsible for a small portion of data. Let $I = 1, 2, ..., n$ be the set of agents-players. For each player, let $P^{(c)}$ be her (in GT literature players are assumed to have feminine gender [12]) finite pure strategies set of the all available *minpaths*. For any *strategy* profile $p_i^{(c)} \in P^{(c)}$, let $u_i(p_i^{(c)})$ be the associated payoff to player i. The objective of each agent is to choose a path that maximizes her payoff. In this study the payoff corresponds to TCP throughput, given by

$$u_i(p_i^{(c)}) = min\left(\frac{1}{RTT\left(p_i^{(c)}\right)}\sqrt{\frac{3}{2PER\left(p_i^{(c)}\right)}}, cp_i\right) \tag{11.17}$$

where $cp_i = min(X_{ij}), \forall(i,j) \in p_i^{(c)}$. The agents are all programmed to play the same pure *incumbent* strategy profile $p_i^{(c)} \in P^{(c)}$ with payoff function given in (11.17). Every time t, let $x_i^{(c)}(t) \geq 0$ be the number of agents who are currently programmed to pure strategy $p_i^{(c)} \in P^{(c)}$, and are responsible to handle traffic for commodity c. Let $x^{(c)}(t) = \sum_{p_i^{(c)} \in P^{(c)}} x_i^{(c)}(t)$ be the total population. In order the flow conservation constraints to be satisfied, the total population should be equal to the summation of all data injected into the network (see e.g., Eq. 11.13) The associated population state is defined as the vector $y^{(c)}(t) = \left(y_1^{(c)}(t), ..., y_n^{(c)}(t)\right)$, where each component $y_i^{(c)}(t)$ is the population share programmed to pure strategy i at time t given by

$$y_i^{(c)}(t) = \frac{x_i^{(c)}(t)}{x^{(c)}(t)} \tag{11.18}$$

The associated *population average* payoff is

$$u^{(c)}(w, w) = \sum_{i=1}^{n} y_i^{(c)} u(e^i, w) \tag{11.19}$$

where $u(e^i, w)$ denotes the payoff when player $i \in I$ plays pure strategy $p_i^{(c)} \in P^{(c)}$, while all other player play according to another profile w.

Suppose that a small group of mutants appear in population $y_i(t)$. Those mutants are all programmed to play some other strategy $p_j^{(c)} \in P^{(c)}$. Let the share of mutants in the postentry population be ϵ, where $\epsilon \in (0,1)$. The strategy of the mutants will be adopted if and only if [26]

$$u[p_j^{(c)}, \epsilon p_j^{(c)} + (1-\epsilon)p_i^{(c)}] > u[p_i^{(c)}, \epsilon p_j^{(c)} + (1-\epsilon)p_i^{(c)}] \quad (11.20)$$

Finally, the probability that a population $y_i^{(c)}$ using pure strategy $p_i^{(c)}$ attaches to pure strategy $p_j^{(c)}$ is given by the following equation [26]

$$Prob_i^{j(c)}(w) = \begin{cases} y_j^{(c)} \phi\left[u(e^j - e^i, w)\right] & if i \neq j \\ 1 - \sum_{j \neq i} y_j^{(c)} \phi\left[u(e^j - e^i, w)\right] & otherwise \end{cases} \quad (11.21)$$

and the corresponding *population dynamics* can be expressed as follows

$$\dot{y}_i^{(c)} = y_i^{(c)} \left(\sum_{j \in P^{(c)}} y_j^{(c)} \left(\phi\left[u(e^i - e^j, w)\right] - \phi\left[u(e^j - e^i, w)\right] \right) \right) \quad (11.22)$$

where $\phi : R \rightarrow [0,1]$ is a strictly increasing function.

11.4.2 Path Selection Algorithm

In this section an algorithm based on EGT for path selection is firstly here presented. Initially, each agent selects randomly a path from the available minpaths set. A rational behavior for an agent would be in case she notices a better throughput to reconsider her routing strategy, and select another path. If the new path offers lower performance then remains to her previous strategy. The times at which each agent revises her strategy are generated at fixed Poisson rates.

Consider, the topology depicted in Fig. 11.2. In case we are interested in transmitting data from node 1 to node 9, an agent employing uniform

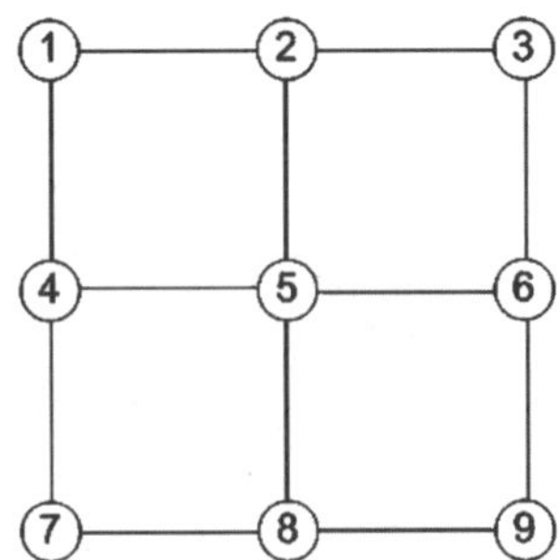

Fig. 11.2. A simple WMN with 9 nodes

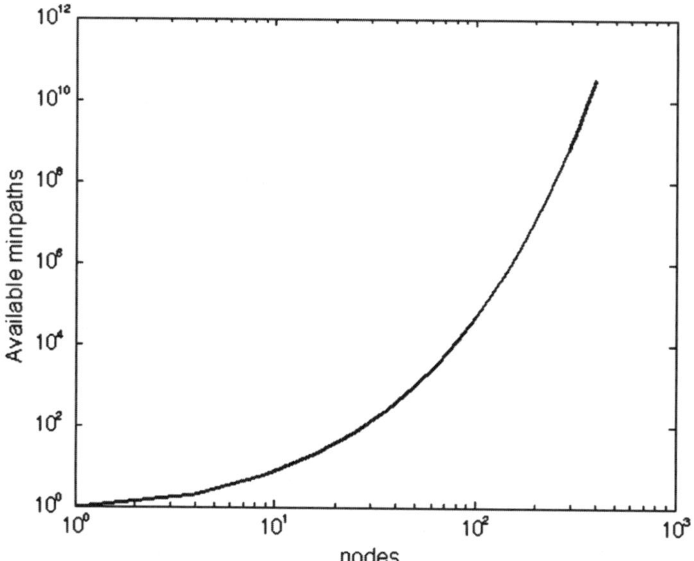

Fig. 11.3. Number of minpaths versus nodes for a nxn WMN

sampling could select one of the six available minpaths (1,2,3,6,9), (1,2,5,6,9), (1,4,5,6,9), (1,4,5,8,9), (1,4,7,8,9).The main problem, however, is that while the number of nodes in the network increases, the total number of the available minpaths increase exponentially. The above is depicted in Fig. 11.3. Hence, it is obvious that the probability an agent to find the maximum throughput path in time is extremely decreased.

To overcome that difficulty an algorithm inspired from cell-cite diversity techniques [21], taking advantage of the spatial inhomogeneity of rain attenuations, is presented here. Diversity schemes are effective techniques to mitigate rain fades for high availability fixed cellular systems operating at Ku and above frequency bands. Their performance is mainly affected by geometrical separation parameters, i.e., the separation distance between stations and the angular separation between paths, respectively.

Using the idea of route diversity, a new metric called P*ath Correlation Coefficient* (PCC) is proposed. PCC_{ij} is a factor indicating the Euclidean distance among paths $p_i^{(c)}$, $p_j^{(c)}$.

Let n be the size of *minpath* $p_i^{(c)}$, and z_i be a $2 \times n$ matrix with each column representing the coordinates of nodes belonging in $p_i^{(c)}$. For paths $p_i^{(c)}$ and $p_j^{(c)}$, PCC_{ij} is defined as

$$PCC_{ij} = \begin{cases} exp\left(-\sum_{k=1}^{n} |z_i(k) - z_j(k)|\right), \; if \; \sum_{k=1}^{n} (z_i(k) - z_j(k)) > 0 \\ -exp\left(-\sum_{k=1}^{n} |z_i(k) - z_j(k)|\right), \; otherwise \end{cases} \tag{11.23}$$

$|PCC_{ij}|$ take values close to one (or close to zero) in case paths $p_i^{(c)}$, $p_j^{(c)}$ are nearby (or distant). Therefore, if paths $p_i^{(c)}$, $p_j^{(c)}$ have small $|PCC_{ij}|$ the probability that both paths suffer from the same tension of rain fades simultaneously is low. Each agent performs the following actions in steps:

1. Initially she selects at random a path $p_i^{(c)} \in P^{(c)}$, where $P^{(c)}$ represents the minpath set.
2. For every $p_j^{(c)} \in P^{(c)}$, the PCC_{ij} values are calculated. To reduce the computational load, these values may be computed a priori and stored at the network nodes.
3. a. An agent selects a new strategy based on the subset of PCC values that are positive and close to zero. If the new strategy offers a better payoff it is adopted.
   b. On the contrary, a strategy from the subset of PCC values that are negative and close to zero is selected. The agent switches to the latter strategy in case it achieves a better payoff.
   c. If neither of the conditions of steps 3a, 3b are satisfied, the agent retains its previous strategy.
4. Having determined a high throughput path, the agents select at random neighboring to that paths (paths with close PCCs). Due to spatial correlation of the physical medium, adjacent paths have similar performance metrics and load could be severed in order to achieve load balancing.
5. As far as the instants following the Poisson distribution, the reader may refer to step 3.

Based on Eq. (11.22), traffic in each path in discrete time is given by

$$y_i^{(c)}(t+1) = y_i^{(c)}(t)\left(\sum_{j\in P^{(c)}} y_j^{(c)}(t)\left(\phi\left[u(e^i - e^j, w)\right] - \phi\left[u(e^j - e^i, w)\right]\right)\right) \tag{11.24}$$

while at the same time capacity and flow conservation constraints (11.13)–(11.14) should be satisfied. Function ϕ is assumed to follow Boltzmann distribution. The employment of Boltzmann distribution in the current approach, is similar to its use in genetic algorithms, where we would like the majority of agents to use the paths with the best performance but we want a fraction of them to choose paths with almost equal performance. In Fig. 11.4, function ϕ for different values of the parameter T (temperature) is depicted.

11.5 Performance Evaluation

11.5.1 Channel Modeling

The performance of the PCC algorithm was tested using a *Matlab* based simulation. The operational frequency of the system was assumed to be 40 GHz. To model rain attenuation, a dynamic rain rate field has been implemented.

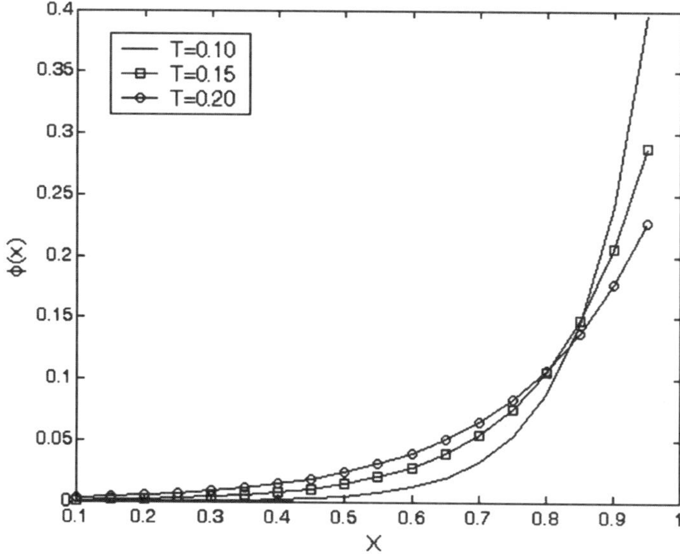

Fig. 11.4. Function ϕ for different values of the parameter T

The physical medium exhibit both spatial and temporal characteristics. For the simulation, the spatial characteristics of rain were simulated using HYCELL, a model for rain fields and rain cells structure developed at ONERA [22,23]. HYCELL is used to produce two-dimensional rain rate fields, $R(x, y)$, over an area corresponding to the size of a typical WMN. So, $R(x, y)$ denotes the rainfall rate at a point (x, y). These rain fields follow the properties of the local climatic conditions. Moreover, the temporal characteristics of the wireless channel, $R(t)$, were simulated using the methodology described in [24] where, $R(t)$ denotes the rainfall rate at time t.

Having implemented the appropriate model for $R(x, y; t)$, the next step is to determine $A_{ij}(t)$, that is, the rain attenuation induced along link (i, j) at time t. This is achieved by integrating the specific rain attenuation $A_0(x, y; t)$ (in dB/km) over the path length L_{ij} of link (i, j) and by taking into account the local properties of the rainfall medium through $R(x, y; t)$.

$$A_{ij}(t) = \int_0^{L_{ij}} A_0(x, y; t)dl \tag{11.25}$$

where

$$A_0(x, y; t) = aR(x, y; t)^b \tag{11.26}$$

In the above, the a, b parameters depend on frequency, elevation angle, incident polarisation, temperature and raindrop size distribution.

11.5.2 Numerical Results

Consider the WMN shown in fig. 11.1. The network consists of N = 25 nodes and A = 40 bidirectional links. The nodes are supposed to be uniformly distributed over a 625 km^2 surface. To keep the analysis tractable, we assume that the system uses BPSK modulation. Therefore the BER along link (i, j) is simply

$$BER_{ij} = 0.5 \exp\left\{-\left(\frac{E_b}{n_0}\right)_j\right\} \tag{11.27}$$

where $(E_b/n_0)_j$ is the bit energy to noise spectral density ratio at the decoder input of node j under clear sky conditions. This ratio depends on the transmitter and receiver power, the antenna gain and the free space loss. Under rainy conditions, A_{ij} [dB] is introduced in the above equation

$$BER_{ij} = 0.5 \exp\left\{-10^{\left[\left(\frac{E_b}{n_0}\right)_j [\mathrm{dB}] - A_{ij}\right]/10}\right\} \tag{11.28}$$

Then, employing a (k, n) = (204, 188) packet FEC (forward error correction) scheme the packet error rate in link (i, j) can be expressed as

$$PER_{ij}(k, n, BER_{ij}) = BER_{ij}\left(1 - \sum_{m=0}^{n-k-1}\binom{n-1}{m} BER_{ij}^m (1 - BER_{ij})^{n-m-1}\right) \tag{11.29}$$

Initially, the effect of the path selection algorithm in the convergence of the routing protocol for different size of topologies is shown in Fig. 11.5. The network is located in area of Athens. We are interested in setting up a flow with data rate 10Mbit/sec. Traffic is controlled by 100000 agents. The parameter T (temperature) is equal to 0.1. Comparing, the proposed algorithm with uniform sampling, we observe that it converges much faster, e.g. about 1100 less iterations are required for a WMN consisting of 100 minpaths.

Then, the impact of climatic characteristics in the convergence of the algorithm is investigated. For that reason, it is assumed that the aforementioned network is established in three different locations of Europe. In Fig. 11.6, the normalized throughput under rain conditions in the most reliable path of the topology is depicted. Compared to Athens and Lisbon, the algorithm in Paris converges faster. In Paris, there are intensive rainfalls (convective type of rain), thus the differences in throughput among paths are high. Therefore, the probability for an agent to switch from a path to another is higher in comparison with Athens and Lisbon where the rain is stratiform.

In Fig. 11.7 the flows paths under rain conditions are depicted. One may observe that the computed flows have the trend to detour in order to avoid paths suffering most from rain fades. Furthermore, load balancing is achieved

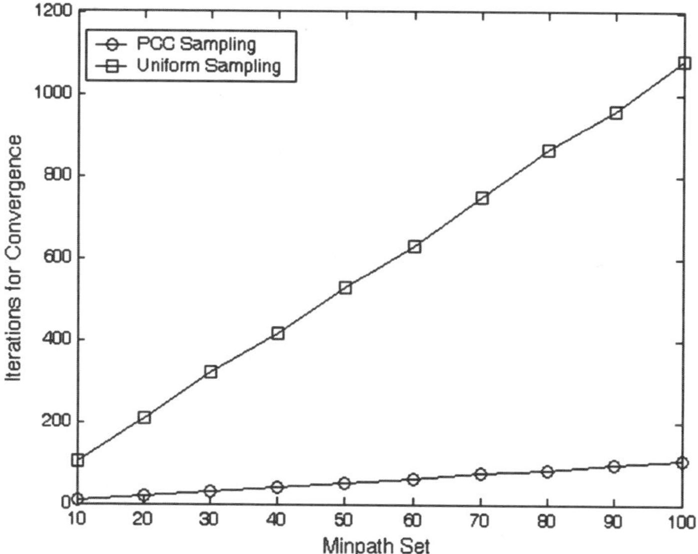

Fig. 11.5. Convergence of PCC algorithm

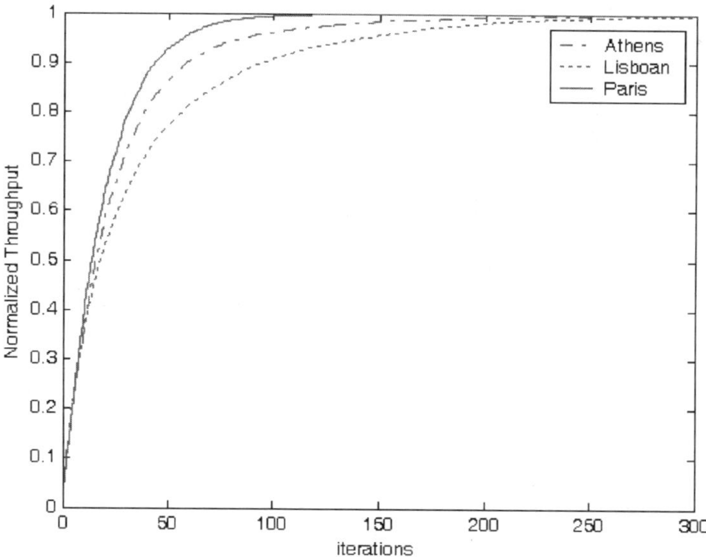

Fig. 11.6. Convergence of PCC algorithm for a WMN located in Athens, Lisbon and Paris

spliting traffic in the four best performing minpaths (1,2,3,4,9,10,15,20,25), (1,2,3,4,9,14,15,20,25), (1,2,3,4,9,14,19,20,25), (1,2,3,4,9,14,19,24,25). Note that the thickness of an edge is proportional to the amount flow on the corresponding wireless link.

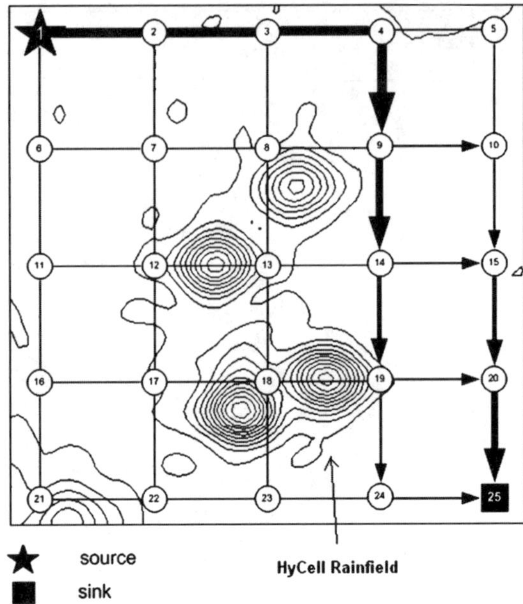

Fig. 11.7. Traffic flows from source node 1 to destination node 25 under rain conditions

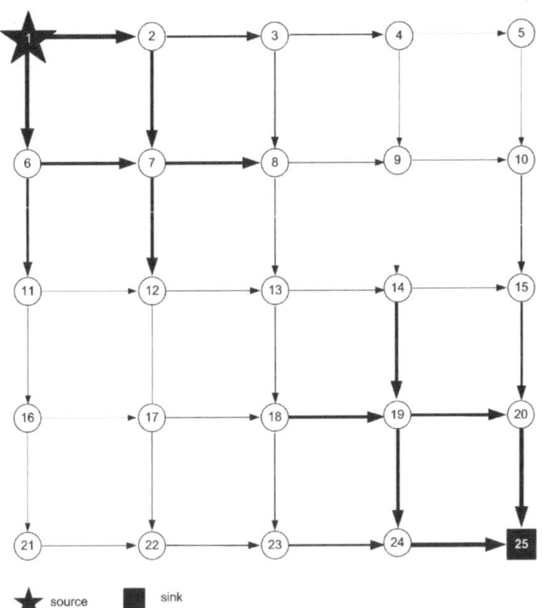

Fig. 11.8. Traffic flows from source node 1 to destination node 25 under clear sky conditions

Finally, data flows in case system operates under clear conditions are depicted in Fig. 11.8. In this scenario the agents gain the same payoff regardless of the selected minpath. Hence, there have no incentive to change their initial strategy. Due to the fact, that, the initials strategies have been chosen at random, traffic has been equally shared among all the available minpaths.

11.6 Conclusions

In this chapter, a new protocol for maximum throughout routing based on evolutionary game theory for wireless mesh network operating above 10 GHz was presented. Rain attenuation has been considered as the major fading mechanism that affects the QoS performance of the WMN. In the current approach, traffic is controlled by a large population of agents, each responsible for a small amount of data. Their aim is to find the maximum throughput path from the source to destination. Due to the dynamic nature of the physical medium, each agent revises his routing strategy continuously by sampling another path. In order to improve convergence an algorithm based on path correlation (PCC) was proposed. This algorithm uses input from the physical medium to find the optimal routing strategy for each agent in minimum iterations. Finally, the performance of the proposed routing scheme is evaluated through comparison with the uniform sampling technique. Its superiority in iterations is obvious in all the tested examples.

Acknowledgements

Markos Anastasopoulos thanks *Propondis foundation* for the financial support.

References

1. Akyildiz, I.F., Wang, X., Wang, W (2005): Wireless mesh networks: a survey. Computer Networks Journal, vol. 47, issue 4, pp. 445–487, March 2005
2. De Couto, D.S.J Aguayo D., Bicket J., Morris R. (2003): A high-throughput path metric for multi-hop wireless routing, in: ACM Annual International Conference on Mobile Computing and Networking (MOBICOM), pp. 134–146.
3. Draves, R, Padhye, J, Zill, B (2004): Comparisons of routing metrics for static multi-hop wireless networks, in: ACM Annual Conference of the Special Interest Group on Data Communication (SIGCOMM), pp. 133–144.
4. Draves, R, Padhye, J, Zill, B (2004): Routing in multi-radio, multi-hop wireless mesh networks, in: ACM Annual International Conference on Mobile Computing and Networking (MOBICOM) pp. 114–128.
5. Mosko, M, Garcia-Luna-Aceves, JJ (2005): Multipath Routing in Wireless Mesh Networks, in WiMesh 2005, Santa Clara; CA.

6. Xiaojing Tao, Kunz, T, Falconer, D (2005): Traffic balancing in wireless MESH networks, in IWCCC 2005, pp. 154–169
7. Mueller, S, Tsang, R, Ghosal, D, Calzarossa, M.C, Gelenbe, E (2004): Multipath routing in mobile ad hoc networks: issues and challenges, in: Lecture Notes in Computer Science. vol. 2965, pp. 209–234
8. Eklund, C, Marks, R.B, Stanwood, K.L, Wang, S (2004): IEEE Standard 802.16: A technical overview of the WirelessMAN air interface for broadband wireless access, IEEE Commun. Mag., vol. 40, no. 6, pp. 98–107
9. ETSI TS 101 999 V1.1.1 (2002): Broadband Radio Access Networks (BRAN); HIPERACCESS; PHY protocol specification.
10. Roughgarden, T, Tardos, E (2002): How bad is selfish routing?, J.ACM, 49(2), pp. 236–259
11. Fiescher, S, Vocking, B (2005): Adaptive Routing with Stale Information, in Proc. 24th Ann. ACM SIGACT-SIGOPS Symp. on Principles of Distributed Computing (PODC)
12. Srivastava, V (2005): Using Game Theory to Analyze Wireless Ad hoc Networks, IEEE Commun. Surveys Tutorials, vol. 7 no. 5, 4th Quarter
13. Felegyhazi, M, Hubaux, J.P, Buttyan, L (2006): Nash Equilibria of Packet Forwarding Strategies in Wireless Ad Hoc Networks, IEEE Trans. Mobile Comp. vol. 5, no. 5
14. Rajgopal Kannan, S. Sitharama Iyengar (2006): Game-Theoretic Models for Reliable Path-Length and Energy-Constrained Routing With Data Aggregation in Wireless Sensor Networks, IEEE J. Select. Areas Commun., vol. 22, no. 6
15. Bertsekas, D.P, Gallager, R.G (1992): Data Networks. Englewood Cliffs, NJ: Prentice-Hall
16. Berstekas, D.P. (1998): Network Optimization: Continuous and Discrete Models. Belmont, MA: Athena Scientific
17. Colbourn, C (1987): The Combinatronics of Network Reliability, New York: Oxford University Press
18. Altman, E, Avratchenkov, K, Barakat, C (2000): A stochastic model for the TCP/IP with stationary random losses, ACM SIGCOMM
19. Lakshman, T.V, Madhow, U (1997): The performance of TCP/IP for networks with high bandwidth-delay products and random losses, IEEE/ACM Trans. Networking, pp. 336–350
20. Mathis, M, Semke, J, Mahdavi, J, Ott, T (1997): The macroscopic behavior of the TCP congestion avoidance algorithm, Computer Commun. Review 27, pp. 67–82
21. Panagopoulos, A.D, Kanellopoulos, J.D (2002): Cell-Site Diversity Performance of Millimeter-Wave Fixed Cellular Systems Operating at Frequencies Above 20 GHz, IEEE Antennas Propag, vol. 1
22. Feral, L, Sauvageot, H, Castanet, L, Lemorton, J. (2003): HYCELL: A new hybrid model of the rain horizontal distribution for propagation studies. Part 1: Modeling of the rain cell, Radio Science, vol. 38, no. 3
23. Feral, L, Sauvageot, H, Castanet, L, Lemorton, J. (2003): HYCELL: A new hybrid model of the rain horizontal distribution for propagation studies. Part 2: Modeling of the rain cell, Radio Science, vol. 38, no. 3
24. Panagopoulos, A.D, Kanellopoulos, J.D (2003): On the rain attenuation dynamics: spatialtemporal analysis of rainfall rate and fade duration statistics, Int. J. Satell. Commun. Network. 2003; 21:595–611

25. RFC 3452 http://www.ietf.org/rfc/rfc3452.txt
26. Weibull, J, W. (1995): Evolutionary Game Theory. MIT press
27. Taylor, P., Jonker, L(1978): Evolutionary Stable Strategies and Game Dynamics, Mathematical Biosciences, 5:455–484
28. Akin, E. (1980): Domination or equilibrium Mathematical Biosciences, 50, 239–250
29. Bjornerstedt, J. (1993): Experimentation, imitation and evolutionary dynamics, Mimeo Department of Economics, Stockholm University

12

Applying Hybrid Multiobjective Evolutionary Algorithms to the Sailor Assignment Problem

Deon Garrett[1], Dipankar Dasgupta[1], Joseph Vannucci[1], and James Simien[2]

[1] University of Memphis, Memphis, TN, 38152, USA
[2] Naval Personnel Research, Studies, and Technology (NPRST), Millington, TN, 38055

Summary. This chapter investigates a multiobjective formulation of the United States Navy's Sailor Assignment Problem (SAP) and examines the performance of two widely-used multiobjective evolutionary algorithms (MOEAs) on large instances of this problem. The performance of the algorithms is examined with respect to both solution quality and diversity, and the algorithms are shown to provide inadequate diversity along the Pareto front. A domain-specific local improvement operator is introduced into the MOEAs, producing significant performance increases over the evolutionary algorithms alone. This hybrid MOEA approach is applied to the sailor assignment problem and shown to provide greater diversity along the Pareto front. The manner in which the local search is incorporated differs somewhat from what is generally reported. Our results suggest that such an approach may be beneficial for practitioners in handling similar types of real-world problems.

12.1 Introduction

Approximately every three years, the United States Navy requires that all sailors rotate into a different job. As a result, at any given time there exists a large group of sailors and jobs awaiting reassignment. Currently, more than 300,000 sailors serve in the Navy; more than 120,000 are rotated each year [17]. Sailors must be reassigned in such a way as to satisfy an array of Naval regulations and personal preferences. The Navy's goal is to find a set of assignments of sailors to jobs that maximizes some criterion of desirability.

In reality, the utility of a possible solution to the SAP is not determined by a single measure. Instead, there are a number of attributes that go into deciding whether a solution is good. Typically, each of these attributes is represented by a distinct objective function. One approach to this type of problem is to attempt to formulate a single metric that encompasses all aspects of solution quality and optimize based on this metric. This single objective formulation has the advantage that the resulting problem can be approached using any of a vast array of classical optimization methods. Constructing the metric requires detailed knowledge of the structure of each objective function.

D. Garrett et al.: *Applying Hybrid Multiobjective Evolutionary Algorithms to the Sailor Assignment Problem*, Studies in Computational Intelligence (SCI) **66**, 269–301 (2007)
www.springerlink.com

In addition, it can be difficult for a decision maker to specify an appropriate set of parameters to arrive at a desirable solution.

A better solution from a usability standpoint is to find a set of solutions to the problem that range across the set of "good" solutions. In multiobjective optimization, this set is defined by the set of optimal nondominated solutions and is called the Pareto front, or Pareto optimal set. Deb [8] refers to this type of approach in which no decisions need be made a priori as an "ideal multiobjective optimizer".

In this work, we examine two well-known approaches to multiobjective optimization using evolutionary algorithms, the Nondominated Sorting Genetic Algorithm (NSGA-II) and the Strength Pareto Evolutionary Algorithm (SPEA2), showing that multiple diverse solutions to the SAP can be found in a single run of the algorithms. Section 12.2 describes the Sailor Assignment Problem in more detail. Section 12.3 gives a brief overview of related work followed by a brief introduction to modern multiobjective evolutionary algorithms in Sect. 12.4. The details of the application of these algorithms to the SAP are described in Sect. 12.5. Section 12.6 describes the experiments and results using SPEA2 and NSGA-II. Section 12.7 details hybrid multiobjective evolutionary algorithms and describes the approach used in this study, and Sect. 12.8 presents the conclusions drawn from the analysis of these results.

12.2 The Sailor Assignment Problem

The SAP is a complex assignment problem consisting of some ninety goals and constraints. In order to ensure that all requirements are met, the Navy employs over 200 detailers full time who are responsible for constructing the sets of job assignments [24]. In the Navy's current system, each detailer is presented with a list of jobs that need to be filled every two weeks. The task of the detailer is to assign each job to a sailor currently undergoing job rotation and do so while adhering to Naval policies and guidelines. A primary objective of the detailer is to ensure that the Navy's personnel requirements are fulfilled, e.g., maintaining a minimum number of trained personnel onboard a submarine, etc. A second objective is to satisfy sailor preferences. Sailors are encouraged to specify their desires in such aspects as geographic location. In the event that sailor preferences cannot be met, morale is lowered and retention may ultimately suffer. The Navy's goal, and thus the objective of the detailers, is to satisfy these competing criteria to the maximum extent possible.

Assuming the existence of a single fitness measure that could determine unambiguously the desirability of a proposed set of assignments, also referred to as a *slate*, the problem could be formally defined as follows. Maximize the following utility function

$$U = \sum_{i=1}^{N} \sum_{j=1}^{M} \mathcal{F}_{i,j} d_{i,j}, \tag{12.1}$$

subject to

$$\sum_{i=1}^{N} d_{i,j} \leq 1, \quad \forall j \in \{1, 2, \ldots, M\} \tag{12.2}$$

and

$$\sum_{j=1}^{M} d_{i,j} \leq 1, \quad \forall i \in \{1, 2, \ldots, N\}, \tag{12.3}$$

where $\mathcal{F}_{i,j}$ denotes the fitness of assigning sailor i to job j and D is an assignment matrix such that

$$d_{i,j} = \begin{cases} 1 & : \text{ Sailor } i \text{ assigned to job } j \\ 0 & : \text{ otherwise} \end{cases}$$

The fitness measure $\mathcal{F}_{i,j}$ encapsulates all information relevant to determining the desirability of the match. The constraints on D ensure that at most one job will be assigned to any sailor and that no job is assigned to multiple sailors. Note that both constraints are inequalities, allowing for the possibility that a given sailor is not assigned a job. The construction of $\mathcal{F}$ is described in [16] where a single-objective genetic algorithm is shown to be effective in solving the problem. In that work, the multiple objectives were scalarized into a single measure which was optimized using a state-of-the-art genetic algorithm.

In [16], it was shown that a state of the art genetic algorithm could successfully find good solutions to the SAP, given a scalar fitness measure defined as a linear combination of four different attributes desired in a solution. However, formulating a single such criterion is problematic, at best. There are several different goals which the Naval detailers seek to satisfy; in many cases, these goals conflict with one another. Thus, a drawback of this approach is that the final solution is strongly dependent upon the weights assigned to each feature in the fitness function. As it is not generally known *a priori* how best to set these weights, the resulting algorithm may be difficult to use in practice.

Figure 12.1 shows an overview of the Web-Based Marketplace (WBM) in which the above system is to be implemented. The complete detailing and assignment process is broken into discrete *windows* – periods of time during which only certain aspects of the process may occur. To begin, a detailer uses the Web-Based Marketplace (WBM), a web-based front end, to enter the application window. During this time, sailors may use the WBM to view and apply for available jobs. The next stage is the command preview window, during which commanders may view and rank the sailors applying for each available job. Finally, the detailer enters the detailing window. At that point, the evolutionary algorithm-based matching system, dubbed GenoSAP, is launched. The MOEA algorithms in GenoSAP produce a set of nondominated solutions that are sent to a module called the Solution Analyzer which facilitates closer examination of the results. The detailer chooses one solution from the set discovered by the MOEA and writes it to the database where it

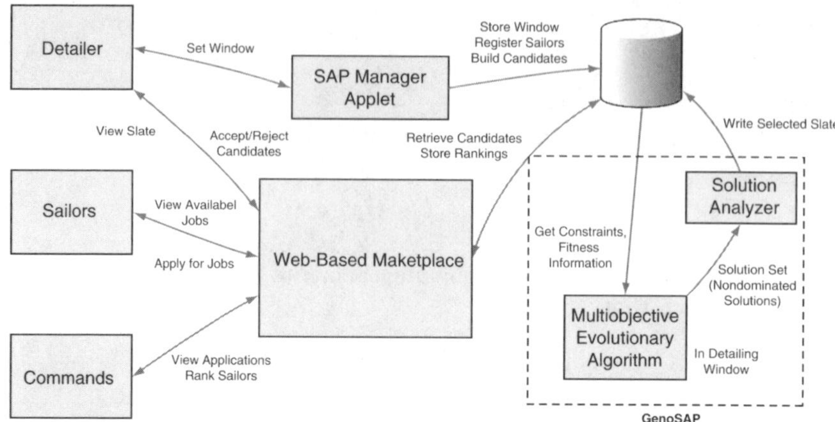

Fig. 12.1. Overview of the sailor assignment process to be implemented in the Web-Based Marketplace

is viewed and manipulated via the WBM. In the following paragraphs, each stage of the process is described in more detail.

The Navy requires that all enlisted sailors rotate jobs regularly. With the arrival of the requisition list every two weeks, the detailer looks to find sailors whose Projected Rotation Date (PRD) falls within the window of availability of each job on the list. Those sailors are examined for a variety of qualifications and constraints to determine if each is a valid match for one of the jobs and a score is computed determining the extent to which the sailor is a valid match. This normalized score, called the *training score*, must be at least 0.7 in order for the sailor to qualify for a particular job. The training score encompasses many factors, including the pay grades of the sailor and the proposed job, the amount of training required for the sailor to be able to perform the duties of the proposed job, and many others. In addition, the monetary cost of assigning the sailor to the proposed job is computed. Called the Permanent Change of Station (PCS) Cost, it encompasses the cost of relocating the sailor and any dependents to the location of the new job, the cost of any required training for the sailor, and other similar factors. Each valid sailor/job match is called a *candidate* and is placed in a list from which the final set of assignments will be drawn.

After all candidates have been identified, the sailors are allowed to rate those jobs for which they are qualified. It is important to note that the sailor's rankings bear no correlation to the needs and desires of the Navy. Therefore, it is likely that the detailer will need to make tradeoffs in assigning a job to that sailor. On one hand, the detailer wishes to assign the sailor a job that will satisfy the sailor's wishes. On the other hand, the detailer has other issues to handle. The final set of assignments must satisfy Naval regulations and must do so at a reasonable cost to the Navy.

It is preferable to assign a sailor a job that was asked for whenever possible. In this work, we will assume that sailors may be assigned only those jobs for which the sailor applied and ranked. This assumption is not required for the proper operation of the multiobjective evolutionary algorithms, and relaxing this restriction would allow the algorithms to sometimes find more complete slates. However, it should be noted that any sailors not assigned a job can then be manually assigned a job that was not applied for; the same as in the current system. By not allowing the algorithm to force unwanted assignments, an additional bias toward satisfying the preferences of the sailors is built into the system.

When each sailor has finished applying for and ranking the available positions, the commands are allowed to view and rate the sailors that have applied for their available positions. Unlike the sailor ranking stage, the command preview is optional. If a command wishes to rate individual sailors, the system should attempt to honor those preferences. However, not all commands will choose to do so, and in the absence of preference information, the system should choose fitter sailors for the position whenever possible.

After the command preview stage, the detailer must construct the set of assignments in accordance with the regulations and preferences of the Navy. The complexity of the detailing process limits the Navy's ability to make effective decisions. Expanding the window from two weeks to three months would allow the Navy to more effectively allocate resources, providing more jobs and sailors with which to arrive at a better solution. However, the three-month job announcement list would quickly overload the detailer's ability to manage the process. An automated system for quickly finding good solutions to the problem is needed in order for this change to occur.

While this problem was successfully attacked using a single-objective genetic algorithm, the inherent problems associated with the scalarization of the objective functions remain. The weights are difficult to set correctly and intuition often fails when attempting to predict the results of changing the weights. Furthermore, if the Pareto optimal front is non-convex, it is well-known that there will exist solutions along the front which are not optimal for any possible weight vector [8]. Thus there can be no possiblility of uniform coverage of the true Pareto optimal set in this case.

In addition, the detailer typically cannot make adequate judgements without greater knowledge than is provided by a single run of the genetic algorithm (GA). For example, suppose that the detailer believes that minimizing cost is the most important factor in constructing a particular set of assignments. Accordingly, the detailer weights the linear combination more heavily in favor of the PCS cost. The GA finds a solution that minimizes the linear combination. There is no information provided concerning the overall structure of the solution space. It may be that a large set of solutions are approximately equivalent with respect to cost, but differ widely with respect to other objectives. In this case, weighting the cost too heavily likely results in poorer solutions than would have been selected by a human detailer, since the human would recognize the situation and select a higher quality solution at the expense of a slight increase in cost.

One possible solution to this problem is to discard the scalarization entirely and attempt to find a Pareto optimal set of solutions covering a wide range of the solution space. Multiobjective Evolutionary Algorithms (MOEAs) have proven themselves a useful tool for many multiobjective optimization problems. One of the primary benefits of MOEAs is that a range of solutions can be found in a single run of the algorithm. In constrast, most classical optimization methods require multiple runs with varying parameter settings in order to find a disparate set of nondominated solutions.

One common practical problem with multiobjective algorithms is the burden placed on the decision maker. There must be a balance between providing so much information that the user cannot cope with the data and providing too few alternatives, hampering one of the goals of multiobjective optimization in general. In multiobjective evolutionary algorithms, this tradeoff is often realized by clustering the final set of alternatives to reduce the number of solutions presented to the decision maker. While this technique does effectively allow the decision maker to handle the information presented, care must be taken to ensure that the most desirable solutions are not lost.

To facilitate processing of the nondominated solutions by the detailer, we instead developed a tool called the Solution Analyzer. This tool provides a set of filters and views that allow the detailer to examine the solution set from many different aspects. For example, the detailer may choose to show solutions in which the PCS cost was below a given threshold and the training score was above another threshold. The set of solutions could be further restricted to only those in which a particular sailor did or did not receive a particular job. By providing these visualization and analysis tools, the burden placed on the decision maker can be minimized without discarding potentially useful solutions.

12.3 Previous Work

In [16], a single-objective genetic algorithm was employed to find very good solutions to the SAP. The GA was compared against the Gale-Shapley algorithm, a deterministic algorithm for finding matches that assures stability for matches based on the scalarized scoring rule used to represent the preferences [15]. The GA was found to find much better solutions with respect to coverage, the number of sailors and jobs not left unmatched, while greatly improving the cost associated with the resulting assignments. In addition, the GA bettered or was competitive with the deterministic algorithm on all other objectives as well.

However, the scalarization of the objective functions used in that study is problematic due to the difficulty in predicting the impact of a change in the weight vector on the solution found. In order to get a broad view of the solution space, the algorithm requires several runs with different parameters settings. Achieving good coverage of the Pareto front thus depends not only upon the ability of the algorithm to find Pareto optimal solutions, but also on the proper selection of parameters by the user. Improper parameter settings could easily bias the coverage of the front to a much smaller region than would be desired.

In this paper, we extend the earlier work to incorporate directly each objective into the optimization problem. Two modern multiobjective algorithms, NSGA-II and SPEA2, are used to solve the resulting problem and the results compared to other methods of generating feasible assignments. The remainder of this section concerns the basic structure of the sailor assignment problem and briefly describes related work by other researchers in solving single and multiobjective variants of related assignment problems.

In addition to our earlier work using evolutionary algorithms, the SAP has been attacked using more traditional optimization techniques. The SAP is a difficult Integer Programming (IP) problem. The difficulty lies not only in the very large numbers of sailors and jobs involved, but also in the nature of the often conflicting goals and objectives. As a result, the Navy has investigated numerous mechanisms to attempt to solve the problem [1, 2, 17, 23]. Due to the difficulty of the resulting integer programming formulation, most mathematical programming techniques cannot adequately handle problem instances of even fairly moderate size. Because of this difficulty, increasing numbers of researchers have turned to metaheuristics such as evolutionary algorithms or tabu search to attack difficult problems such as the SAP.

The SAP bears some similarities to the so-called airline rostering problem. The goal in this problem is to find a way to assign flight crews to tours of duty called *pairings* for commericial airlines. Like the SAP, the Airlines Crew Rostering Problem (ACRP) involves multicriteria optimization while simultaneously meeting several hard and soft constraints. Unlike the SAP, in the ACRP, all tours must be staffed by a crew. Each crew is assigned multiple pairings until all pairings have been assigned. In contrast, sailors are assigned a single job for the duration of their tour of duty and preventing the assignment of multiple jobs to a single sailor is considered a hard constraint.

In [27], genetic algorithms are used to solve a multiobjective formulation of what is referred to as the nominal Airline Crew Rostering Problem (ACRP). In addition to minimizing the costs incurred by the airline, their work attempted to maximize the satisfaction of the crew resulting in a dual-objective problem. The approach taken was to utilize a heuristic procedure to generate a set of solutions considered very good with respect the crew preferences, then use a domain specific genetic algorithm to search for low cost solutions given the initial population of highly satisfactory solutions.

One commercial tool for generating crew rostering schedules [20] utilizes a series of heuristics to first generate feasible schedules and then iteratively optimize these schedules in an effort to improve solution quality. This approach is limited in that decisions must be made on which aspect of solution quality to emphasize. In addition, the authors report run times of up to 14 hours on problems consisting of a few thousand crews. While such numbers are difficult to compare to other techniques due to differences in computing power, one may at least infer than an algorithm may be permitted to perform a rather intensive search while remaining within the limits of available run times.

Other approaches to the ACRP [5] formulate the problem as an integer linear programming problem and use additional heuristics to solve the problem. However, as a multiobjective problem, these approaches suffer from the requirement of a scalar objective function. While the specific details of the ACRP differ from the SAP, the problems share many aspects. To the best of the author's knowledge, no multiobjective formulation of the ACRP has been approached using MOEAs.

Perhaps the best-known assignment problem is the ubiquitous Quadratic Assignment Problem [6]. In the QAP, the goal is to find an optimal assignment between elements of a set of facilities and elements of a set of locations. Each potential facility/location match incurs a cost and the objective is to minimize the total cost of assigning all facilities to a location. Several researchers have applied genetic algorithms to the QAP, usually combined with a heuristic improvement operator [12, 14]. In most reported studies, evolutionary algorithms have performed rather poorly without additional domain specific information such as these local search techiques embedded into the algorithms.

The canonical QAP is an unconstrained, single-objective problem. As such, most GA approaches utilize a permutation encoding representing the fact that valid solutions represent a mapping from facilities to locations that is bijective. In contrast, the mapping from sailors to jobs in the SAP forms an injective map. Note that in practice, even this restriction may not hold, since any given solution may leave one or more sailors without a job. In addition, the constraints on which jobs a sailor is allowed to receive make a permutation encoding all the more problematic, since most randomly generated permutations would be infeasible. For the same reasons, a binary encoding is unsuitable for the sailor assignment problem. A mutation operator which simply flipped a bit would likely produce an infeasible individual, as would crossover applied at arbitrary points on the string. Because of these issues, we employ a domain specific integer encoding scheme described in Section 12.5.

12.4 Multiobjective Evolutionary Algorithms

Some of the most promising approaches to many multiobjective optimization problems arise from evolutionary techniques. One of the primary benefits touted by practioners is the ability of multiobjective evolutionary algorithms to cover the Pareto front in a single run of the algorithm. Many classical approaches to multiobjective optimization rely on multiple runs of a single algorithm with slight modifications in parameter settings. In contrast, because

EAs require a population of solutions for their basic operation, it is hoped that the algorithms may provide a uniform coverage of the front in a single run.

Any multiobjective EA has two basic goals. First the algorithm must push the initial population in the direction of Pareto optimal solutions. Ideally, the algorithm would terminate with a set of nondominated solutions such that no possible solution could dominate any member of the Pareto front. This set is called the true Pareto optimal set, or true Pareto front. In practice, we often wish only to find solutions that are very good and nondominated with respect to one another. This requires that the algorithm routinely improve the quality of the initial randomly generated solutions until some level of acceptability has been met.

Several MOEAs have been proposed in the literature with varying degrees of success. The first multiobjective evolutionary algorithm, Schaffer's Vector Evaluated Genetic Algorithm [29], VEGA, operated much as a single-objective evolutionary algorithm. However, the objective function used to evaluate solutions was chosen from the multiple objectives defined by the problem. As a result, different elements of the population evolved with respect to different fitness functions. Unfortunately, this algorithm fails to find useful trade-off solutions, finding mostly solutions near the extremes of the Pareto front. More recent research has prompted a flurry of activity among MOEA researchers resulting in a series of new algorithms based on Pareto dominance and incorporating features such as elitism. Of these algorithms, the Nondominated Sorting Genetic Algorithm by Deb and others (NSGA-II) [9] and the Strength Pareto Evolutionary Algorithm (SPEA2) of Zitzler et al. [30] have been widely studied and found to be effective across a range of common test functions as well as combinatorial optimization problems.

SPEA2 is an elitist multiobjective genetic algorithm using Pareto dominance to determine the fitness of candidate solutions. In addition, an archive construction mechanism ensures that the archive size remains constant while ensuring that the salient structure of the Pareto front remain intact. In order to construct the archive, all individuals in the multiset union of the previous archive and the population are compared to one another to determine Pareto dominance. The *strength* of an individual is defined to be the number of other individuals that are dominated by the given individual. The raw fitness of an individual is computed as the sum of the strengths over all dominating individuals. Therefore, nondominated solutions are assigned a raw fitness of zero and the fitness is to be minimized.

Problems could potentially arise with this strategy whenever many individuals are mutually nondominated, as each would share the same fitness value. As a way to differentiate between such individuals, a crowding factor is included into the fitness and is equal to the distance in objective vector space from an individual to its kth nearest neighbor. Because the goal is to uniformly spread the solutions across the Pareto front, the algorithm prefers individuals with a larger crowding distance.

Like SPEA2, NSGA-II is an enhancement of an earlier algorithm which was developed to address several issues with the original NSGA. In NSGA-II, solutions are segregated according to their *nondomination rank*. The nondomination rank is the number of "layers" which must be peeled away before the solution becomes nondominated. The basic nondominated sorting algorithm consists of finding the set of nondominated solutions, or the nondominated front, assigning them rank 1, then removing them from the population. This results in some solutions becoming nondominated which were dominated in the original population. These individuals will have been dominated only be elements of the first nondominated front, and are assigned nondomination rank 2. The process continues until the population is empty. Like SPEA2, NSGA-II includes a density estimation component to differentiate between solutions at the same nondomination rank.

The details of the algorithms differ, but the basic ideas are the same. Both algorithms seek to reward individuals that are currently nondominated and very different from the other nondominated solutions in the current set. Both algorithms consider the dominance relation to be the primary component of the fitness of an individual. Because Pareto dominance provides only a partial ordering on the sets, many solutions will be equally fit. To provide useful selection pressure in these cases, the algorithms prefer solutions that are more distant from other solutions, thus providing more meaningful diversity along the Pareto set approximation.

We considered M-PAES [22], a multiobjective memetic algorithm utilizing a strict Pareto local search technique to improve individuals during the evolutionary process. M-PAES performed very slightly worse in our preliminary testing than the preceding algorithms. Much of the literature on so-called "many-objective" optimization [13, 21], optimizing problems with more than two or three objectives, seems to indicate that pure Pareto based approaches may be less effective on these problems. As the number of objectives increases, the ability of the dominance relation to adequately discriminate between points is degraded. Instead, methods which combine objectives in some way or which utilize additional preference information [7, 10] are preferred. Thus, we did not consider M-PAES for further study.

12.5 Multiobjective Formulation of the SAP

12.5.1 Encoding

One of the principle decisions to be made when adapting a genetic algorithm to a particular problem is how to encode solutions to the problem in a manner amenable to genetic search. Table 12.1 briefly notes the major constraints present in the SAP and how each is handled in the system. As discussed in Section 12.3, the conventional permutation encoding used in typical assignment problems is not satisfactory due to the large number of constraints in

Table 12.1. Handling constraints relevant to the sailor assignment problem. Each constraint is labeled either "Hard" or "Soft", denoting in the usual sense whether the constraint may be violated by feasible solutions

Issue	Resolution
No sailor may be assigned more than one job. (Hard)	Encoding mandates at most one job per sailor. (Section 12.5.1)
Only certain jobs are acceptable for each sailor. (Hard)	Maintain list of acceptable jobs and ensure than only items from that list appear in solutions.
No job assigned to multiple sailors. (Hard)	Dummy job introduced to break conflicts; repair procedure to handle dummy assignments. (Section 12.5.2)
Each sailor should be assigned a job. (Soft)	Repair procedure (Section 12.5.2) tries to reassign as many unassigned sailors as possible

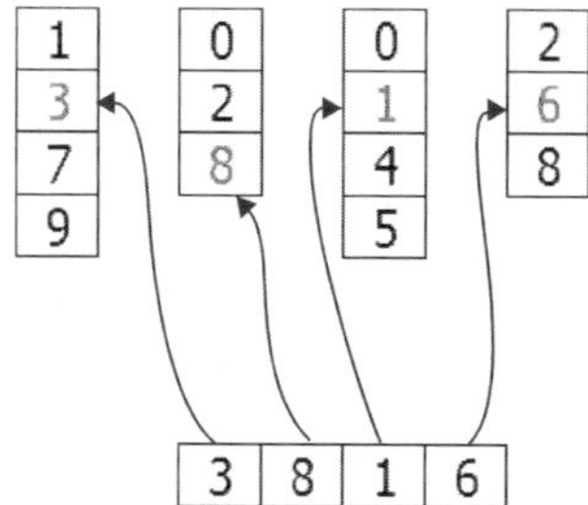

Fig. 12.2. Example of the chromosome encoding for the SAP. The chromosome is shown along the bottom. Each element of the chromosome is a job number chosen from a list of acceptable jobs for each sailor. Note the possibility of multiple sailors receiving the same job

the SAP. A binary encoding encounters the same problems and likely some additional ones.

Instead, we utilize an integer encoding in which each gene is chosen from a subset of possible numbers, precisely those representing valid jobs for that sailor. In this way, we ensure that only those jobs for which a sailor is qualified may be assigned to the sailor. Any respectful [28] crossover operator will respect these constraints, thereby ensuring that recombination will not assign a sailor to a job for which he is not qualified. One simple mutation operator, assigning a randomly selected sailor a new job, is easily made to maintain these constraints as well by selecting the new job from the list of acceptable jobs.

Figure 12.2 shows an example of a chromosome given a problem instance of four sailors and ten jobs. Each sailor is given an identifying number in the closed interval [0..3]. Similarly, each job is assigned a number in [0..9]. Decoding a chromosome, C, is simply performed by assigning sailor i to job

$C[i]$. Note in Figure 12.2 that a random crossover or mutation could easily result in a violation of another constraint–that of uniqueness. Equations 2 and 3 from Section 12.1 ensure that feasible solutions assign at most one sailor to each job and one job to each sailor. In reality, no job should be assigned to multiple sailors nor should multiple jobs be assigned to a single sailor. The proposed encoding prevents violations of the latter constraint, but allows the former to occur.

12.5.2 Genetic Operators

To provide valid solutions to the SAP, violations of the uniqueness constraint must be detected and somehow removed. There are three events which could potentially result in this type of violation: randomizing the chromosome, performing crossover, and performing mutation. While most of the common crossover operators are easily formulated for this encoding, they can obviously produce infeasible children from two feasible parents. "Disruptive" operators, such as uniform crossover, may be expected to violate more constraints than operators such as one-point crossover. To see why, note that both parents must be feasible, i.e., have no two sailors assigned to the same jobs. Under one-point crossover, the offspring are unlikely to assign two sailors the same job unless the locations of the sailors in the chromosome are physically separated by some distance.

Because crossover can and often does violate the constraint that no job be assigned to multiple sailors, a repair procedure is employed to help ensure that the algorithm maintains feasible, high-quality solutions. The repair method traverses the chromosome in a random order. Whenever a duplicate job is found, it is changed to a dummy job (with job ID-1 in our implementation). Following this step, any sailor assigned a dummy job is assigned a job from his job list if one is available.

Not every violation can be repaired with this method. When computing fitness, one may choose to penalize the solution for unassigned sailors. This is typically unnecessary since an unassigned sailor receives no training score, sailor rank score, or command rank score. However, PCS cost could present a problem since receiving zero PCS cost would be considered advantageous. In reality, the Navy does not allow enlisted personnel to remain unemployed. Instead, the sailor would need to be manually assigned to a remaining unfilled job. Because this manual process would typically cost more than an automated assignment, we consider the PCS cost of an unassigned sailor as equal to the maximum cost of any of the sailor's acceptable jobs. The result is that an unassigned sailor is guaranteed to receive the lowest possible scores for all four components of fitness, thereby removing the need for an explicit penalty function.

As mentioned previously, mutation consists of randomly reassigning a chosen sailor probabilistically. Suppose that sailor i is selected for mutation. There exists a list of acceptable jobs for sailor i from which the new job must be drawn. It may be that no job in that list is available (all are assigned to other sailors). In that case, a conflict arises over the selected job, and one of the sailors contending for the job must be reassigned. The mutated sailor cannot be reassigned, as we have just determined that all feasible jobs are already taken by other sailors. But to reassign the second sailor could result in the same problem. Thus, simply finding a feasible assignment of sailors to jobs is in general NP-hard. It should be noted the search depth required is likely to be limited in most cases, since often only a small number of sailors need be shuffled around to find a feasible set of assignments. Thus, although this repair might be done in polynomial time in most cases, it is likely to be rather expensive on average. Therefore, we have employed a simpler repair mechanism, described below, which operates in constant time while still providing adequate performance in most cases.

Mutation of a candidate solution begins as described above – a sailor i is selected at random to undergo mutation. Each job j in the selected sailor's list is considered in a random order. If no other sailor is currently assigned job j, then sailor i is assigned job j and the process is halted. If another sailor i_2 is already assigned job j, the job list for i_2 is checked to see if any of those jobs are currently unassigned. If so, sailor i_2 is assigned a new job, freeing job j for assignment to sailor i. If none of sailor i_2's jobs are available, then the attempt to assign job j to sailor i fails, and a new attempt is made for the next job in sailor i's list.

It is clear then, that given a specified mutation rate P_m, the effective mutation rate, the percentage of attempted mutations that actually effect some change in the individual, is less than P_m, since only some of the attempted mutations are completed successfully. As the proper setting of the mutation rate is an important aspect of any evolutionary algorithm, it is very important to understand the relationship between the specified mutation rate and the effective mutation rate on the SAP. Clearly, this relationship is intricately linked to the amount of contention for jobs, which in turn depends critically on the ratio of jobs to sailors in a particular instance of the problem. As this ratio increases, the likelihood of finding a job for any given sailor which is not currently assigned to another sailor increases, thereby increasing the percentage of mutations that can be completed successfully.

12.5.3 Initialization

Randomization of a newly created chromosome utilizes the same process as the mutation operator described above. Each sailor receives a randomly selected job if one is available. If no acceptable job is available, the randomization procedure marks the sailor unassigned. To prevent bias toward assigning lower numbered sailors over higher numbered sailors, the order in which the chromosome is traversed is randomized.

12.5.4 Objective Functions

As mentioned in Section 12.2, the SAP consists of finding solutions that maximize the total training score, the sailor preference score, and the command preference score, and minimize the PCS cost. Each of these metrics is summed over all sailor-job matches to compute the score for the match as a whole. The total is then divided by the number of sailors to yield mean values over all sailors. These mean values are the fitness vector which is used by the algorithms to determine Pareto dominance.

While multiobjective evolutionary algorithms typically do not require normalization of objective functions, visualizing trade-offs becomes easier if all objectives are similarly scaled. In addition, converting the problem to a homogeneous problem, i.e., one in which all objectives are to be maximized (or minimized) often simplifies the implementation of the algorithms. Therefore, we normalize each instance of the SAP as described below.

While the training score for a sailor-job match is ostensibly already normalized to a real-valued number in the interval [0, 1], the actual range of values falls within [0.7, 1], due to constraints placing a minimum on the fitness of a sailor to perform the required duties. Each sailor applies for a fixed subset of the jobs available to him, and the relevant measurements are known for each of these jobs *a priori*. Given the ideal situation in which each job was applied for by only a single sailor, the global minimum and maximum for each objective can be computed by simply assigning each sailor to the job providing the extreme value for each objective. Note that typically, this will not be a single job, and thus the result will not be a feasible solution. This *ideal vector*, the vector consisting of the optimal values of each objective independent of all others [8], provides a baseline with which to compare the quality of the solutions found by the algorithms. Similarly, the corresponding vector of worst possible fitness values provides a lower bound used to scale the objective function values.

The PCS cost is expressed in actual dollars, most of which fall between $0 and $20,000. For each candidate, the PCS cost is divided by the largest PCS cost of any candidate matching the sailor. After the mean is computed over all matched candidates, the PCS score is subtracted from 1.0 to yield a maximization problem.

The sailor and command preferences are ordinal ranks with one caveat. While sailors are required to apply for and rank all jobs which they are interested in receiving, there is no such requirement for commanders. If a commander so wishes, he may rank one or more of the sailors that have applied for the job he oversees. However, if this ranking is not performed, then the system should prefer those sailors with higher training scores. Note that in computing the command preference score, all unranked sailors are treated as equal. The bias toward selecting more fit sailors arises from the training score aspect of the fitness vector.

Such ordinal ranking schemes introduce two problems. First, ignoring rankings of zero, the better solutions are those with lower numbers. We need to convert this to a maximization problem to match the framework defined above. In addition, a naïve linear scaling would likely imply that the difference in fitness of two solutions would depend only on the absolute difference in their ranks. In practice, it seems likely that the difference between ranks one and two should be more pronounced than the difference between ranks six and seven, for example.

The ability of commanders to specify a ranking of zero leads to other problems. Obviously, a solution should be rewarded for assigning sailors to commanders who ranked those sailors with low rankings (one or two, for example). However, any positive number is preferred over a ranking of zero. Thus the function determining payout based on rank is nonmonotonic. In order for the sum of the score to be meaningful, we must convert this function into a monotonic function that can be maximized. Therefore, we define the rank score as follows.

$$Rank_{score} = \begin{cases} 0.0 & : \quad Rank = 0 \\ \sqrt{Rank}^{-1} & : \quad \text{otherwise} \end{cases}$$

The resulting function can be more easily interpreted, since higher values of the function are always preferred over lower ones.

Throughout the remainder of the paper, the terms *training score*, *PCS Cost*, *Sailor Rank*, and *Command Rank* will be used to refer to both the sum of the raw numbers as well as to these normalized quantities described above. Which meaning is intended should be clear from context. Whenever the discussion involves the quantities that are manipulated by the evolutionary algorithms, the latter meaning is intended. The raw scores will be used occasionally to compare results on realistic test data.

To compute the fitness of a solution, we then iterate over each sailor-job match and find the normalized training score, PCS cost, sailor rank score, and command rank score. Summing each of these four values over all sailors in the match and dividing by the number of sailors yields four objective functions in the interval [0, 1], each a maximization problem. Of course, one would not expect there to be a single solution with a fitness vector consisting entirely of ones. The existence of such a solution implies that there is but a single globally optimal solution as opposed to a set of mutually nondominated solutions.

However, it should be noted that while in principle, these values are scaled such that the objective function values range from 0 to 1, in practice the extreme values are almost guaranteed not to occur. The reason is that multiple sailors are in contention for the same jobs. The scaling procedure tacitly assumes that each sailor can receive those jobs which maximize or minimize the values of each objective function. However, uniqueness constraints make it exceedingly unlikely that all such matches can be simultaneously attained. Thus, the actual upper and lower bounds for each objective are unknown for any problem instance of even relatively benign size. As a result, not only will we not see a single solution near 1.0 in all objectives, we will most likely never see a single solution near 1.0 in *any* objective.

12.6 Experiments and Results

Due to limitations on the amount of real data available, a random problem generator was developed to allow testing on problem instances of controlled size and difficulty. These random instances contain training scores and PCS costs produced in accordance with the observed distributions of these values in the actual data. In previous work [16], it was found that one of the most important factors governing the difficulty of a problem instance was the contention for jobs. Given a fixed number of sailors, the more total jobs available for the sailors to choose from, the easier the problem became for classical optimization methods. In a real world scenario, it is unlikely that there will exist many more available jobs than sailors to fill them. This observation implies that procedures such as the Gale-Shapely stable marriage algorithm will have great difficulty finding high quality solutions under realistic assumptions.

To test the evolutionary algorithms, we thus produced a set of sample problems which contain a reasonable ratio of sailors to jobs. Each problem was generated according to a specified number of sailors, a specified number of jobs, and the mean and standard deviation of a normally distributed random variate governing how many jobs each sailor applied for. Table 12.2 shows the parameters of each sample problem used in this study.

Thirty runs of each algorithm were performed using a population size of 100, binary tournament selection, uniform crossover with probability 0.9, and bit-flip mutation with probability 0.005. Each algorithm was allowed 100,000 fitness function evaluations. In each trial, the final Pareto set approximation was recorded. The mean and variance of the solutions in this set with respect to each objective are then recorded. Tables 12.4 and 12.5 summarize the distribution of the mean values and standard deviations along each objective over the 30 trials.

Table 12.3 and Figure 12.3 shows the results of the SPEA2 and the NSGA-II algorithms on a single run on each instance of the SAP. Note that the figures and table reflect different views of the results of the same experiments – the table showing summary statistics of the average Pareto

Table 12.2. Randomly generated SAP instances used in this work

Sailors	Jobs	Applications
1000	1100	$7{\pm}3$
2000	2100	$5{\pm}1$
3000	3500	$5{\pm}1$
4000	4500	$6{\pm}1$

Table 12.3. Characterization of the Pareto front found by the SPEA2 and NSGA-II algorithms on sample runs on each of four data sets. The numbers reported for each objective are the mean and range of values found during the run. The subscripts on CHC denote the weight vector in use, where the weights refer to training score, PCS cost, sailor choice, and command choice, in that order. Bold rows correspond to the runs represented graphically in Figure 12.3

Alg.	Size	Training Score		PCS Cost		Sailor Rank		Command Rank	
		Avg	Rng	Avg	Rng	Avg	Rng	Avg	Rng
SPEA2	**1000**	**0.62**	**0.07**	**0.67**	**0.07**	**0.42**	**0.08**	**0.37**	**0.08**
	2000	0.58	0.03	0.64	0.04	0.46	0.04	0.39	0.05
	3000	0.57	0.03	0.65	0.02	0.45	0.02	0.42	0.04
	4000	0.57	0.03	0.64	0.03	0.40	0.03	0.35	0.05
NSGA-II	**1000**	**0.59**	**0.13**	**0.67**	**0.12**	**0.39**	**0.12**	**0.35**	**0.12**
	2000	0.56	0.06	0.63	0.08	0.43	0.07	0.35	0.13
	3000	0.56	0.06	0.63	0.05	0.43	0.04	0.39	0.09
	4000	0.54	0.05	0.62	0.06	0.38	0.08	0.33	0.09
CHC_{1000}	**1000**	**0.80**	**0.05**	**0.58**	**0.04**	**0.31**	**0.03**	**0.22**	**0.04**
	2000	0.71	0.04	0.56	0.03	0.36	0.04	0.26	0.04
	3000	0.68	0.13	0.56	0.03	0.36	0.02	0.27	0.03
	4000	0.68	0.02	0.57	0.02	0.33	0.02	0.24	0.03
CHC_{0100}	**1000**	**0.49**	**0.05**	**0.86**	**0.02**	**0.32**	**0.03**	**0.23**	**0.04**
	2000	0.49	0.03	0.77	0.04	0.36	0.03	0.25	0.06
	3000	0.50	0.03	0.74	0.02	0.36	0.02	0.28	0.04
	4000	0.49	0.01	0.75	0.02	0.33	0.02	0.25	0.03
CHC_{0010}	**1000**	**0.49**	**0.05**	**0.58**	**0.02**	**0.67**	**0.11**	**0.20**	**0.05**
	2000	0.49	0.03	0.56	0.03	0.60	0.09	0.26	0.04
	3000	0.50	0.03	0.56	0.02	0.58	0.09	0.28	0.03
	4000	0.49	0.02	0.57	0.03	0.57	0.12	0.26	0.02
CHC_{0001}	**1000**	**0.50**	**0.04**	**0.58**	**0.03**	**0.30**	**0.03**	**0.66**	**0.04**
	2000	0.49	0.03	0.56	0.04	0.36	0.02	0.59	0.04
	3000	0.49	0.02	0.56	0.02	0.35	0.02	0.58	0.04
	4000	0.49	0.02	0.57	0.02	0.33	0.02	0.53	0.03
CHC_{equal}	**1000**	**0.66**	**0.03**	**0.74**	**0.02**	**0.47**	**0.03**	**0.55**	**0.03**
	2000	0.62	0.03	0.70	0.03	0.49	0.03	0.57	0.06
	3000	0.63	0.02	0.69	0.02	0.49	0.01	0.63	0.03
	4000	0.62	0.02	0.70	0.02	0.46	0.02	0.56	0.03

fronts found over ten trials with the figures providing a graphical representation of the complete Pareto front found during a single run. Table 12.3 shows, for each objective, the mean values present in the final Pareto front along with the range of values, reported as $maximum - minimum$, in the Pareto front. For comparative purposes, the average results found by running separate trials of the CHC genetic algorithm [11], each using the same number of evaluations as the multiobjective MOEAs, are also shown. These results show the best solution found using CHC and treating all objectives other

Table 12.4. Mean and standard deviation of the mean objective function values found over 30 different Pareto set approximations

Alg.	Size	Training Score		PCS Cost		Sailor Rank		Command Rank	
SPEA2	1000	0.59	1.6e-2	0.68	9.8e-3	0.43	1.5e-2	0.35	1.5e-2
	2000	0.59	1.3e-2	0.65	9.2e-3	0.46	1.5e-2	0.40	2.0e-2
	3000	0.60	1.2e-2	0.66	1.2e-2	0.46	1.1e-2	0.43	1.9e-2
	4000	0.58	1.1e-2	0.77	1.0e-2	0.43	1.2e-2	0.38	2.1e-2
NSGA-II	1000	0.59	8.2e-3	0.67	7.8e-3	0.41	8.2e-3	0.37	1.3e-2
	2000	0.58	6.6e-3	0.65	9.0e-3	0.45	7.2e-3	0.39	1.0e-2
	3000	0.57	7.0e-3	0.64	6.9e-3	0.43	7.6e-3	0.40	1.2e-2
	4000	0.56	6.0e-3	0.64	6.9e-3	0.40	6.8e-3	0.36	9.6e-3

Table 12.5. Mean and standard deviation of the standard deviation along each objective function found over 30 different Pareto set approximations

Alg.	Size	Training Score		PCS Cost		Sailor Rank		Command Rank	
SPEA2	1000	5.9e-2	5.6e-3	5.3e-2	4.1e-3	6.6e-2	7.1e-3	9.1e-2	7.3e-3
	2000	4,8e-2	7.6e-3	4.4e-2	5.6e-3	5.1e-2	4.6e-3	8.0e-2	6.2e-3
	3000	6.2e-2	6.9e-3	5.8e-2	5.2e-3	6.4e-2	5.6e-3	1.0e-1	7.4e-3
	4000	6.3e-2	5.9e-3	5.7e-2	5.5e-3	6.9e-2	5.4e-3	9.9e-2	9.9e-3
NSGA-II	1000	3.3e-2	4.1e-3	3.2e-2	3.0e-3	3.5e-2	3.9e-3	4.9e-2	3.9e-3
	2000	2.0e-2	2.9e-3	2.0e-2	2.4e-3	1.8e-2	2.6e-3	3.0e-2	3.2e-3
	3000	1.4e-2	1.9e-3	1.4e-2	1.9e-3	1.4e-2	2.0e-3	2.4e-2	3.5e-3
	4000	1.5e-2	2.0e-3	1.4e-2	2.1e-3	1.4e-2	1.6e-3	2.2e-2	2.5e-3

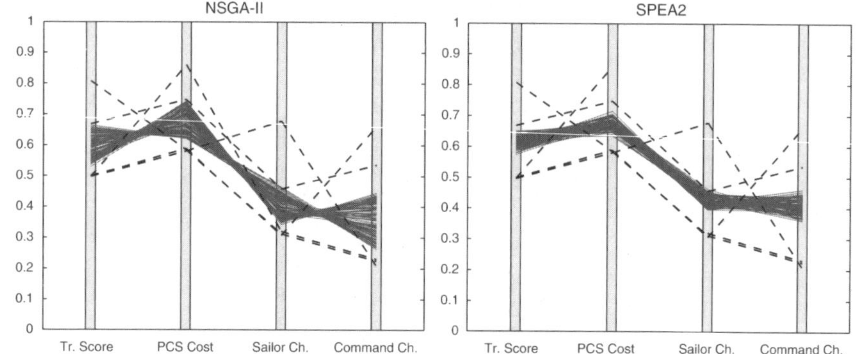

Fig. 12.3. Diversity of solutions found in the Pareto front using both the NSGA-II and the SPEA2 on 1000 sailor instances of the SAP. The axes are scaled from 0.0 to 1.0 and all objectives are to be maximized. Note that the global optima for each objective is not known and will not in general approach 1.0

than the current one as wholly unimportant. For example, row three of the table shows that if we allow CHC to simply maximize the training score on the 1000 sailor instance of the problem, the algorithm finds solutions with a mean training score of 0.80 on average.

We would expect that CHC focusing only on a single objective would find solutions that are better with respect to that objective than those found by the Pareto-conscious algorithms. Therefore, those solutions found by CHC will be nondominated. The hope would be that the Pareto-optimal set found by the MOEAs will contain solutions that are similar to those found by four different runs of CHC. Of course, the MOEAs also need to find solutions that lie "in the middle" as well, but uniform coverage of the true Pareto front should contain solutions that are at least somewhat similar to those found by CHC.

The obvious conclusion to be drawn from the table is that for any single objective, the best solution found by the MOEAs is far inferior to that found by CHC when using only that objective to compute the fitness of the individual. This is not entirely unexpected, since the MOEAs must concern themselves with not only the extreme solutions but also with finding a uniform approximation across all regions of the Pareto optimal set. However, the magnitude of the differences in the quality of the solutions found by CHC and the MOEAs is disconcerting. While beating CHC on any single objective may be an unrealistic ambition, the algorithm must be at least competitive in order to leverage the advantages of multiobjective optimization.

We note here that CHC was chosen as a single objective optimizer due to its superior performance in a previous study [16]. In addtion, we tested a simulated annealing algorithm and found the performance to be competitive with CHC; simulated annealing performed slightly better for short runs, but was slightly outperformed for the longer runs used in this study.

Figure 12.3 shows visually the diversity of solutions in the Pareto front obtained via NSGA-II and SPEA2 on a 1000 sailor instance of the problem. Note that both algorithms find similar solutions on average, but the variation in solutions found by the NSGA-II is significantly greater than that of the SPEA2 on this problem. While, the NSGA-II does provide some variation, all solutions are still concentrated in a reasonably constrained region of the solution space. Ideally, we would like to see a greater range of solutions for both algorithms.

The darker dashed lines in Figure 12.3 show the average solutions found by CHC using one of five distinct weight vectors to construct the linearly weighted fitness function. Each graph contains five such lines, corresponding to the following weight vectors: (1, 0, 0, 0), (0, 1, 0, 0), (0, 0, 1, 0), (0, 0, 0, 1), and (0.25, 0.25, 0.25, 0.25). Which line corresponds to which vector should be clear upon examination. The line with the largest value with respect to training score (the first objective) corresponds to weight vector (1, 0, 0, 0).

Of course, a single run does not provide a statistically valid sample. Table 12.3 and Figure 12.3 are included as representative samples providing a snapshot of typical algorithm performance. A more significant picture of the overall performance of the algorithms is provided in Tables 12.4 and 12.5. These tables show the average characteristics of the Pareto fronts discovered by the algorithms over 30 trials. Table 12.4 shows the mean value in each objec-

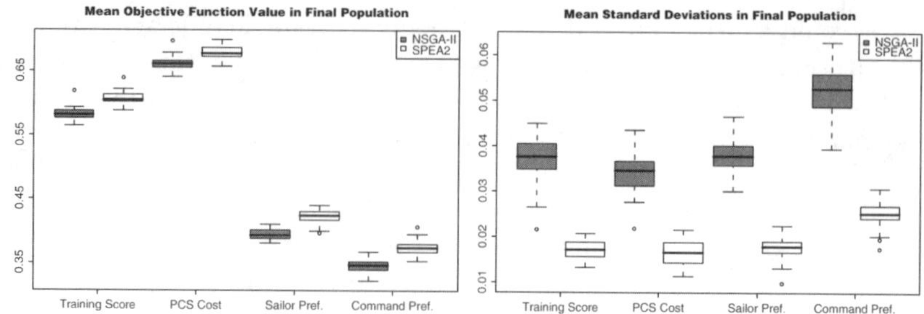

Fig. 12.4. Relative performance of SPEA2 and NSGA-II on a 1000 sailor instance of the SAP. Note that SPEA2 finds better solutions, but cannot maintain diversity along the front as well, while NSGA-II maintains more diversity but cannot approach the Pareto front as well as SPEA2

tive present in the final Pareto set approximation, while Table 12.5 shows the average spread along each objective, as measured by the standard deviation.

Figure 12.4 shows graphically the comparison between the SPEA2 and NSGA-II algorithms. The correct interpretation of Figure 12.4 may not be apparent. On the left, the plot is comparing the mean objective function values found in the final Pareto set approximation. On the right, the comparison is over the standard deviation along each objective function. Thus, the former represents the ability of the algorithms to converge toward the true Pareto front. The latter represents the maintainence of diversity along the discovered Pareto set approximation. Ideally, an algorithm would exhibit the following behavior. On the left, the boxes for the algorithm would be very high and thin, indicating that the algorithm found very high quality solutions in each objective with little variation in the quality between runs. On the right, the algorithm would be shown again with boxes that were very high and thin. This would indicate that the diversity of solutions found by the algorithm was large, and that the difference in diversity between runs was small. For the algorithms considered here, the boxplots show that SPEA2 finds better solutions on average with respect to any single objective, but cannot find as many meaningful alternatives. Note again that Figure 12.4 shows the mean and standard deviation of the mean and standard deviation found over 30 trials of the algorithms. Thus, it cannot generally be inferred that one algorithm is always better than another due to the smoothing effects of the mean.

Although Figure 12.3 shows only a single trial of the algorithms, Figure 12.4 and Tables 12.4 and 12.5 show that every trial results in very similar performance characteristics. Thus we may treat the runs shown in Figure 12.3 as a representative sample. We can then easily see that the multiobjective algorithms are having a difficult time producing any solutions competitive with CHC on any single objective, and the solutions found by CHC using a homogeneous weight are typically reasonably close to the best

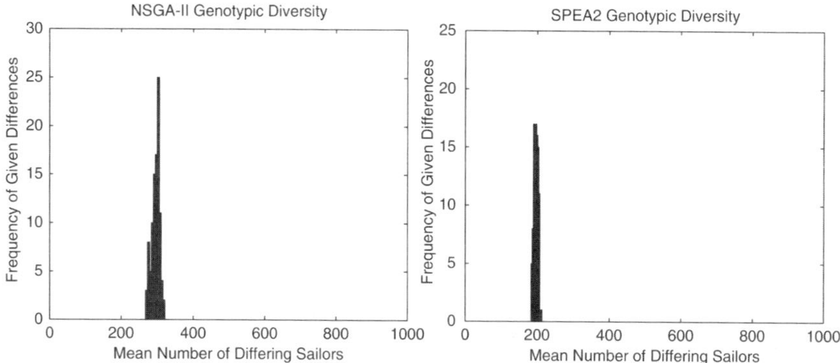

Fig. 12.5. Histogram of mean distance between solutions from the Pareto set approximation in terms of number of sailors assigned to different jobs

that the MOEA has to offer. In addition, the spread of solutions is quite small in the objective space.

Spread in objective space is only one measure of the diversity of solutions in the final Pareto set approximations. Another possible measure is the diversity of the solutions in the parameter space. For the SAP, the question is how many sailors are assigned different jobs in any two solutions. A large spread in the parameter space would mean that the detailer would have many options for each sailor, although the total training score, PCS cost, etc., may be very similar. Figure 12.5 shows a measure of diversity in the parameter space. For each trial, the solutions in the final Pareto set approximation are compared to each other to determine the number of sailors receiving different jobs. These differences are then averaged over all pairs of solutions to arrive at the frequency with which two solutions differ by a given number of sailors. Over all trials, the computed frequency is shown in histogram form in Figure 12.5. Given a 1000-sailor instance of the problem, NSGA-II finds solutions that differ in about 300 sailors on average, compared to an average difference of 200 sailors for SPEA2.

Thus, the MOEAs find solutions that are quite similar to one another in both the objective space and the decision space. Since one of the primary goals of any multiobjective algorithm is to find disparate solutions along the Pareto front, the performance of the two MOEAs tested here appears questionable. The spread of solutions in the decision space is certainly significant. However, as we will see later, these differences can be made much greater. In addition, in the objective space, the solutions are clustered in a small region. CHC using a vector of equal weights is able to find a solution which dominates a large percentage of the solutions found by the MOEAs. Given this result, any diversity maintained by the algorithms comes at too great a cost. Essentially every choice found by the algorithms is worse than the solution found by CHC.

12.7 Hybrid Multiobjective Evolutionary Algorithms

Because the solutions found by the MOEAs are clustering in regions of relatively low quality, we may be able to introduce diversity while simultaneously moving closer to the true Pareto-optimal front by pushing individual solutions in a direction known to be closer to the true front. In general, it is difficult to determine how changes in the decision variables will affect the objective functions. However, by considering one objective at a time, we can certainly move toward the extreme solutions on the true Pareto-optimal front. By moving selected points toward these solutions, we also introduce diversity into the population. Not only does the population become more diverse, but in some sense we can guarantee useful diversity. To achieve the stated goal, a hybrid algorithm utilizing conventional multiobjective evolutionary techniques along with a hill-climbing component allowing local improvements was developed to help produce a set of nondominated solutions both closer to the true Pareto-optimal front and more varied along the Pareto-set approximation.

The benefits of hybridization are well known in the genetic algorithm community for many single objective problems [25, 26]. However, hybrid multiobjective evolutionary algorithms have not received the same degree of attention as of yet. Knowles' M-PAES [22] is perhaps the best known hybrid MOEA. M-PAES is based on the Pareto Archived Evolution Strategy (PAES) of the same author. PAES is a $(1 + 1)$-Evolution Strategy, which is, like any other $(1+1)$-ES, a hill climber. Unlike most hill climbing algorithms, however, PAES uses Pareto dominance to determine the utility of a proposed move in the space. PAES maintains an external archive of nondominated solutions discovered during the run. The novelty of PAES lies in how offspring are compared to parents to determine which is better. As a $(1 + 1)$-ES, PAES maintains a single parent, generating a single child via Gaussian mutation with zero mean and a specified mutation strength. If the child dominates the parent, then the child replaces the parent and is added to the archive. If the child is dominated by the parent, then the child is discarded and another is generated via a new mutation. If the parent and child are mutually nondominated, each is compared to the archive. If the child is dominated, it is discarded. If the child dominates an archive member, then each such archive member is deleted, the child replaces the parent and is added to the archive. If the child and archive are mutually nondominated, then a density estimation technique is employed to determine the winner.

The extension to the Memetic PAES (M-PAES) produces a $(\mu + \lambda)$-ES involving crossover. Unlike PAES, however, M-PAES utilizes two distinct external archives. The first serves the same purpose as in PAES. However, a second archive is maintained for the purpose of determining improving moves during each local improvement stage. This second archive is composed of solutions from the first which do not dominate the current solution undergoing improvement. The authors report performance competitive with several other

MOEAs, although their work predates the more effective algorithms SPEA2 and NSGA-II.

Ishibuchi and Murata proposed one of the earliest hybrid MOEAs [18]. In their algorithm, each individual is evaluated using a randomly selected weight vector. Selection, recombination, and mutation are then performed as in a single-objective evolutionary algorithm. The resulting offspring are then improved by a local search operator using the same weight vector as the original evaluation of the individual.

Subsequently, Jaszkiewicz [19] proposed additional hybrid MOEAs using more sophisticated local search operators such as simulated annealing and also incorporating mating restrictions so that parents are similar to one another in terms of their objective function values. In comparisons with Ishibuchi and Murata's algorithm, Jaszkiewicz reported better results. However, as stated in [22], the test problems used in that study were convex and could be exploited by the mating restriction scheme.

More recently, proposed algorithms have attempted to utilize gradient information in multiobjective optimization. Bosman and de Jong [3], and Brown and Smith [4] have proposed algorithms which attempt to compute a gradient that may be followed leading to the true Pareto front. However, while such gradient descent approaches may be useful for many numerical optimization problems, combinatorial optimization problems and problems in which the structure of the fitness landscape is not well-behaved may present problems for these algorithms.

For this study, we chose a local search operator reminiscent of VEGA and that of Ishibuchi and Murata. However, unlike those studies in which the local search operator utilizes randomly selected weight vectors to optimize every newly created solution, we choose a small set of weight vectors designed to push solutions toward the extremes of the current Pareto set approximation. The goal of such an operator is to seed the archive with extreme values which then affect the trajectory of the evolutionary algorithm to find better solutions.

Each time a new individual is generated, it is subjected to a fixed number of evaluations of the hill climber with some (very low) probability. Unlike in many other approaches, our method was to allow the hill climber to find single points that are very good with respect to a single objective. This point is then re-inserted into the population where it affects the evolution of other individuals, providing a continuum of solutions along the Pareto front. To test this hypothesis, a next ascent hill climber was incorporated into NSGA-II and SPEA2.

In a conventional unconstrained optimization problem, a next ascent hill climber is defined by the neighborhood operator. From any point in the decision space, the neighborhood operator defines the set of points reachable in one "step" from the current point. The algorithm visits each of these points in turn, testing to determine if moving to the new point results in an improvement in the objective function. The first such improving move is taken, and the neighborhood operator is applied to this new point to iterate the algorithm. The algorithm terminates when no neighboring point provides an improvement in the objective function value.

Defining the hill climber for the SAP involves two questions: what is the neighborhood operator and what is the objective function. The neighborhood operator conventionally defines a small perturbation in the decision space. In the SAP, such an operator would attempt to find a different job for a single sailor such that the resulting assignment remains feasible. However, as discussed above in dealing with the mutation operator, such a function results in a fairly weak form of search, since the uniqueness constraints prevent us from exploring many moves that involve jobs already given to other sailors.

A better neighborhood operator needs to allow more of the possible neighboring moves to provide feasible solutions. To accomplish this, the operator is allowed to perform a single layer of conflict resolution. This means that to generate neighboring points, a sailor is chosen and the operator selects another possible job for that sailor. If no other sailor has the new job, then the operator modifies the assignment and adds the resulting solution to the set of neighboring points. If another sailor has previously been assigned the selected job, the algorithm attempts to find another job for this latter sailor such that both sailors can be reassigned resulting in a feasible solution. If no such available job can be found, then the algorithm moves on to the next possible neighboring point. Empirical evidence suggests that this one-step resolution operator provides roughly double the number of feasible neighborhood points of the naïve mutation operator. The hill climbing algorithm is sketched in Listing 1.

The drawback of a larger neighborhood embedded in a local search operator is that examining the neighborhood of a point requires greater computation time. Let $SAP(n, m, k)$ denote a randomly generated SAP instance with n sailors and m jobs, where each sailor is eligible for k different jobs. The degree to which the two sets k jobs available to two different sailors overlap determines the likelihood that a conflict will be encountered, requiring the conflict resolution step. The size of the neighborhood is in turn determined by the relative frequencies of the one-swap versus the two-swap moves, hereby denoted 1X and 2X.

We note that the probability that a single 1X operation is sufficient is equal to the probability that a randomly selected job is already taken by another sailor. As there are n sailors selecting k of m jobs each, we note that each job is selected by $\frac{nk}{m}$ sailors on average. The probability that a randomly selected sailor applied for any given job is thus $\frac{k}{m}$. Suppose that we choose a sailor and a job at random. We wish to find the probability that another sailor already has the selected job. $\frac{k}{m}$ denotes the probability that any single sailor applied for the job. However, any such sailor applied for k jobs in all, and thus the probability that the sailor actually is assigned the job in question is given by $\frac{k}{m}\frac{1}{k} = \frac{1}{m}$. As there are n sailors in all, the total probability of any other sailor having been assigned the job in question is thus $\frac{n}{m}$. Therefore, upon selecting

Listing 1 Local Improvement Operator

```
repeat
  Generate random sailor order
  for each sailor i do
    Get list of available jobs
    Generate random job order
    for each job j do
      if job j not taken then
        Evaluate new solution
        if solution improved then
          Accept move
        else
          Reject move
        end if
      else
        Find sailor i' holding job j
        Get available jobs for i'
        Generate random job order
        for each job j' in job list do
          if j' not taken then
            Assign sailor i' to job j'
            Assign sailor i to job j
            Evaluate new solution
            if solution improved then
              Accept move
            else
              Reject move
            end if
          end if
        end for
        if No available j' then
          Move to next sailor
        end if
      end if
    end for
  end for
until No improvement found
```

a random sailor and a random job to assign to the sailor, the job will be taken by another sailor with probability $\frac{n}{m}$.

If none of the 1X neighbors are feasible, the algorithm must begin to look in the 2X neighborhood for feasible neighbors. Because exactly one sailor will be assigned each of the possible jobs, the 2X neighborhood consists of looking at potentially all of the 1X neighbors resulting from applying the 1X operator to each of the $k-1$ sailors currently assigned to the $k-1$ jobs we originally tried to assign to the first selected sailor.

As a result, we can compute the expected neighborhood size as follows. Define X to be the number of neighbors evaluated given a particular sailor and a job to assign to the sailor. Then

$$\mathbf{E}[X] = \left(1 - \frac{n}{m}\right)|\mathcal{N}_{1X}(x)| + \left(\frac{n}{m}\right)|\mathcal{N}_{2X}(x)|,$$

where $|\mathcal{N}_{1X}(x)|$ is the number of solutions that must be examined to take a 1X move from solution x.. Because the $1 - \frac{n}{m}$ term ensures that the 1X move is feasible, $|\mathcal{N}_{1X}(x)| = 1$, since we need only to accept the feasible move.

On the other hand, after we learn that we must apply the 2X operator, there exists more than one possible move. We must reassign the conflicting sailor, but there are $k-1$ possible ways to perform this reassignment. Of those $k-1$ possible jobs, the number that are not already taken by other sailors is only $(k-1)\left(1 - \frac{n}{m}\right)$. Thus, $|\mathcal{N}_{2X}(x)| = (k-1)\left(1 - \frac{n}{m}\right)$. Additionally, these calculations assume that we have selected a single job to assign to the selected sailor. In reality, there are $k-1$ possible jobs we could choose from for each sailor, and n possible sailors to choose from. Thus, the actual expected number of neighbors we may need to examine is given by

$$\begin{aligned}\mathbf{E}[|\mathcal{N}(x)|] &= n\sum_{i=1}^{k-1}\left(1 - \frac{n}{m}\right)1 + \left(\frac{n}{m}\right)(k-1)\left(1 - \frac{n}{m}\right) \\ &= n(k-1)\left(1 - \frac{n}{m}\right) + n(k-1)^2\left(\frac{n}{m}\right)\left(1 - \frac{n}{m}\right),\end{aligned}$$

which is $O(nk^2)$. For typical problems of a few thousand sailors with perhaps 10 feasible jobs per sailor, the total number of possible moves from a given points is on the order of 10-50 thousand. While not overwhelmingly large, a hybrid evolutionary algorithm attempting to perform a complete local search on every individual would require relatively long computation times. This has implications on the effectiveness of multistart techniques for multiobjective optimization such as multiple runs of simulated annealing or tabu search. The time spent exploring the neighborhood as well as the need to find and maintain multiple solutions seems to indicate that such techniques are less desirable.

Because the SAP is a multiobjective problem, there is in general no obvious answer to the question of an appropriate objective function for the hill climber. A typical approach would be to consider an improving move to be one such that the resulting solution dominates the previous one. From this point of view, all mutually nondominated points form a plateau along which the hill climber cannot move, since no point is better than any other. This prevents the hill climber from improving the spread of solutions along the Pareto set approximation, eliminating one of the benefits of the hybridization. Instead, each point selected to undergo local improvement is paired with a randomly selected objective function chosen from the set of objective functions inherent to the problem. In the case of the SAP, each improvement phase operates exclusively on either training score, PCS cost, sailor rank score, or command rank score.

As a result, during the evolutionary run, certain points are occasionally chosen to undergo local improvement via the next ascent hill climber. Each such point is pushed toward the extreme solution on the Pareto front corresponding to the dimension associated with the chosen objective function. Because the point is then placed back into the population, it can affect the evolution of other points in the population. Crossover serves to spread the beneficial components of these solutions throughout the population where it helps to fill in the gaps between the extreme values found by the hill climber and the median values making up the population of the MOEA as a whole. As a result, the algorithm is able to spread more solutions across promising regions of the Pareto set approximation, providing additional choices for the decision maker.

One potential drawback of the hill climbing scheme described above is that the algorithm is biased toward solutions near the extremes of the Pareto front. While it is generally useful to ensure that these solutions are maintained, it is usually necessary to also provide solutions near the center of the trade-off surface. These solutions, not too bad in any objective, are often the favored solutions to the problem at hand. To help prevent the loss of these solutions, the hill-climber is also employed periodically using a weight vector treating each objective equally.

With no further modifications, one would expect the small number of possible weight vectors to adversely affect the diversity present in the population. However, two factors help to overcome this tendency in our implementation. First, the probability of applying the local search operator is very small – approximately 15 to 25 applications per 100,000 generated offspring. This is necessary for the SAP due to the relatively large cost of performing the local search. However, another implication is that the other genetic operators have a much greater chance to restore diversity between local search applications.

A second effect is due to the repair procedure invoked on each offspring. Crossover tends to violate a reasonably large number of constraints, in that it will often assign two sailors the same job, as is evident from closer examination of Figure 12.2. As a result, the repair procedure described above is employed to bring the offspring back into compliance with all constraints. While the repair procedure is not directed in the sense of a particular weight vector, it is certainly biased toward better solutions, in the trivial sense that solutions with few unassigned sailors typically dominate those with many. Combined with the relatively large number of changes made to the genome by the crossover/repair operators, this bias tends to provide a fairly diverse sample of possible solutions which helps to offset the loss of diversity caused by the local search operator.

One problem is that these "intermediate" solutions sometimes exhibit slightly more unassigned sailors than those very near one of the solutions found via local search. The immediate cause of this is the repair procedure, which is often unable to completely eliminated unassigned sailors. Two possible solutions exist. The simplest is to allow the repair procedure to perform a broader search. However, the cost of the repair procedure should be kept low, as it must be performed for every generated offspring. Another possible solution is to perform very short bursts of some form of local search on these repaired solutions to attempt to shuffle sailors around to allow for fewer unmatched sailors. One advantage of this method is that it could be applied sporadically to keep the search cost low. We hypothesize that such an approach might be fruitful, but further research must be done to ultimately determine the utility of such a scheme.

To ensure a fair comparison between the naïve MOEAs and their hybrid counterparts, the total number of fitness function evaluations was held constant. For all tests, each algorithm was allowed 100,000 fitness function evaluations before the algorithm was terminated. For the hybrid MOEAs, the total number of evaluations must be shared between the local search operator and the evolutionary aspect of the algorithms. As an example, for the 1000-sailor instance, the local search operator described below required an average of about 20,000 evaluations to conclusively reach a local optimum. Because the hybrid algorithm is limited to a total of 100,000 evaluations, the evolutionary aspects of the hybrid algorithm would have very little chance to affect the quality of solutions found. Thus, the local search operator was terminated after a fixed number of evaluations of the fitness function, regardless of local optimality. For the results reported here, the maximum number of evaluations allowed was set to three times the length of the chromosome, or three times number of sailors. In addition, because the focus of the local search is to simply seed the population with good solutions, the probability of undergoing the local improvement stage was set such that approximately 15-25 individuals were selected for the local improvement stage during the entire run. Figure 12.6 shows the Pareto fronts found by the two hybrid MOEAs on the 1000 sailor instance.

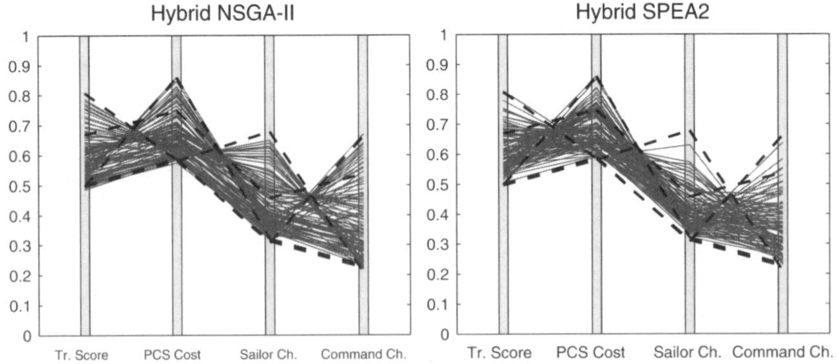

Fig. 12.6. Diversity of solutions found in the Pareto front using both the NSGA-II and the SPEA2 with hybridization on 1000 sailor instances of the SAP.

Table 12.6. Mean and standard deviation of the mean objective function values found over 30 different Pareto set approximations

Alg.	Size	Training Score		PCS Cost		Sailor Rank		Command Rank	
SPEA2	1000	0.60	2.9e-2	0.68	2.7e-2	0.42	3.0e-2	0.36	3.3e-2
	2000	0.59	1.9e-2	0.67	1.9e-2	0.46	2.2e-2	0.42	3.9e-2
	3000	0.59	2.6e-2	0.66	2.7e-2	0.48	3.3e-2	0.43	3.9e-2
	4000	0.59	4.5e-2	0.67	3.6e-2	0.44	4.3e-2	0.41	5.2e-2
NSGA-II	1000	0.60	9.9e-3	0.68	9.8e-3	0.43	1.1e-2	0.38	1.3e-2
	2000	0.59	1.0e-2	0.65	9.7e-3	0.46	9.1e-3	0.43	1.6e-2
	3000	0.60	9.1e-3	0.66	7.7e-3	0.47	9.9e-3	0.46	1.3e-2
	4000	0.60	7.1e-3	0.67	7.6e-3	0.44	1.1e-2	0.43	1.7e-2

Table 12.7. Mean and standard deviation of the standard deviation along each objective function found over 30 different Pareto set approximations

Alg.	Size	Training Score		PCS Cost		Sailor Rank		Command Rank	
SPEA2	1000	5.3e-2	1.3e-2	4.5e-2	1.2e-2	5.4e-2	1.5e-2	7.5e-2	1.9e-2
	2000	3.3e-2	1.1e-2	3.6e-2	8.6e-3	3.5e-2	1.2e-2	5.6e-2	1.4e-2
	3000	4.2e-2	1.5e-2	4.6e-2	1.4e-2	5.2e-2	1.3e-2	7.6e-2	2.0e-2
	4000	4.3e-2	2.1e-2	4.2e-2	1.6e-2	4.7e-2	1.6e-2	8.1e-2	2.1e-2
NSGA-II	1000	7.9e-2	5.3e-3	7.0e-2	3.3e-3	8.4e-2	4.6e-3	1.2e-2	7.5e-3
	2000	6.0e-2	3.7e-3	5.8e-2	2.9e-3	6.2e-2	5.1e-3	1.0e-2	6.1e-3
	3000	7.5e-2	4.6e-3	7.0e-2	4.4e-3	7.6e-2	5.1e-3	1.3e-1	6.7e-3
	4000	7.6e-2	3.3e-3	7.0e-2	4.2e-3	7.8e-2	5.1e-3	1.2e-1	6.7e-3

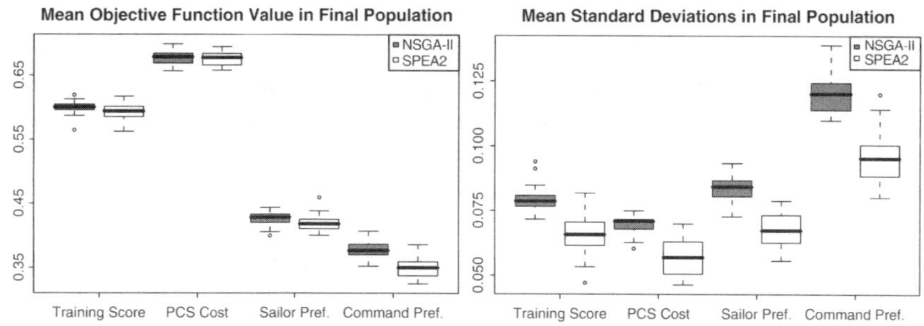

Fig. 12.7. Relative performance of the hybridized SPEA2 and NSGA-II on a 1000 sailor instance of the SAP. Hybrid NSGA-II typically finds slightly better solutions than the hybrid SPEA2, while also finding more diversity along the Pareto set approximation

Note that Figure 12.6 shows only a single run of the algorithms, in the same fashion as in Figure 12.3. In the same manner, we present the performance of the algorithms averaged over 30 trials in the following figures and tables. Table 12.6 shows the mean objective function values, while Table 12.7 shows the standard deviations in each objective, both averaged over 30 runs of the algorithms. Figure 12.7 shows graphically the performance of the hybridized

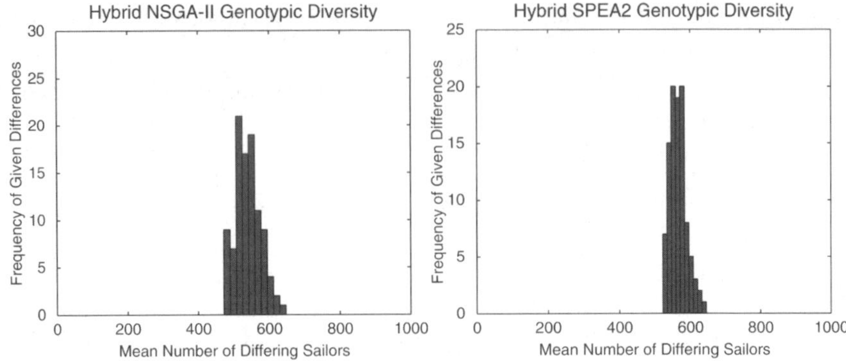

Fig. 12.8. Histogram of mean distance between solutions from the Pareto set approximation in terms of number of sailors assigned to different jobs

algorithms in terms of the mean and deviation along each objective function in the final Pareto set approximations. Comparing Figure 12.7 with Figure 12.4 shows that the hybridization produces very similar values for the mean in each objective function. However, the spread along each objective is much larger in the hybrid versions. Additionally, Figure 12.8 shows that in the decision space, the spread of the located solutions is significantly larger than those found by the MOEAs without the local search operator.

12.8 Conclusions and Future Work

In this chapter, we have presented a hybrid multiobjective evolutionary algorithm approach to solving the United States Navy's Sailor Assignment Problem. The proposed approach has been shown to provide solutions competitive in any single objective to solutions found by using very good single objective evolutionary algorithm. However, unlike the single objective formulations, the algorithm also provides a continuum of solutions in the interior portions of the trade-off surface. The insights gained from this application could potentially be useful to practitioners working on other difficult multiobjective problems.

Two well-known MOEAs, NSGA-II and SPEA2, were compared according to empirical performance on the SAP. NSGA-II was shown to provide greater diversity of solutions along the Pareto set approximation, while SPEA2 was typically able to find slightly better solutions. Neither algorithm was able to find solutions which were significantly better than those found by a CHC, a very effective single objective evolutionary algorithm utilizing a weight vector. The addition of a local search operator resulted in a dramatic improvement in the performance of both algorithms, NSGA-II in particular.

A local search operator was constructed and integrated into the MOEAs in such a way as to emphasize the importance of finding diverse solutions across the Pareto set approximation. The resulting hybrid evolutionary algorithms are able to leverage the points near the extremes in each objective to help find a continuum of solutions along the Pareto front. In addition, because the focus of the local search is on diversity rather than proximity to the true Pareto front, fewer evaluations of the are required. This allows the ability of the evolutionary algorithms to perform a global search to potentially find better solutions than would have been found by either an evolutionary algorithm or a local search algorithm alone.

The hybridization technique employed in this work is not the only possible method. Further work is needed to determine whether other local search operators or hybridization mechanism would provide additional improvements in performance. In addition, the development of improved domain specific genetic operators could potentially provide further improvements. In particular, a crossover operator which better maintains feasibility would reduce the need to repair individuals. Such an operator could potentially provide benefits both in solution quality as well as computation time.

Many previous approaches to the SAP have considered only a single objective. While our approach currently incorporates four distinct objectives, there are several additional components which require further investigation. Among the most important of these additional objectives are job priority and integration of what are known as AIP jobs, or Assignment Incentive Pay jobs, into the system. The former refers to the fact that not all jobs are equally important to fill. An effective algorithm should bias solutions toward filling the most important jobs. The latter refer to jobs that are typically difficult to fill. A bidding system provides the opportunity for sailors to apply for these jobs in return for some specified additional compensation. Minimizing the cost associated with these AIP payouts is another goal of the Navy. Integrating these additional goals into the algorithm increases the dimensionality of the objective space. Further work is needed to determine what effect this has on the performance of the algorithms.

Acknowledgements

This work was supported by the Navy Personnel Research, Studies, and Technology Department (James Simien) under the auspices of the U.S. Army Research Office Scientific Services Program administered by Battelle (Delivery Order 0495, Contract No. DAAD19-02-D-0001).

The authors wish to thank Professor Fernando Niño of the Computer and Industrial Engineering Department, National University of Columbia and Rodrigo Silva of the Department of Computer Science, University of Memphis for their valuable comments and suggestions.

References

1. A. Ali, J. Kennington, and T. Liang. Assignment with en route training of navy personnel. *Naval Research Logistics*, 40:581–592, 1993.
2. T. Blanco and R. Hillery. A sea story: Implementing the navy's personnel assignment system. *Operations Research*, 42(5):814–822, 1994.
3. P. Bosman and E. de Jong. Exploiting gradient information in numerical multi-objective evolutionary optimization. In Una May O'Reilly, editor, *Proceedings of the Genetic and Evolutionary Computation Conference (GECCO '05)*, 2005.
4. M. Brown and R. E. Smith. Effective use of directional information in multi-objective evolutionary computation. In E. Cantu-Paz et al., editors, *Proceedings of the 2003 Genetic and Evolutionary Computation Conference (GECCO '03)*, pages 778–789. Springer-Verlag, 2003.
5. Paola Cappanera and Giorgio Gallo. A multicommodity flow approach to the crew rostering problem. *Operations Research*, 52(4):583–596, July–August 2004.
6. Eranda Cela. *The Quadratic Assignment Problem: Theory and Applications.* Kluwer Academic Publishers, 1998.
7. D. Cvetković and I. Parmee. Use of preferences for ga-based multi-objective optimization. In W. Banzhaf et al., editors, *Proceedings of the 1999 Genetic and Evolutionary Computation Conference (GECCO '99)*, pages 1504–1509. Morgan-Kaufmann, 1999.
8. Kalyanmoy Deb. *Multi-Objective Optimization using Evolutionary Algorithms.* John Wiley & Sons, 2001.
9. Kalyanmoy Deb, A. Pratap, Sameer Agarwal, and T. Meyarivan. A fast and elitist multiobjective genetic algorithm: NSGA-II. *IEEE Transactions on Evolutionary Computation*, 6:182–197, 2002.
10. N. Drechsler, R. Drechsler, and B. Becker. Multi-objective optimization based on relation favour. In E. Zitzler et al., editors, *First International Conference on Evolutionary Multi-Criterion Optimization*, pages 154–166. Springer-Verlag, 2001.
11. L. Eshelman. The CHC adaptive search algorithm. In *Foundations of Genetic Algorithms I*, pages 265–283. Morgan Kaufmann, 1991.
12. Harald Feltl and Günther Raidl. An improved hybrid genetic algorithm for the generalized assignment problem. In *2004 ACM Symposium on Applied Computing*, pages 990–995, 2004.
13. P. Fleming, R. Purshouse, and R. Lygoe. Many-objective optimization: An engineering design perspective. In C. Coello-Coello et al., editors, *Third International Conference on Evolutionary Multi-Criterion Optimization*, pages 14–32. Springer, 2005.
14. Charles Fleurent and Jaques Ferland. Genetic hybrids for the quadratic assignment problem. *DIMACS Series in Discrete Mathematics and Theoretical Computer Science*, 16, Quadratic Assignment and Related Problems:173–187, 1994.
15. D. Gale and L. S. Shapley. College admissions and the stability of marriage. *The American Mathematical Monthly*, 69(1):9–15, 1962.
16. J. Deon Garrett, Joseph Vannucci, Rodrigo Silva, Dipankar Dasgupta, and James Simien. Genetic algorithms for the sailor assignment problem. In *Proceedings of the 2005 Genetic and Evolutionary Computation Conference (GECCO '05)*, 2005.
17. Allen Holder. Navy personnel planning and the optimal partition. *Operations Research*, 53(1):77–89, January-February 2005.

18. H. Ishibuchi and T. Murata. Multi-objective genetic local search algorithm. In T. Fukuda and T. Furuhashi, editors, *Proceedings of the 1996 International Conference on Evolutionary Computation*, pages 119–124, 1996.
19. A. Jaszkiewicz. Genetic local search for multiple objective combinatorial optimization. Technical Report RA-014/98, Institute of Computing Science, Poznan University of Technology, December 1998.
20. Hamid Kharraziha, Marek Ozana, and Sami Spjuth. Large scale crew rostering. Technical Report CRTR-0305, Carmen Systems, September 2003.
21. J. D. Knowles and D. Corne. Memetic algorithms for multiobjective optimization: Issues, methods and prospects. In N. Krasnogor, J. E. Smith, and W. E. Hart, editors, *Recent Advances in Memetic Algorithms.* Springer, 2004.
22. Joshua D. Knowles and David Corne. M-PAES: A memetic algorithm for multiobjective optimization. In *Proceedings of the 2000 Congress on Evolutionary Computation (CEC-2000)*, pages 325–332. IEEE Press, 2000.
23. Mark W. Lewis, Karen R. Lewis, and Barbara J. White. Guided design search in the interval-bounded sailor assignment problem. Technical Report HCES-04-04, Hearin Center for Enterprise Science, The University of Mississippi, April 2004.
24. Lee McCauley and Stan Franklin. A large-scale multi-agent system for navy personnel distribution. *Connection Science*, 14(4):371–385, December 2002.
25. Peter Merz and Bernd Freisleben. A genetic local search approach to the quadratic assignment problem. In *Proceedings of the Seventh International Conference on Genetic Algorithms*, 1997.
26. P. Moscato. Memetic algorithms: A short introduction. In D. Corne, F. Glover, and M. Dorigo, editors, *New Ideas in Optimization*, pages 219–234. McGraw-Hill, 1999.
27. Walid El Moudani, Carlos Alberto Nunes Cosenza, Marc de Coligny, and Félix Mora-Camino. A bi-criterion approach for the airlines crew rostering problem. In Eckart Zitzler, Kalyanmoy Deb, Lothar Thiele, Carlos A. Coello-Coello, and David Corne, editors, *First International Conference on Evolutionary Multi-Criterion Optimization*, pages 486–500. Springer-Verlag, 2001.
28. N. Radcliffe. Forma analysis and random respectful recombination. In R.K. Belew and L.B. Booker, editors, *Proceedings of the Fourth International Conference on Genetic Algorithms.* Morgan Kaufman, 1991.
29. J. D. Schaffer. Multi-objective optimization with vector evaluated genetic algorithms. In J. J. Grefenstette, editor, *Proceedings of an International Conference on Genetic Algorithms and Their Applications*, pages 93–100, 1985.
30. Eckart Zitzler, Marco Laumanns, and Lothar Thiele. SPEA2: Improving the strength pareto evolutionary algorithm. Technical Report 103, Computer Engineering and Networks Laboratory (TIK), Swiss Federal Institute of Technology (ETH), Zurich, Switzerland, May 2001.

13

Evolutionary Techniques Applied to Hardware Optimization Problems: Test and Verification of Advanced Processors

Ernesto Sanchez and Giovanni Squillero

Dipartimento di Automatica e Informatica - Politecnico di Torino, Torino – Italy
edgar.sanchez@polito.it, giovanni.squillero@polito.it

Summary. In this chapter, a software-based methodology to automatically generate test programs is described. The methodology is based on an evolutionary algorithm able to generate test programs for microprocessor cores, and may be used for different processors since their instruction set architecture is described appropriately, and because a feedback can be defined, computed, and used to drive the test program generation process. The usefulness of the methodology is backed up by the presentation of three different cases of study: the first one tackles the verification of the DLX/pII processor; the second one generates post-silicon verification programs for the Pentium® 4; and the third one evolves a test set for the PLASMA processor. The gathered experimental results demonstrate the algorithm versatility and efficiency.

13.1 Introduction

Evolutionary algorithms (EAs) do have a significant exploitation in the field of the electronic Computer Aided Design (CAD). Several optimization problems are effectively tackled resorting to evolutionary heuristics, since no exact methodology appears usable. This chapter describes the successful usage of an evolutionary tool for devising programs suitable for test and verification of microprocessor cores. The reported cases of study experimentally demonstrate the effectiveness of evolutionary computation applied to very complex problems.

In the last years, the performance of processor and microprocessor cores has impressively increased due to mainly four aspects: huge quantities of resources, high work frequencies, low power and elevated parallelism. Even though all these issues are correlated, physical resources, work frequencies, and low power are directly backed by advances in technology, while parallel execution depends on the evolution of the processor architecture.

In early 70s the first commercial microprocessor was released: the *4004* by Intel® was built with about two thousand transistors and the clock frequency

E. Sanchez and G. Squillero: *Evolutionary Techniques Applied to Hardware Optimization Problems: Test and Verification of Advanced Processors*, Studies in Computational Intelligence (SCI) **66**, 303–326 (2007)
www.springerlink.com

reached about 100kHz. Currently, synthesis technologies allow logic densities of about 90 million transistors per square centimeter, and taking into account the increasing pace of technology advances that reach at least a 10% reduction of the feature-size a year, it is likely that in the next years, transistor densities go beyond 120 million transistors per square centimeter. Additionally, Very-Large-Scale Integration (VLSI) circuits achieve clock rates in the GHz range and the power consumption per transistor is ever decreasing, thanks to operation voltages below 1 volt. In fact, evaluating at a glance some facts on modern processors, clock frequencies and transistors have been incredibly increased. For example the number of transistor of an Intel® Core™ 2 Duo is about 291 millions, 126,000 times the *4004*, while the clock frequency is about 20,000 times higher.

The evolution of the processor architecture, on the other hand, is being characterized by a high regularity of development, from the initial Von Newman machines up to today's speculative or hyper-threaded processors, processor features have been supported by the advantages in technology. Initial processors were distinguished by an in-order sequential execution of instructions, in fact the preliminary instruction cycle was very rigid and the parallel instruction execution was not possible. Soon after, the evolution of instructions allowed executing extremely complex operations in order to perform a series of multifaceted functions, such as the *LOOP* instruction present in the x86 architecture that executes in a row a register decrement, a comparison with zero, and depending on the comparison results a jump to a label. However, complex instructions were not as frequently used as processor architects thought; instead, a small set of very simple instructions were very frequently used. This led to a paradigm change in processor architecture.

The parallelization tackled the execution stage of the instruction cycle: pipelined microprocessors allowed executing some instructions in parallel. Thus, the bottleneck became the instruction dispatch not the instruction execution. Later, the introduction of superscalar processors that exploit not only parallel execution but also parallel issue, overcame this situation.

These advanced architectures are very complex to conceive and design, but even more complex to verify and test. Design verification is defined as the *proof* that the intent of a design is preserved in its implementations; in other words design verification is the process of corroborating that all modeled behaviors of a design are consistent with another model, usually described at a higher level of abstraction. Processor testing is related to the final design steps, where an already produced microprocessor chip is checked to reveal possible production errors, i.e., defects in the production process. In both cases, the processor under verification or test must be shown to execute instructions correctly, and to operate within the timing constraints given by the specification. Although different, verification and test are strictly related activities, because they both consist in checking the result of a step in the design process (the circuit architecture, or the produced chip) against the specifications, aiming at detecting possible errors (design errors in verification, production errors in test).

As maintained by the SIA05 roadmap [1], today most of the integrated circuit (IC) manufacturing cost is brought about by testing and verification processes. The increasing difficulty to generate appropriate testing and validation stimuli and the expensive elaboration times required to fault simulate an IC, raised these costs up to near 70%. Moreover, test and verification methodologies do not progress at the same pace the manufacturing technology does, contributing to increase the cost gap.

A fundamental step in test and verification of processor cores is the execution of carefully manufactured programs. These programs, usually called *test programs*, are composed of valid sequences of assembly instructions.

The quality of test and verification processes directly relies on the processor features excited by the test programs used. It should be pointed out that in these contexts the quality of a program is measured by its coverage capacity: in terms of production defects in the case of test, and in terms of design errors in the case of verification. Furthermore, code size, and execution time may be also considered relevant measures of test programs.

Considering the problem of test program generation, it is necessary to tackle the high parallelization on instruction execution made possible by modern architectures. Therefore, a test program must be able to test all features of the processor, but it makes no sense either trying to test individual instructions, or trying to execute all possible sequences of instructions.

Manual generation of test programs does not seem to be an efficient solution tackling modern processors, due to the number of specific cases and instruction interactions to fulfill. Manually written programs are useful to check some specific behavior that is known to be critical and difficult to cover by test programs built with other techniques.

A general solution for a high-quality test program is to devise automatic generation methodologies. An automatic approach able to generate test programs should be characterized by:

- high flexibility regarding the target microprocessor, in order to allow the maximum applicability of the method
- syntactically correct generation of assembly programs depending on the specific singularities of the target processor
- high versatility with respect to the evaluation system in order to allow tackling different problems such as test or verification
- the ability of driving the generation process exploiting coverage metrics, for example fault coverage, as feedback values.

In the following, the evolutionary algorithm that has been developed to automatically generate test programs is described. Additionally, the algorithm usefulness is backed up by the presentation of 3 cases of study tackling different flavored problems.

13.2 μGP

Alan Turing was probably the first to suggest that a computing machine could manipulate itself just as well as any other data. However, the pioneering attempts of generating programs in an automatic way date back to 1958, when Friedberg tested randomly mutated code against a given problem trying to cultivate a valid solution [2] [3]. Probably the required computational effort was the most limiting factor, and these experiments showed limited success. Better results were obtained in the 1980s limiting the scope and targeting simple mathematical *expressions* instead of full Turing-complete programs, as in most Genetic Programming (GP) applications.

One of the first successful attempts to cultivate a test program was reported in the DATE conference in 2001 [8]. Test programs were built concatenating macros, and a genetic algorithm was eventually exploited to choose the optimal values of the constants. The approach was experimented tackling an Intel 8051 and attained a quite good result. However the methodology was not extensible to more complex microprocessors: the usage of a set of fixed macros prevented the stimulation of several interactions between instructions inside a more complex processor architecture such as a pipelined one.

Learning from the experience a new evolutionary algorithm for evolving programs was developed by the same group. The new tool, called μGP (MicroGP), was originally devised to overcome the limitations of [8] in terms of versatility and applicability. μGP was expressly designed for generating syntactically correct assembly programs. It is able to handle variable size individuals, fully exploit assembly peculiarities (e.g., different addressing modes, instruction set asymmetries), and generate complex program structures (e.g., loops, subroutine calls and interrupt handlers).

The first working prototype of μGP was presented in 2002, but its development was discontinued only in 2006, and in 4 years of continuous development several new features were added to the original project. In 2006 the μGP was completely redesigned and rewritten in a object-oriented language (μGP^3). However, the new tool is still in a preliminary phase and will not be discussed here.

In more details μGP is composed of clearly separated blocks: an *evolutionary core*, an *instruction library*, and an *external evaluator*. The evolutionary core cultivates a population of individuals. It uses self-adaptation mechanisms, dynamic operator probabilities, dynamic operator strength, and variable population size. The instruction library is used to map individuals to valid assembly language programs. It contains a highly concise description of the assembly syntax or more complex, parametric fragments of code. The external evaluator evaluates the assembly program exploiting a simulator or other tools and eventually provides the necessary feedback to the evolutionary core.

Programs are represented internally as directed graphs. Graphs are composed of nodes, each one mapped to a macro (a generic fragment of code,

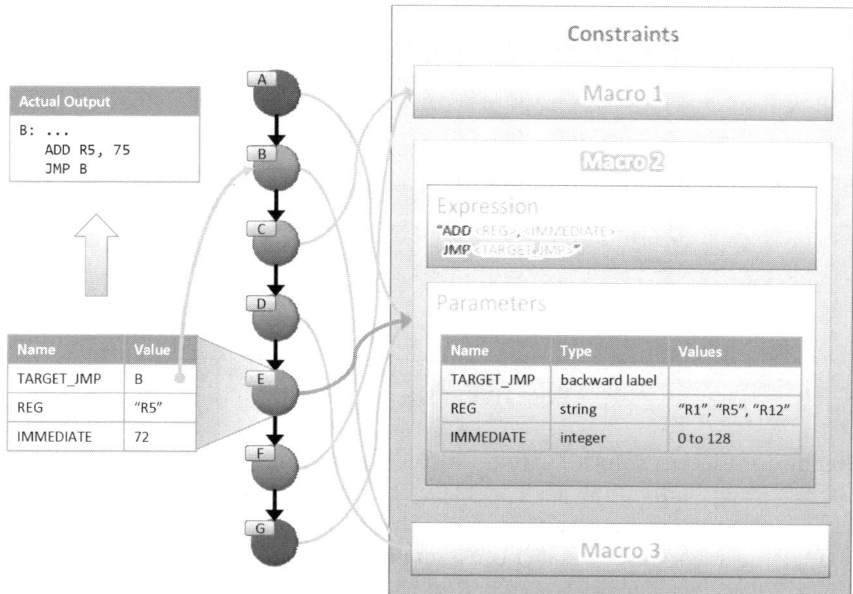

Fig. 13.1. Test program example

possibly with parameters). Figure 13.1 represents a fragment of a possible test program.

Test programs are generated by an evolutionary algorithm implementing a pseudo $(\mu + \lambda)$ strategy modifying graph topologies and mutating parameters inside nodes. A population of μ individuals is cultivated, each individual representing a test program. In each step, λ' new individuals are generated applying λ genetic operators. Each genetic operator may produce a variable number of new individuals. Parents are selected using tournament selection with tournament size τ (i.e., τ individuals are selected randomly and the best one is picked), while operators are selected using a roulette wheel mechanism on their activation probabilities. Finally, the best μ programs in the population of $(\mu + \lambda')$ are selected to survive to the next generation (see Fig. 13.2).

The fitness evaluation is intentionally external to μGP. The *external evaluator* is requested to compute a fitness value for each program, using any required tool. To further extend the usability, μGP was enhanced to handle a fitness value $F = (f_1, f_2, \ldots f_T)$ composed of T terms of strictly decreasing importance, i.e., $f^1 > f^2$ if $\exists i \leq T : ((j < i \Rightarrow f^1_j = f^2_j) \wedge f^1_i > f^2_i)$. The parameter T must be selected at the beginning of the evolution process.

Analyzing the graphs, μGP is able to detect if two individuals are identical. In such a situation, the fitness values of the *clones* may be scaled down by user defined quantities.

Moreover, since in a test-program generation problem maintaining a high degree of diversity in the population is critical, μGP is able to estimate the

```
evolution() {
  P ← μ random individuals;
  while not stopping conditions {
    repeat λ times {
      GO ← select one genetic operator
      A ←select parents in P
      O ←generate offspring from A, applying GO (A)
      P ←P ∪ O
    }
    sort P on individual fitness
    keep the best μ individuals in P and discard the others
  }
}
```

Fig. 13.2. μGP pseudocode

amount of unique genetic material introduced in the population by each individual. Then, individuals that increase the population diversity are favored.

The amount of novelty is measured considering the whole population as a single message and calculating Shannon's Entropy. Then, a measure called *Delta Pseudo Entropy* is associated to each individual. The approach used to favor certain individuals is based on Riccardo Poli's fitness hole [5] and can be easily tuned to fit the problem needs.

Last but not least, to allow the reuse existing code, μGP has the ability to analyze a set of existing programs and to translate them back to a possible internal representation. Since μGP has the ability to generate almost any fragment of code, translating generic programs (phenotypes) back to their originating graphs (genotype) led to several, non-trivial practical problems. Moreover, the mapping from genotypes to phenotypes is problem-dependent: depending on the *instruction library*, a single fragment of code may correspond to different sets of macros (the representation may be non-univocal), or a given fragment of code may be even impossible to generate.

The existing code is fully *assimilated* by μGP: the evolutionary core is able to modify the code, or mix fragments of codes originally in different programs exploiting a process that is akin to sexual recombination.

Almost all parameters of μGP are either endogenous or automatically self-adapted parameters by the algorithm. A more detailed description of μGP can be found in [4] and [6].

13.3 Cases of Study

The effectiveness and feasibility of the test program generation methodologies based on the evolutionary algorithm described above have been attested in very different problems ranging the whole design cycle of processor cores. In the next sections three cases of study are presented tackling diverse processors that range form a VHDL description of a DLX/pII at RT level, up to the first silicon of a Pentium® 4.

During the complete design cycle of microprocessor cores, these devices require to be constantly tested. At earlier stages, verification process aims at guaranteeing that the delivered model conforms to initial specifications. Similarly at final stages, the produced processor must undergo manufacturing tests to assurance the final processor correctness. In both cases verification and testing, functional tests based on test programs are being used to support the whole testing process.

The methodologies described in the following are based on a step-by-step approach: test program generation is performed each time a new microprocessor model is delivered. Moreover, being a cumulative process, that tackles high-level descriptions, the approach minimizes the use of explicit fault simulation to drive the test program generation, greatly reducing computational efforts. In fact, fault simulation is only used at the end of the generation to complete the set of test programs.

Additionally, exploiting the evolutionary algorithm described before, a set of programs, such as the functional programs developed by designers or tests automatically-generated in early phases of the processor life cycle, can be automatically extended to a complete test set reaching the necessary coverage on all chosen coverage metrics. Reusing existing material cuts down the time required to generate a set of test programs.

First Case Study: Pre-Synthesis Processor Verification

µGP was exploited for the automatic program generation along with the program assimilator: the module able to translate existing programs and fragments of code to graphs. The proposed approach was able to enhance existing test sets, adding new content for completing them or targeting different metrics. Previous verification sets were not merely included in the new set of programs for the verification, but assimilated and used as a starting point for the new test program cultivation task.

A DLX/pII microprocessor [13] was targeted. The DLX/pII is a small microprocessor with a 5-stage pipeline. The core under verification is described in VHDL at the RT-level with 4,558 statements. The description contains 3,695 branches that can be activated, and 193 conditional statements based on 1,764 different expressions. The design contains 8,283 logic bits that can be potentially toggled. Due to its size and complexity, the DLX/pII is a typical example of a small microprocessor core designed for a SoC.

The initial instruction library for the DLX/pII consists of 98 macros, containing the information to map a large amount of the microprocessor instructions and operands.

As stated by Piziali in [14], the real success of a simulation-based verification process relies on the adequacy of the initial verification route-map, called *functional verification plan*. Roughly speaking, the functional verification plan is composed of three aspects: *coverage measurement*, defining the verification problem, the different metrics to be used and the verification progress;

stimulus generation, providing the required stimulus to thoroughly exercise the device obeying to the directives given; and *response checking*, describing how to demonstrate that the behavior of the device conform to the specifications.

Within the microprocessor design life cycle, each time a new stage is reached, a new processor model is delivered and it must be verified. Comparing the behavior of the new model with the previous one is required, but all the new features must also be checked. A set of test programs able to excite the required functionalities is eventually exploited to unveil possible design errors.

The methodology is particularly suited for a multi-stage design strategy: as the microprocessor core is modified and detailed in the different design stages, the test programs are modified to handle the increasing complexity of the model. This approach allows to improve the quality of the generated set, and to significantly reduce the computational effort to generate new test programs.

Verification Plan Driven by Code Coverage Metrics

Code coverage metrics directly derive from metrics used in software testing. These metrics identify which code structures belonging to the circuit description are exercised by the set of stimuli, and whether the control flow graph corresponding to the code description has been thoroughly traversed. The structures analyzed in code coverage metrics range from a single line of code to if-then-else constructs.

Performing a verification process guided by code coverage metrics allows achieving ample design verification limiting the redundant efforts. Code coverage metrics act as heuristic measures for quantifying the verification completeness and identifying inadequately exercised design aspects.

As supported by many authors, while a high coverage figure increases the confidence in the design correctness, it is not possible to select one single code coverage metric as the most reliable [15] [16]. A complete coverage on any particular metric would never guarantee a 100% flaw free design, nor thorough code exploration. Therefore, the current verification trend is to combine multiple coverage metrics together to obtain better results.

The five considered coverage metrics were: *statement coverage* (the percentage of executable statements in the model that have been exercised during the simulation); *branch coverage* (the percentage of Boolean expressions evaluated to both true and false); *condition coverage* (the percentage of Boolean subexpressions evaluated to both true and false); *expression coverage* (as condition coverage, but calculated against concurrent signal assignments); *toggle coverage* (the percentage of bits that toggle at least once from 0 to 1 and at least once from 1 to 0 during the simulation).

Automatic Completion of Existing Functional Test Sets

An existing test set composed of 12 programs was used as a starting point for the experiments. The test set included six functional test programs written during the design process. The set also includes six application programs implementing simple algorithms (e.g., 32-bit multiplication).

After the assimilation of the initial set, μGP was asked to complete it, adding new content in order to maximize the 5 coverage metrics mentioned previously. μGP was set to generate new test programs targeting one metric at a time.

The final test set is composed of 31 programs: the 12 initial ones and 19 automatically generated by μGP. Results are summarized in Table 13.1.

It can be easily noticed that it is quite easy to saturate the statement coverage. Thus, while a program that executes all statement is required in any verification test set, reaching complete statement coverage is not probably sufficient.

On the other hand, the toggle coverage is a very hard metric and it does not saturate easily. Toggle coverage is clearly related to the data loaded into registers and not only to the control flow of the evaluation.

As a result of the completion process, the obtained set of programs outperforms the previous one. This shows the effectiveness of the evolutionary algorithm enhanced with the assimilation process.

For the sake of comparison, a test set was devised using a purely random approach. The random instruction generator exploits the μGP instruction library to devise syntactically correct fragments of code and the external evaluator used by μGP to simulate them. Differently from the evolutionary experiments, however, the approach is merely cumulative: coverage figures are not used as a feedback to optimize candidate programs, but a program is added to the test set whenever it increases at least one coverage figure.

About 20,000 random programs were generated and evaluated in about one week. The final test set contains about 230 K lines of code in about 2,000 random programs.

Table 13.2 shows the cumulative results obtained by the random experiment related to each code coverage metrics. The row size reports the total instruction lines that compose the set of random programs.

Table 13.1. Automatic Completion of Designers Functional Test Set

	Initial [%]	Enhanced [%]	Size [inst]	CPU Time [h]
Statement coverage	68.91	99.69	268	20
Branch coverage	52.49	99.10	118	33
Condition coverage	43.01	80.31	148	27
Expression coverage	26.93	53.71	226	37
Toggle coverage	30.34	85.29	285	95

Table 13.2. Pure Random Approach verification set

	Random
Statement coverage [%]	99.30
Branch coverage [%]	98.59
Condition coverage [%]	79.79
Expression coverage [%]	48.75
Toggle coverage [%]	56.84
Size [lines]	230 K

As already remarked, the statement, branch and condition coverage can be saturated easily, although µGP shows slightly higher performances. Reaching high expression coverage is more difficult and the random approach is significantly worse than µGP.

It is interesting to see that toggle coverage is a random resistant metric. 20,000 random programs are only able to attain a 57%, while the evolutionary approach surpasses 85%. This fact indicates that a feedback-based mechanism is required to toggle all register bit in a design, and that suitable test programs are slowly evolved and tweaked to increase their effectiveness. Indeed, experiments targeting the toggle coverage were the longest.

Second Case Study: Post-Silicon Processor Verification

As sketched in the introduction, methodologies exploiting the feedback from a simulator to generate test programs minimize human intervention; however, most of the approaches belonging to the state of the art of test program generation techniques are hardly used to tackle high-end processors. The fact may be easily explained: leading-edge microprocessors models are not freely available to researchers, and even when models are accessible, the computational effort required to simulate them is excessive.

Tackling high-end processors, verification is performed using a set of carefully chosen programs to provide wide coverage of application and benchmarks that cover all market segments. The set of programs must range from micro-benchmarks up to desktop applications. As described in [18], performance validation is becoming a necessary stage in the general verification process, because delivering a functional processor is not enough to the current processor marketplace. Initially, pre-silicon performance verification is performed using performance simulators to assert that the processor implementation achieves the desired levels of performance and that this result does not degrade during the design convergence step. Secondly, post-silicon performance verification is performed resorting to the first silicon; the basic micro-architecture features are measured and compared to the expected performance specifications and to the previous processor family implementation.

Most modern high-performance processors have special on-chip hardware that can monitor some of their internal behavior. This hardware allows identifying the causes of performance degradation, and typically includes event detectors and counters, qualification of event detection and counting by privilege mode and event characteristics and support for event-based sampling. The case study presented in this section exploits μGP using an interesting approach to automatically generate test programs that takes advantage of native monitor hardware of microprocessors. The automatic program generator is run on the microprocessor under test; candidate programs are executed (not simulated), and feedback is eventually extracted from the appropriate performance counters to drive subsequent optimization steps. Thus, the microprocessor under test is exploited twice: for running the test program generator and for evaluating candidate programs.

The obtained programs, derived from the presented approach, are suitable test programs that can be added to the set of test programs that stress the processor micro-architectural features for the post-silicon performance verification.

Moreover, the proposed test program generation architecture could be implemented by the final user to generate test programs able to excite critic conditions of the microprocessor to perform a qualitative incoming test.

Specifically, the approach has been tested on a Pentium® 4 microprocessor, and again, an instruction randomizer was used to corroborate the method suitability.

As stated before, with modern processors functional test programs must be targeted at specific micro-architectural features and generating them by hand is a challenging task. Engineers must understand the micro-architectural implementation of the design as well as how to write a full chip functional test to excite and observe the possible design faults.

Today several microprocessors (e.g., PowerPC™, Intel® Pentium®, AMD Athlon™, Sun UltraSPARC™, etc) include hardware performance counters that allow monitoring events like cache misses, pipeline stalls and floating-point operations. While the performance counters were mainly introduced to help identifying the causes of performance degradation, they can also be exploited for the test program generation process. Thus, a possible solution for automatic test program generation for high-end processors is to rely on a trial-and-error automatic approach able to exploit internal performance counters.

The basic structure of the proposed approach is presented in Figure 13.3. The target processor is used twice: for hosting the evolutionary algorithm, and for evaluating program candidates. Thus, μGP is executed on the target microprocessor and candidate test programs are not simulated on a model of the target microprocessor, but rather executed by the target microprocessor itself. The feedback information regarding each candidate program resorts to the microprocessor performance counters. Thus, extracting performance counters information, it is possible to excite better the specific unit under evaluation.

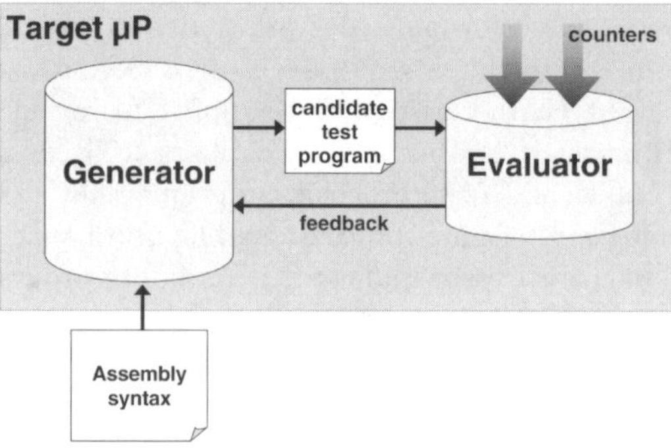

Fig. 13.3. Proposed architecture

A major advantage of this solution lies in the removal of the simulation overhead; moreover, there is no more the need for a simulation model of the target processor. However, since performance counters monitor the microprocessor activity at the lowest possible level, new problems may arise.

The Pentium® 4 Microprocessor

Performance monitoring was introduced in the IA-32 architecture in 1993 with the Pentium® microprocessor as a set of model-specific performance-monitoring counters. These counters permit a selection of processor performance parameters to be monitored and measured. The information obtained from these counters can then be used for tuning system and compiler performance [19].

The counters architecture of the Pentium® 4 microprocessor represents a definite enhancement and overcomes many of the limitations of its predecessors [20]. The Pentium® 4 microprocessor supports 48 event detectors, 18 event counters and 18 counter configuration control registers, enabling the concurrent collection of a large set of event counts. Identifiable events range from page misses to fetch operations requested by the branch prediction unit. The Pentium® 4 microprocessor also provides several instruction-tagging mechanisms that enable counting of non-speculative performance events (e.g., events generated by instructions that retire).

Pentium® 4 microprocessor performance monitors are used as a proxy for the creation of certain micro-architectural events. The maximization/minimization of the appropriate counters is used to stress specific micro-architectural features and excite subtle corner cases. The generated test programs could be added as new content to improve existing validation suites such as those presented in [18]. In addition, such a kind of programs can be used to perform regression testing on future processor models.

In this case study the events are the misprediction of branches and the execution from the trace cache. While the concept of mispredicted branches deserve no further explanation, it may be useful to sketch the basic concepts behind the trace cache.

IA-32 instructions have a variable number of bytes and many different options, thus they are complex to decode. The instruction decoding logic of the Pentium® 4 convert the IA-32 instructions into simpler micro-operations (μops). The trace cache is located between this instruction decode logic and the out-of-order execution core. Typically the instructions are decoded once and the corresponding μops placed in the trace cache, then μops are used repeatedly from there like from a normal instruction cache. The IA-32 instruction decoder is only used when the machine misses the trace cache and needs to go to the L2 cache to get and decode new IA-32 instruction bytes [21]. The case study focuses on maximizing/minimizing the percentage of cycles in which the trace cache is able to deliver μops to the execution unit.

Accessing Performance Counters

There are different approaches to extract information from the performance counters. The simplest one is the time-based sampling, and consists in periodically accessing counters through the *RDMSR* instruction (read machine-specific register) or the *RDTSC* (read time stamp counter). This approach has a limited overhead, but there is no mechanism to isolate the instructions that caused key performance events. The intrusiveness of time-based sampling may be tuned by modifying the sample ratio.

A different approach is the event-based sampling, and consists in generating an interrupt whenever a specified event occurs. The interrupt handler is eventually able to identify (with a certain approximation) what instruction caused the event. Event-based sampling is more intrusive than the former approach, but the overhead may be reduced using a set of internal registers for counting the events and hooking the interrupt routine on the overflow condition. However, there is no guarantee that all the counted events originated from the same instruction. In event-based sampling, the instruction that caused the performance event cannot be identified with absolute precision due to the deep pipelines of the Pentium® 4 microprocessor, the superscalar execution, and the latency between the counter overflow and the actual interrupt.

A variant of the previous approach is the precise event-based sampling. Instead of issuing an interrupt at the level of instructions, the Pentium® 4 brand handle the event at the level of μop. Using this mechanism, precise event-based sampling may unambiguously identify instructions that cause key performance events. Indeed, as in imprecise event-based sampling there is no guarantee that all the counted events originated from the same instruction. Precise event-based sampling only supports a limited number of non-speculative performance events.

Experimental Setup

A prototype of the proposed approach was implemented and tested on a Pentium® 4 microprocessor running the Linux operating system. The prototype exploits µGP and some different tools to access the Pentium® 4 performance counters.

The instruction library for µGP describing the IA-32 consists of 1,711 lines (239 macros). To access the performance counters, three different tools have been tested: *VTune*, *Brink and Abyss* (BaA), and *OProfile* (OP). The first is a commercial product distributed by Intel® and it was merely used to check the final results. The last two are distributed as source code under the GNU Public License and are very easily controllable with scripts and command line switches. The interfaces between the tools and µGP cumulatively require about 300 lines of ANSI C, and a few scripts.

In the experimental evaluations, candidate programs were created as assembly programs; then they were assembled using NASM (Netwide Assembler), linked using GNU ld and eventually run. The environment exploited the operating system in order to perform several tasks (e.g., accessing the file system), and relied on external tools, such as the assembler and the linker.

This approach can be easily implemented and does not require any additional hardware, but since the performance counters monitor the microprocessor activity at a low level, it is not easy to distinguish between activity caused by the program under exam and activity caused by concurrent tasks and the operating system itself. Additionally, candidate test programs are usually very small and the percentage of activity caused by their executions may be negligible compared to the standard startup code inserted by the linker.

Imprecise event-based sampling is not a solution to the first problem since, as mentioned before, while the source of the last event can be determined with acceptable precision, there is no guarantee about all the previous ones. The number of events required for a counter to trigger the overflow interrupt must be carefully tuned, and may not be reduced under a certain threshold or the microprocessor would spend most of its time executing the interrupt handlers.

To reduce these issues, the evaluation of the candidate program was performed at the highest priority supported by the operating system (multitasking was not disabled because it was required by the performance monitoring tools). Additionally, the fragments of code devised by µGP were inserted in a loop, so the time consumed by the microprocessor for executing the code under exam become relevant with respect to the fixed parts.

Experimental Results

The first experiment targets the ratio of mispredicted branches over the total branches. Only non-speculative (retired) instructions are considered.

To excite the branch-prediction logic, each functional test is iterated 1 million times. Table 13.3 details the programs in terms of number of instructions (column "#INST"), and the attained mispredicted ratio measured

Table 13.3. Non-speculative mispredicted ratio

Program	$\#INST$	Sampling Type	
		Time	Event
Random [max]	278	6.01	5.93
Random [min]	426	0.10	0.00
Random [avg]	353.87	1.63	1.33
Random [std]	91.91	2.11	2.19
μGP (maximizing)	442	49.34	49.55
μGP (minimizing)	266	0.02	0.01

both using time-based sampling (column "Time") and event-based sampling (column "Event"). Remarkably, there are only small discrepancies between the two sampling methodologies.

Devising each test program required about 12 hours on a desktop PC equipped with a Pentium® 4 microprocessor; during this time, μGP generated and evaluated about 5,000 candidates. The population evolved through 200 generations, creating about 20 new individuals in each one (parameter λ). Different sizes of the population were experimented, ranging from 30 individuals to 200 (parameter μ), but no significant effects have been recorded.

For the sake of comparison, 5,000 programs were generated using an instruction randomizer. Internal weights were adjusted in order to produce a number of instructions and a percentage of conditional branches similar to the evolved tests. Table 13.3 details the programs attaining the maximum and the minimum values in rows "Random [max]" and "Random [min]" respectively. Finally, rows "Random [avg]" and "Random [std]" show the average mispredicted branch ratio and the standard deviation.

The two programs evolved by μGP are shown in rows "μGP (maximizing)" and "μGP (minimizing)".

A random fragment of code consisting in 10% conditional branches may show an already low mispredicted ratio when iterated several thousand times. μGP is able to further reduce this quantity to few hundredths. The experiments were interrupted because the measured values have the same order of magnitude of the noise introduced by the operating system and the concurrent tasks.

On the other side, when maximizing the ratio is the target, random programs can only attain a 6% figure; the automatically generated test program is able to increase this value up to nearly 50%.

The ability of a program to increase the branch prediction rate is very useful as the approach that will be taken is, by definition, random. The generated functional tests include very interesting combinations of instructions and branches that are likely to excite corner-case events. The resulting functional test could cover unique flaws that would be hardly detected by manually-written targeted tests.

Table 13.4. Trace cache deliver-mode ratio

Program	#INST	Sampling Type	
		Time	Event
Random [max]	6	99.32	98.13
Random [min^T]	5	85.39	
Random [min^E]	53		81.46
Random [avg]	32.13	91.68	88.58
Random [std]	21.52	4.38	4.75
µGP (maximizing^T)	36	99.49	
µGP (maximizing^E)	40		93.94
µGP (minimizing^T)	5	48.13	
µGP (minimizing^E)	55		23.33

The second set of experiments targets the ratio of clock cycles in which the trace cache is delivering µops to the execution unit instead of decoding or building traces. Table 13.4 summarizes the experiments; its columns are similar to the previous table.

It is interesting to note that values gathered with time-based sampling are strongly different from the ones measured using event-based sampling. This is due to the stronger impact of counter-accessing overhead and task switching on trace-cache behavior.

Due to the differences recorded using the two sampling methodologies, most experiments have been duplicated. Columns annotated with the letter "T" (e.g., min^T) refer to experiments using the time-based sampling, while columns annotated with the letter "E" (e.g., maximizing^E) using the event-based one. Non-relevant quantities, i.e., the deliver-mode ratio recorded with one type of sampling for tests optimized using the other one, are not reported.

Devising a test program required again about 12 hours with the same setup of the previous experiment. It is important to notice that, since all experiments have been duplicated, Table 13.4 shows 4 different µGP experiments.

Executing random code, the microprocessor spent in trace-cache deliver mode between 80% and 99% of clock cycles. Comparing with an almost saturated value, µGP is still able to increase it. When minimizing, on the other hand, generated functional tests yield a deliver-mode ratio of only 23% (measured with event-based sampling) or 48% (with time-based sampling), showing again the ability of the method to significantly improve the quality of the generated programs with respect to random generation.

The metric for the trace cache will likely produce tests which are not biased to any specific solutions. The ability of a program to execute from the trace cache is an inherent feature of the design. If a test program is generated by an architect to exercise the cache it will target the specific features that have been implemented while if a tool targets the feature it can likely cover multiple cases.

Third Case Study: Post-Production Processor Testing

Traditionally, tests have been applied using external Automatic Test Equipments (ATEs) [22]. However, technological progress is pushing up the complexity and operating frequencies of low-end microprocessor cores. It has become apparent that parametric testing alone was not sufficient to achieve the high quality goals required. Moreover, even though ATE effectiveness on applying parametric test is unquestionable, the costs for an ATE able to run at-speed functional tests are becoming prohibitive for manufacturers of moderate quantities of units [1]. As a consequence, the test community headed to alternative solutions.

To overcome these problems, industries are trying to reduce the use of expensive ATEs. One interesting strategy is to perform part of the test plan resorting to the so called Software-Based Self-Test (SBST) [23], where the test consists in a mere set of assembly programs and does not rely on any special test point to force values or observe behaviors during test application. Such programs could be loaded in RAM (e.g., resorting to a DMA or other mechanisms), and executed to test the core. A minimum effort is needed to extract test results.

Clearly, an SBST test is executed at-speed, and requires a very simple ATE and little hardware overhead to download and upload the test information. Furthermore, SBST is a suitable solution for stand-alone microprocessors as well as for those embedded in a SoC.

In this section, the evolutionary algorithm is exploited devising effective test programs using a test strategy based on SBST. The resulting methodology is almost automated and takes benefit from the information available in the different design stages, from the ISA to register transfer level (RT-level), and eventually netlist. The key idea is to build a set of test programs through the whole design cycle of the processor, using the microprocessor high-level descriptions delivered during the design process. Indeed, the most of the efforts and analysis in the methodology presented in this chapter are focused on automatic generation of test programs for microprocessors at RT-level; in fact, a systematic exploitation of the available code coverage metrics at this design abstraction level reports interesting results.

Additionally, the framework set up allows a detailed analysis of the effectiveness of programs devised for maximizing different metrics, and their usability for test is also reported.

Methodology Description

The proposed methodology is based on a step-by-step approach that allows generating and collecting test programs through the whole design cycle. Test programs generation is performed each time a new microprocessor model is delivered; thus, it is not necessary to wait up to the end of the design cycle to start the generation of programs. Moreover, being a cumulative process that

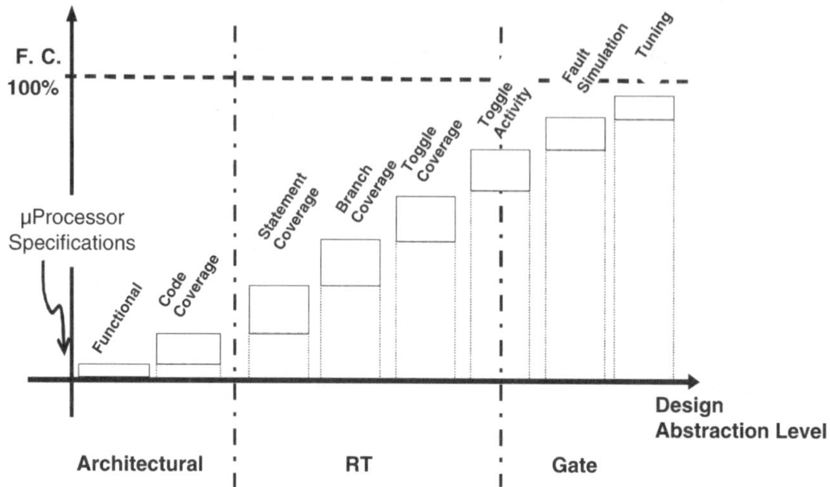

Fig. 13.4. Qualitative description of the Methodology

tackles high-level descriptions, the approach minimizes the use of explicit low-level fault simulation to drive the test program generation, greatly reducing computational efforts. In fact, fault simulation is only used at the end of the generation to complete the set of test programs. Figure 13.4 is a graphical representation of the possible progress of the final fault coverage (FC%) at gate-level attained by the test programs devised in the different phases.

To determine if a logic circuit is good, patterns or test vectors are applied to the circuit inputs, and then the circuit responses are compared with the expected outputs. If the responses match, the circuit can be considered as good. However, the results of the test process are directly correlated whit the quality of the patterns [22]. The performance measure of the test patterns is commonly called the fault coverage (FC). This percentage value represents the patterns ability to detect specific faults into the logic circuit. A fault is a model of one or more possible physical defects. In most of the cases, when a FC value is provided, the single stuck-at fault is chosen as the fault model. However, other fault models can be adapted, such as the delay fault model.

Figure 13.4 also provides a general microprocessor design life cycle. The whole cycle has been split in three successive macro phases. Each phase is characterized by a different kind of design abstraction model: architectural, RT-level, and gate-level.

A pure simulatable model, like the one embedded in an Instruction Set Simulator (ISS), characterizes the first phase; architectural simulatable models such as high level programs possibly catching some peculiar aspects of the internal design, may be found too. In the second phase, more structured models such as RT-level descriptions are available; hybrid models, where some blocks have been detailed down to gate-level, but others are still at the

RT-level, could be included in this phase. Behavioral simulators drive this section of the process. Finally, in the last phase of the design, before the manufacturing stage, the processor netlist is assumed available.

Figure 13.4 gives a qualitative idea of the FC% that could be attained by generating test programs using different microprocessor models; these percentages reflect the potential performance of test programs generated on each stage. As estimated, test programs generated in the early stages provide poorer results in terms of FC%; however, these collected programs are a good initial point regarding the entire process. In particular, the case study tackles herein shows that test programs generated in the second part of the design cycle, when RT-level and hybrid descriptions are available, are able to reach an important contribution of fault coverage regarding the complete set of programs.

The descriptions available in the final phase lead to a more expensive test programs generation process; consequently, this phase should mainly be devoted to completion and tuning regarding FC%.

As outlined before, the main goal of the generation process is to gather, during the whole design process, a set of programs able to reach a high FC%. However, devising test programs suitable for testing at every phase of the design cycle of the processor is not a trivial task.

First of all, at initial stages of the design cycle, functional programs do not target fault coverage but special corner cases assessed to support the design process. Additionally, there is not a very strong correlation between simulatable models metrics and final processor faults mainly because the observability lacks. Thus, the FC% results attained using these programs are low.

At medium levels of the design cycle, it means at RTL, a common practice is to generate test programs using code coverage values as driving information, as explained recently. However, every coverage metric stresses different aspects of the microprocessor description, and as stated before choosing only one metric is not a clever decision.

Finally, at final levels of design, automatic generation of test programs requires considerable computational efforts mainly due to the huge quantities of information that must be computed to simulate a single processor run.

Thus, the step-by-step methodology proposed herein suggests to take advantage of the available metrics in modern simulators at every phase of the design cycle to drive the program generation.

Architectural stages deserve no longer considerations, since does not exist a well-known relationship between coverage metrics at this level and FC%. Conversely, test programs generated at this stage may provide valuable regarding special cases.

Regarding to RTL, it seems to be convenient to firstly face the metrics regarding the program flow (i.e., statement coverage, branch coverage and condition coverage), and in a refinement phase the use of the metrics observing internal information (i.e., expression coverage and toggle coverage).

This must be of course a gradual process, for example, the first step could start maximizing the statement coverage of the microprocessor description; then, when a satisfactory program or set of programs has been obtained, the next metric to deal with is the branch coverage; but, this new program generation must not start from scratch; instead a kind of coverage grading could be devised. The performance metric of the new program set will not be the total branch coverage, but the branch coverage without the branches already covered by the first program set. Once again, when a satisfactory value of branch coverage is reached, a new metric is targeted, and so forth.

As described previously, a set of test programs had been collected through the processor design cycle; however, it is possible that the maximum FC% has not been reached by the gathered set of programs. Then a completion and tuning processes must be performed at the final stages of the design.

The first step is to perform a fault grading using the set of accumulated programs. Thus, an updated scenario about the remaining faults is presented. At this point, a new test program generation process can be implemented using as feedback value the FC% regarding the uncovered faults. As supposed, this process could demand excessive amount of time, then it is not possible to perform this program generation loop from scratch.

The resulting programs of this process are also collected as test programs; their performance depends on the available time to carry out the generation; usually heavy time restrictions pushed by the time to market guide this generation.

At the last stage, an expert engineer performs the final tuning phase. Then, hand-written programs will tackle specific uncovered faults. This practice is again very expensive and very time demanding because it relies on a skilled engineer.

Case Study

As a case study, the proposed methodology was implemented tackling a 32-bit pipelined processor. The core is available in two different descriptions: the first is a RT-level microprocessor description and the second one is the microprocessor netlist. μ GP was exploited as automatic generator, following the step-by-step method.

It must be noticed that when a FC% value is provided, the single stuck-at fault is chosen as the fault model.

The processor chosen is called PLASMA. This is a free microprocessor core that conforms the MIPS I™ architecture [24], and supports interrupts and all MIPS® user mode instructions except unaligned load and store operations. The original core is implemented in VHDL with a three-stage pipeline.

PLASMA Descriptions

The original PLASMA core is an RT-level microprocessor description available in VHDL. Table **13.5** is a general description of the core at this level.

Table 13.5. PLASMA: RT-level description characteristics

PLASMA	RT-level description
Active Statements	720
Active Branches	379
Active Conditions	40
Active Expressions	43
Nodes at RT-level	1,436

Table 13.6. PLASMA: gate-level description characteristics

PLASMA	gate-level description
Gates	36,991
Flip-Flops	1,466
Nodes	17,290
Faults	98,140

Once synthesized, the PLASMA gate-level implementation consists in about 37K gates (1,466 are flip-flops). The complete fault list consists of 98,140 permanent single-bit stuck-at faults (95,810 of them are detectable). The following table summarizes the characteristics of the gate-level description.

Exploiting both, the microprocessor descriptions and the automatic test program generator, a sequence of experiments were launched obeying the proposed rules of the step-by-step method.

Apart of µGP, the elements used to devise the experiments are the following:

- External evaluators: two CAD tools were required to complete the µGP generation loop:
  - Modelsim v5.8b by Mentor Graphics at RT-level.
  - Faultsim 1999.10-TG4.1-2150 by Synopsys for gate-level fault simulations.
- The microprocessor core: two PLASMA descriptions previously mentioned.
- The µGP instruction library: including about 60 macros and devised in two programmer working days.

In addition, to interface the µGP code with the external evaluators, a few scripts accessing the different metrics were required. All experiments were run on a Sun Enterprise 250 with two 400 MHz UltraSPARC-II CPUs and 2Gb of RAM.

For the completeness of the method, test programs were devised using both manual and automatic methods considering the actual microprocessor design process. At the end of the process when gate-level stuck-at FC% was targeted, the observability points used were the buses around the register bank.

Table 13.7. Test program generation results

Generation process	Processor description	Metric	Size[# inst]	Time [days]	FC [%]
Manual	ISA	Functional	57	7.0	27.12
μ GP	RT-level	Statement Coverage	1,325	5.0	41.92
μ GP	RT-level	Branch Coverage	344	2.1	44.31
μ GP	RT-level	Condition Coverage	1,187	2.4	44.81
μ GP	RT-level	Expression Coverage	2,017	2.7	46.18
μ GP	RT-level	Toggle Coverage	226	2.9	46.28
μ GP	gate-level	Toggle Activity	514	8.1	70.36
μ GP	gate-level	Fault simulation	260	21	83.57
Manual	gate-level	Tuning FC	720	28	91.97

Experimental Results

Table 13.7 shows the results obtained trough the complete test program generation process. It is necessary to highlight again that the process was performed as a cumulative collection of test programs.

Table 13.7 reports the results of the complete step-by-step test program collection process. The first column, called Generation process describes the method used for program generation of each step. Manual means that engineers have generated those programs, while μ GP denotes automatically generated programs.

The second column illustrates the level of abstraction of the microprocessor description, ranging from the instruction set architecture to gate-level. Taking advantage of μ GP; in this collecting approach, automatic test program generation can be applied at all description levels; however, manual generation is allowed at the beginning and at the end of the process, only. In the specific case study, the automatic generation of test programs starts at the RT-level of microprocessor description.

The Metric column evidences the target metrics used to generate test programs. These metrics were described previously as the suitable ones for each microprocessor design stage. Since no PLASMA architectural models are available, it was not possible to face this phase.

Along with the information about the final fault coverage (FC [%]) obtained in the cumulative process, for each set of test programs the total programs size, expressed as number of instructions, and the time to develop the appropriate set, in days, are presented in the columns 4 and 5.

The final resulting test-set is composed of 15 test programs: the first one, at ISA level, was developed targeting quite simple functionalities. At RT-level, 8 programs compose the entire set, and all of them were created exploiting the automatic tool; out of the 8, 3 programs were devised targeting the statement coverage, 1 the branch coverage, 1 the condition coverage, 2 the expression coverage, and the last one aimed at maximizing the toggle coverage. At gate-level,

3 programs exploit the µGP potentiality, among these, 2 maximize the toggle activity, but 1 faces directly the fault coverage. A skilled engineer finally constructed the 3 remaining programs.

The whole test-set elaboration process requires about 80 working days; however, as sketched before, test programs generated up to the logic design macro stage were also exploited in the verification process.

Acknowledgements

Authors wish to thank Matteo Sonza Reorda and Massimiliano Schillaci for their important support and significant comments.

References

1. International Technology Roadmap for Semiconductors – ITRS 2005 Edition, http://www.itrs.net/Links/2005ITRS/Home2005.htm/
2. R. M. Friedberg, "A Learning Machine: Part I", IBM Journal of Research and Development, Vol. 2, No. 1, pp 2–13, 1958
3. R. M. Friedberg, B. Dunham, J.H. North, "A Learning Machine: Part II", IBM Journal Research and Development, Vol. 3, pp. 183–191, 1959
4. G. Squillero, "MicroGP — An Evolutionary Assembly Program Generator", Genetic Programming and Evolvable Machines, vol. 6, no. 3, 2005, pp. 247–263
5. R. Poli, "A Simple but Theoretically-Motivated Method to Control Bloat in Genetic Programming", EuroGP 2003, pp. 204–217
6. F. Corno, E. Sanchez, G. Squillero, "Evolving Assembly Programs: How Games Help Microprocessor Validation", IEEE Transactions on Evolutionary Computation, Special Issue on Evolutionary Computation and Games, Dec. 2005, vol. 9, pp. 695–706
7. L. Chen, S. Dey, "Software Based Self Test methodology using a embedded Software Tester", IEEE transaction on computer aided design of integrated circuits and systems, pp 369–380 (2001)
8. F. Corno, M. Sonza Reorda, G. Squillero, M. Violante, "On the Test of Microprocessor IP Cores", DATE01: IEEE Design, Automation & Test in Europe, 2001, pp. 209–213
9. F. Corno, G. Cumani, M. Sonza Reorda, G. Squillero, "Evolutionary Test Program Induction for Microprocessor Design Verification", ATS02: IEEE Asian Test Symposium, 2002, pp. 368–373
10. F. Corno, G. Cumani, M. Sonza Reorda, G. Squillero, "Fully Automatic Test Program Generation for Microprocessor Cores", DATE03: IEEE Design, Automation & Test in Europe, 2003, pp. 1006–1011
11. F. Corno, M. Sonza Reorda, G. Squillero, "Automatic Test Program Generation for Pipelined Processors", SAC03: ACM Symposium on Applied Computing, 2003, pp. 736–740
12. F. Corno, G. Squillero, M. Sonza Reorda, "Code Generation for Functional Validation of Pipelined Microprocessors", ETW03: IEEE European Test Workshop, 2003, pp. 113–118

13. D. A. Patterson, J. L. Hennessy, Computer Architecture — A Quantitative Approach (2nd edition), Morgan Kaufmann, 1996.
14. A. Piziali, Functional Verification Coverage Measurements and Analysis, Kluwer Academic Publishers, 2004
15. S. Tasiran, K. Keutzer, "Coverage Metrics for Functional Validation of Hardware Designs", IEEE Design & Test of Computers, vol. 18(4), 2001, pp. 36–45
16. J. L. Chien-Nan, C. Chen-Yi, J. Jing-Yang, L. Ming-Chih, J. Hsing-Ming, "A novel approach for functional coverage measurement in HDL Circuits and Systems", ISCAS2000: The 2000 IEEE International Symposium on Circuits and Systems, 2000, pp. 217–220
17. D. A. Patterson, J. L. Hennessy, Computer Architecture — A Quantitative Approach (2nd edition), Morgan Kaufmann, 1996
18. R. Singhal, K. S. Venkatraman, E. R. Cohn, J. G. Holm, D. A. Koufaty, Meng-Jang Lin, M. J. Madhav, M. Mattwandel, Nidhi Nidhi, J. D. Pearce, M. Seshadri. "Performance Analysis and Validation of the Intel® Pentium® 4 Processor on 90 nm Technology". Intel Technology Journal, Volume 8, issue 1, 2004. pp 33–42
19. Intel® Corporation, IA-32 Intel® Architecture Software Developer's Manual Volume 3: System Programming Guide, 2004, http://developer.intel.com/design/pentium4/manuals/
20. Brinkley Sprunt, "Pentium® 4 Performance Monitoring Features", IEEE Micro, July-August 2002, pp. 72 82R
21. G. Hinton, D. Sager, M. Upton, D. Boggs, D. Carmean, A. Kyker, P. Roussel, "The Microarchitecture of the Pentium® 4 Processor", Intel® Technology Journal, 1st quarter 2001, http://www.intel.com/technology/itj/q12001/articles/art_2.htm
22. V. Agrawal, M. Bushnell, Essentials of Electronic Testing for Digital, Memory and Mixed-Signal VLSI Circuits, Kluwer Academic Publishers, 2000
23. L. Chen, S. Dey, "Software Based Self Test methodology using a embedded Software Tester", IEEE transaction on computer aided design of integrated circuits and systems, pp 369–380 (2001)
24. MIPS TECHNOLOGIES. (2002) MIPS32™ Architecture For Programmers Volume I: Introduction to the MIPS32™ Architecture. Revision 1.90. http://www.mips.com/

Printing: Krips bv, Meppel
Binding: Stürtz, Würzburg